REALITY AND RATIONALITY

REALITY AND RATIONALITY

Wesley C. Salmon

EDITED BY

Phil Dowe and Merrilee H. Salmon

OXFORD

UNIVERSITY PRESS

2005

OXFORD
UNIVERSITY PRESS

Oxford University Press, Inc., publishes works that further
Oxford University's objective of excellence
in research, scholarship, and education.

Oxford   New York
Auckland   Cape Town   Dar es Salaam   Hong Kong   Karachi
Kuala Lumpur   Madrid   Melbourne   Mexico City   Nairobi
New Delhi   Shanghai   Taipei   Toronto

With offices in
Argentina   Austria   Brazil   Chile   Czech Republic   France   Greece
Guatemala   Hungary   Italy   Japan   Poland   Portugal   Singapore
South Korea   Switzerland   Thailand   Turkey   Ukraine   Vietnam

Copyright © 2005 by Oxford University Press, Inc.

Published by Oxford University Press, Inc.
198 Madison Avenue, New York, New York 10016

www.oup.com

Library of Congress Cataloging-in-Publication Data
Salmon, Wesley C.
   Reality and rationality / Wesley C. Salmon; edited by Phil Dowe and Merrilee H. Salmon.
      p.   cm.
   Includes bibliographical references and index.
   ISBN-13 978-0-19-517784-8;   978-0-19-518195-1 (pbk.)
   ISBN 0-19-517784-3;   0-19-518195-6 (pbk.)
      1. Science—Philosophy.   I. Dowe, Phil.   II. Salmon, Merrilee H.   III. Title.

Q175.S23414 2005
501—dc22   2004050051

9 8 7 6 5 4 3 2 1

Printed in the United States of America
on acid-free paper

Acknowledgments

M ost chapters in this book are edited versions of previously published material, with minor deletions and additions. The editors gratefully acknowledge permission of the original publishers, as named below, to use these materials.

Chapter 1 is based on "Empiricism: The Key Question," in *The Heritage of Logical Positivism*, ed. Nicholas Rescher (Lanham, Md.: University Press of America, 1985), 1–21.

Chapter 2 is based on "Carnap, Hempel and Reichenbach on Scientific Realism," in *Logic, Language, and the Structure of Scientific Theories: Proceedings of the Carnap-Reichenbach Centenniel*, ed. Wesley Salmon and Gereon Wolters. (Pittsburgh: University of Pittsburgh Press; Konstanz: Universitätverlag Konstanz, 1994), 237–254.

Figure 2.1 in chapter 2 is from *Experience and Prediction* by Hans Reichenbach (Chicago: University of Chicago Press, 1938); with kind permission of the publisher.

Chapter 4 is based on "Inquiries into the Foundations of Science," in *Vistas in Science*, ed. David L. Arm (Albuquerque: University of New Mexico Press, 1968), 1–24.

Chapter 5 is based on "Bayes's Theorem and the History of Science," in *Minnesota Studies in the Philosophy of Science Volume V: Historical and Philosophical Perspectives of Science*, ed. R. H. Stuewer (Minneapolis: University of Minnesota Press, 1970), 68–86.

Chapter 6 is based on "Rationality and Objectivity in Science, or Tom Kuhn Meets Tom Bayes," in *Minnesota Studies in the Philosophy of Science Volume XIV: Scientific Theories*, ed. C. Wade Savage (Minneapolis: University of Minnesota Press, 1990), 175–204.

Chapter 8 is based on "Dynamic Rationality: Propensity, Probability, and Credence," in *Probability and Causality*, ed. James H. Fetzer (Dordrecht: Reidel, 1988), 3–40; with kind permission of Kluwer Academic Publishers.

Chapter 9 is based on "Religion and Science: A New Look at Hume's Dialogues," *Philosophical Studies* 33 (1976): 143–176; with kind permission of Kluwer Academic Publishers.

Chapter 10 is based on "Consistency, Transitivity and Inductive Support," *Ratio* 7, 2 (1965): 164–169; with kind permission of Blackwell Publishers, Ltd., Oxford.

Chapter 11 is based on "Partial Entailment as a Basis for Inductive Logic," in *Essays in Honor of Carl G. Hempel,* ed. Nicholas Rescher (Dordrecht: Reidel, 1969), 47–82; with kind permission of Kluwer Academic Publishers.

Chapter 12 is based on "Confirmation and Relevance," in *Minnesota Studies in the Philosophy of Science Volume VI: Induction, Probability, and Confirmation,* ed. G. Maxwell and R. M. Anderson, Jr. (Minneapolis: University of Minnesota Press, 1975), 3–36.

Editors' Preface

Wesley Salmon (1925–2001) has been aptly described as one of the most influential philosophers of science of the second half of the twentieth century. His work falls into three main areas: explanation and causality, space and time, and inductive reasoning. *Causality and Explanation* (New York: Oxford University Press, 1998) nicely anthologized his work in the first area. This collection attempts to do the same for the third.

"Reality," the first part of the book, presents Salmon's view that scientific realism is true and is appropriately defended by empirical arguments. The second part, "Rationality," lays out his characteristic inductivist and Bayesian account of how empirical reasoning best operates. Both parts contain introductions especially prepared by the author and articles previously published throughout his working life, chosen and edited by him for this collection. Two of the chapters are not published elsewhere. He also wrote new appendices for several of the chapters. In addition to the works included here, he had intended to add a third part, entitled "Reactions," in which he responded to works of several other philosophers. He said that when the book was finished, he would feel that it presented a fair picture of his contributions on the topics of reality and rationality. Unfortunately, he did not live to complete the last part. Nevertheless, we believe that the chapters that make up this book form a lucid and cohesive account of the problems treated therein and that bringing these articles together provides a chance for readers to recognize important connections that would not otherwise be apparent.

In the fall of 2000, the Center for Philosophy of Science at the University of Pittsburgh held a workshop entitled "Induction/Probability and Causation/ Explanation" in honor of Wes's retirement from teaching. In his moving closing comments, he said that he felt his work on these areas was largely completed and that he was happy to retire while knowing that the philosophical topics he cared most about were being left in the "younger and more capable hands" of his former students and colleagues. In that spirit, we would like to dedicate the book to all those students and colleagues—as well as future students of his work.

The editors would like to thank Paul Humphreys, Patrick Maher, James Woodward, and Peter Ohlin for their helpful comments and support.

Author's Preface

Two perennial philosophical questions motivate this collection of essays. First, is science a rational enterprise? Second, does science yield objective information about our world, in particular, about the existence and nature of aspects that we cannot observe directly with our normal unaided sense organs? Although these questions have been addressed by many philosophers, two are at the front of my mind as I undertake these inquiries. The first is Thomas S. Kuhn, whose book *The Structure of Scientific Revolutions* (1962) led many philosophers to question the rationality of science. Although Kuhn has often been accused of denying that science is rational, he rejects the charge, maintaining instead that mature physical science is the best example of rationality we can appeal to. The problem, according to Kuhn, is to discover just what such rationality consists in. I want to join him in this enterprise. The second philosopher is Bas C. van Fraassen, whose 1980 book, *The Scientific Image*, argues that we cannot have knowledge of unobservable entities and processes. His *constructive empiricism* does not deny their existence but rather the possibility of our knowing anything about them. His position is therefore agnostic regarding many entities to which accepted scientific theories seem to refer. I want to show, to the contrary, that such knowledge *is* possible and how it is in fact achieved.

The two questions just mentioned intersect substantially, and the locus of intersection is in the area of scientific confirmation and inductive inference. I begin with the realism issue. Its consideration leads us directly into a careful examination of scientific confirmation and inductive reasoning. These in turn lie at the core of the rationality issue. Although it might seem logically preferable to begin with induction and confirmation, I believe it is heuristically desirable to start with the question of realism. We see in the first chapter that this question provides strong motivation for tackling confirmation, induction, and the rationality of science.

In *The Structure of Scientific Revolutions* (1960, 1972), Kuhn maintained that choices among scientific theories are based on considerations that go beyond observational data and logic, even if logic is taken broadly enough to include scientific confirmation. In order to evaluate this claim we need to consider explicitly—*as Kuhn did not*—the nature of scientific confirmation. My conjecture is that Kuhn was thinking of scientific confirmation in terms of a simple hypothetico-deductive model. I argue that a Bayesian approach provides a more satisfactory conception of confirmation—one that allows us to construe "logic" in still broader terms, enabling us to incorporate within logic considerations that probably seemed to Kuhn to fall outside its scope. This approach allows us at least partially to bridge the gap between Kuhn's historically oriented philosophy and that of such logical empiricists as Rudolf Carnap, Carl G. Hempel, and Hans Reichenbach.

Empiricism is a major theme in van Fraassen's *Scientific Image*; indeed, it takes as its point of departure the thesis that empiricism and realism are incompatible doctrines. I claim, to the contrary, that they are fully compatible. In order to see how this can be, we need to look carefully at the meaning of empiricism—we must, in fact, see what constitutes its *key question*. Twentieth century empiricists have, to a large extent, focused their attention on the meanings of theoretical *terms*. This strikes me as a mistake. We should ask, instead, how—if at all—it is possible to confirm theoretical *statements*, that is, statements about unobservable entities. The key question that thus emerges is whether inductive logic contains the resources to provide legitimate inferences from data about observables to conclusions about unobservables. I offer an affirmative answer.

This book is divided into two parts, Reality and Rationality. The first part addresses the problem of scientific realism, the second deals with issues concerning scientific confirmation.

Part I: Reality

The first chapter in this collection, "Realism and Empiricism: The Key Question," attempts to articulate the question clearly, to place it in historical context, and to set out the fundamentals of the affirmative response. At the most basic level, I claim, reasoning from observables to unobservables involves inferences that are both causal and analogical. In the second chapter, "Scientific Realism in the Empiricist Tradition," I offer a critical examination of the approaches to realism of two philosophical giants: Rudolf Carnap and Hans Reichenbach. This discussion leads immediately to the third chapter, "An Empiricist Argument for Realism," which tries to show how the considerations that convinced serious physical scientists of the reality of atoms and molecules in the early years of the twentieth century provide a philosophically sound argument for realism that does not exceed the bounds of empiricism. This argument differs substantially from the many defenses of realism that have been offered in response to van Fraassen's book.

Part II: Rationality

Van Fraassen's basic answer to my defense of realism is to doubt that the sort of inductive logic I envision even exists. To answer this challenge I turn to a discussion of the Bayesian approach to confirmation and to the basic considerations involved in my response to Kuhn. Any appeal to Bayes's theorem is bound to raise questions about the nature of prior probabilities; my answer is embodied in the fourth chapter, "Plausibility Arguments in Science."

When I first looked at Kuhn's *Structure*, I was shocked by his rejection of the distinction between the *context of discovery* and the *context of justification*. Because this distinction seems to me to have fundamental importance, I offered the reply contained in the fifth chapter, "Discovery and Justification." It seems to me likely that Kuhn's problem with the discovery/justification distinction might have arisen from a puzzlement about the role of plausibility considerations in the evaluation of theories. Reflection on the history of science shows, I believe, that plausibility arguments play an essential role; yet, by their very nature, they *seem* to pertain to the discovery rather than the justification of theories. My strategy is simply to identify plausibility assessments with the prior probabilities, which, as Bayes's theorem shows, constitute an indispensable part of the context of justification. This approach goes a long way in overcoming Kuhnian worries about considerations affecting theory choice that go beyond observation and logic—including confirmation. In the sixth chapter, "Rationality and Objectivity in Science," I bring these ideas to bear directly on Kuhn's problems concerning theory preference and theory choice. The seventh chapter, "Revisions of Scientific Convictions," also deals with prior probabilities, in this case as a response to van Fraassen's discussion of Bayesianism in his *Laws and Symmetry* (1989). The eighth chapter, "Dynamic Rationality" defines various grades of rationality and addresses the question of the relationship between rationality and objectivity within Bayesian confirmation theory.

Chapters 4 through 8 show how Bayes's theorem helps us to deal with certain profound and problematic issues in contemporary philosophy of science. The ninth chapter, "Hume's Arguments on Cosmology and Design," is a sort of encore, showing how the same theorem can be applied to the thesis that scientific evidence can be offered, in the form of the *design argument*, for the existence of God.

The remaining three chapters in this part take a deeper look at confirmation and induction—that is, at the foundations of rationality. The tenth chapter, "The 'Almost-Deduction' Theory," turns from Bayes's theorem to the theorem on total probability to undermine what might be called "the almost-deduction notion of induction" and to expose certain dangers that lie within it. The eleventh chapter, "The 'Partial-Entailment' Theory," considers the intuitively appealing idea of founding inductive reasoning on a relation of partial entailment in much the same way as deductive reasoning can be seen as founded on full entailment. This idea, which seems to have appealed heuristically to Carnap as well as many other philosophers, turns out to be unable to bear the weight. Incidentally, the original version of this chapter, which was published long before the notorious Popper-Miller 'theorem' on the impossibility of inductive support, shows clearly the defect in their 'demonstration'. The twelfth chapter, "Confirmation and Relevance," discusses ramifications of the fundamental distinction between two different senses of confirmation, namely, high degree of confirmation (or high probability) and incremental confirmation (or increase in degree of probability). As this chapter demonstrates, confusion of these two senses has led to serious difficulties in confirmation theory.

Contents

PART I

REALITY

Introduction

It is widely agreed that we live in a "postpositivist" era; these days the word "positivist" functions chiefly as an ill-defined term of abuse. Logical positivism is dead.[1] One cannot deny, however, that it was a potent philosophical force in the earlier part of the twentieth century; many conferences, books, and articles have endeavored to assay that influence. One conference, which generated such a book, was held in Pittsburgh in 1983.[2] The first chapter in the present collection was my contribution to that conference.

1. Realism and Empiricism: The Key Question

Logical empiricism is the direct descendent of logical positivism. For that reason I chose it as the focus of my paper for the conference on "The Heritage of Logical Positivism." It was also the dominant approach to philosophy of science in the pre-Kuhnian period. As its name implies, logical empiricism adopts an empirical approach to scientific knowledge, freely employing the tools of modern logic. After a brief nod to achievements in formal logic, I direct attention to empiricism and the problems it encounters. One of the most difficult questions concerns the possibility of knowledge of unobservable entities, that is, the problem of scientific realism. This issue was also a major point of disagreement between logical positivists and logical empiricists. The key question, as I see it, is "whether inductive reasoning contains the resources to enable us to have observational evidence for or against statements about unobservable entities and/or properties." I argue for an affirmative answer in part I. The main purpose of the first chapter is, however, just to state the question explicitly and urge its careful consideration. As I say immediately after articulating it, "The most surprising thing about the *key question* is how seldom it has been raised explicitly."[3]

3

2. Scientific Realism in the Empiricist Tradition

Rudolf Carnap and Hans Reichenbach were both born in 1891; in 1991 this dual
centennial was celebrated at many conferences in Europe and America. One
such meeting, sponsored jointly by the University of Pittsburgh and the
University of Konstanz, was held in Konstanz (see Salmon and Wolters 1994).
On that occasion the University of Konstanz conferred an honorary doctorate
upon Carl G. Hempel, the most distinguished living representative of the
logical positivist/logical empiricist movement. This is the paper I presented in
his honor on that occasion.

Prior to 1933, when Hitler came to power, Carnap was a member of the
Vienna Circle of logical positivists, Reichenbach was the leader of the Berlin
group of logical empiricists, and Hempel was closely associated with both
groups. Subsequently these philosophers became the most influential leaders
of logical empiricism.[4] Over a period of several decades they struggled with
problems concerning the existence and nature of unobservable objects and the
meanings of theoretical statements and terms. Although, in my opinion,
Reichenbach's position was the most tenable of the three, his exposition of that
viewpoint was far from clear. Attempting to sort out the various arguments
involved in the discussion among these three figures is both historically and
philosophically illuminating.

3. An Empiricist Argument for Atomism

Taking atoms and molecules as prime examples of unobservable entities, I asked
myself why contemporary scientists believe that there are such things. This
approach to the fundamental issue of scientific realism reflected my general
dissatisfaction with arguments offered by philosophers for or against the
existence of unobservables. I was first led to address these questions in
connection with concerns about the nature of theoretical explanation in the
sciences.

John Dalton advanced his atomic theory in 1808, and the molecular-kinetic
theory of gases had been elaborated in great detail before the end of the
nineteenth century. Nevertheless, at the turn of the twentieth century
knowledgeable physical scientists did not agree about the actual existence
of atoms and molecules. During roughly the first decade of the twentieth
century—1908 is an especially important year[5]—an almost universal consensus
did emerge. As Mary Jo Nye (1972) made clear, the work of Jean Perrin on
Brownian motion and Avogadro's number played a crucial role.[6] Avogadro's
number can be viewed as *the link* between the microcosm and the macrocosm;
given various macroquantities we can, with its aid, calculate numerous
microquantities and vice versa. With the assurance that Avogadro's number is,
indeed, the number of molecules in a mole of any given substance, we can
establish many facts about unobservable entities.

Using a variety of historical sources, including Perrin's own account (Perrin
[1913] 1916), I trace some of the most significant trials and tribulations of the

atomic/molecular theory of the constitution of matter from Dalton to Perrin, and I offer a philosophical reconstruction and analysis of Perrin's decisive argument. This argument seems to me to be cogent, and I commend it to those who are interested in scientific realism.

To this paper I have added an appendix dealing with a rather specific issue about scientific realism. It deals with a truly 'far out' form of antirealism; it is Ian Hacking's extragalactic antirealism. I argue that, contra Hacking, we have adequate evidence to believe in the existence of at least some types of entities located beyond the confines of our own galaxy.

Although all three chapters in part I have strong historical slants, they deal with issues that are very much alive today. Taken together they seem to me to provide the best contemporary answers to our questions about scientific realism.

Realism and Empiricism

The Key Question

What is our chief inheritance from logical positivism? The answer is simple—
logical empiricism. Rudolf Carnap and Hans Reichenbach were two of the
most prominent early exponents of logical empiricism. Indeed, the volume devoted to Carnap in *The Library of Living Philosophers* (Schilpp 1963) was originally
planned as a volume on logical empiricism, which would have dealt with Carnap
and Reichenbach. This plan had to be abandoned because of Reichenbach's
untimely death in 1953.

Given the degree to which formal logic has flourished in the twentieth century, I think we can say that the logical part—at least the *deductive* logical part—of
logical empiricism is alive and well. The empiricism part is another story, and that
is what I intend to discuss in this chapter.

1. Some Historical Background

Reichenbach saw a sharp dividing line between logical positivism and logical
empiricism, which he drew in *Experience and Prediction* (1938)—a book that he
regarded as a refutation of logical positivism. As he saw it, the distinctive feature of
his approach was his thoroughgoing probabilism. Two important consequences,
which represented differences with some of the early Vienna positivists, emerged.
First, Reichenbach rejected the notion that there are any protocol sentences or
sense data reports that are absolutely certain. He took physical object statements
as the fundamental type of evidence statement, and he regarded them as fallible.
Impressions, or sense data, are not objects of immediate awareness; instead, they
are theoretical entities of psychology. Second, Reichenbach was a scientific realist;
he maintained that we can have probabilistic empirical knowledge of unobservable
entities. Throughout his philosophical career—after an early period as a Kantian—
he remained a vocal advocate of empiricism. Indeed, I think it is fair to say that
Reichenbach's fundamental motivation was the desire to provide a purely empirical philosophical foundation for modern science. He maintained to the very end

his conviction that scientific realism and empiricism are compatible with one another—though one might wonder, in the light of his remarks about "extension rules of language" (1951, 55–58), whether his realism had not by that time been completely eviscerated.

Carnap's attitude toward empiricism was rather different. In his famous article "Empiricism, Semantics, and Ontology" (1950a), he attempted to retain empiricism by avoiding metaphysical commitment. Although this article was chiefly concerned with the status of abstract entities in logic and mathematics, it contains enough obiter dicta concerning the natural sciences to enable one to conclude that he was willing to embrace instrumentalism in order to preserve empiricism: for example, "The acceptance or rejection of abstract linguistic forms, just as the acceptance or rejection of any other linguistic forms in any branch of science, will finally be decided by their efficiency as instruments" (40). Later, however—on the basis of arguments offered by Carl G. Hempel (1963)—he was persuaded of the tenability of scientific realism (Carnap 1963, 960).[1] The fact that Hempel's argument was based on considerations of inductive systematization will be relevant to my main theme.

Carnap devoted many years of concentrated effort to the foundations of inductive logic. In the last analysis, he maintained, the reasons for accepting axioms of inductive logic "are based upon our intuitive judgments concerning inductive validity." Moreover, "It is impossible" to give a purely deductive justification of induction" and "the [aforementioned] reasons are a priori" (1963, 978). In this context, it appears, Carnap abandoned empiricism. (For further details see Salmon 1967a.) When considering twentieth-century scientific philosophy, it seems significant to note that Bertrand Russell—who cannot properly be classed as a logical empiricist, but who surely provided a good deal of inspiration for members of that movement—also rejected empiricism. In his last major epistemological work, *Human Knowledge, Its Scope and Limits* Russell (1948) also employs "postulates of scientific inference," which cannot be justified empirically. He argues, it will be recalled, that empiricism leads to solipsism—a view that can neither be refuted by logical argument nor sincerely embraced (for further details, see Salmon 1974).

Reichenbach maintained that it is possible to provide a deductive justification of induction, but whether that program can be carried through successfully is a vexed question into which I shall not enter in this chapter. Carnap and Russell, as just noted, relinquish empiricism in order to furnish suitable foundations for scientific inference. Karl Popper, the vocal anti-inductivist, seems to have taken a metaphysical commitment to realism as a surrogate for induction. Actually the views of Popper and Russell on scientific inference bear striking resemblances. Russell rejects induction and introduces synthetic postulates as a supplement to deduction. Popper seems to have adopted something like a postulate of uniformity of nature as a basis for his notion of scientific inference (see Salmon 1974).

Empiricism, it seems, is a tough row to hoe; the difficulties arise at three main junctures. First, there is the question of whether we can have empirical knowledge of ordinary, middle-sized physical objects. In claiming that empiricism leads to solipsism, Russell offered a negative answer to this question. In my opinion, however, the physicalism of Carnap, Hempel, Reichenbach, and many other logical empiricists provides a tenable answer. I shall not pursue the issue further in this chapter. Second, there is the question of whether induction is admissible within an empiricist framework. Carnap, Popper, and Russell all give negative answers. For the present discussion, however, I shall assume that some rudimentary form of inductive reasoning is legitimate. Third, there is the question of scientific realism. As I said at the outset, Reichenbach claimed that he had an affirmative answer to this question, but we shall have to see whether this claim can be sustained. Whether realism is compatible with empiricism is the main issue to be addressed in this chapter. This issue takes on considerable contemporary importance in connection with Bas C. van Fraassen's (1980) constructive empiricism as articulated in *The Scientific Image*. Van Fraassen is quite sincere (pace Russell) in advocating empiricism, but the price he feels required to pay is complete agnosticism regarding the existence of unobservable entities. The scorecard for empiricism is given in table 1.1:

It will be noted that Reichenbach is the only philosopher among those mentioned who gives an unequivocal affirmative answer on all three items, and we have already found reason to wonder whether he actually made good on these claims.

2. Posing the Question

I have refrained, so far, from giving any explicit characterization of empiricism; it is time to rectify that situation. Traditionally, empiricism has come in two forms—concept and statement. In maintaining that all of our ideas must originate in experience, Locke, Berkeley, and Hume upheld the doctrine of concept empiricism.

Operationism is, I think, the twentieth-century version of concept empiricism. In contrast, the logical empiricists have advocated one form or another of statement empiricism. Each form of empiricism has an associated criterion of cognitive meaningfulness. Hume's doctrine that ideas must be analyzed in terms of impressions, and the operationist principle that every scientific term must be operationally definable, are expressions of concept empiricism. The verifiability

TABLE 1.1 Items Admissible in an Empiricist Framework

	Yes	No
Ordinary, middle-sized physical objects	Carnap Hempel Popper Reichenbach Van Fraassen	Russell
Rudimentary induction	Reichenbach Van Fraassen	Carnap Popper Russell
Unobservable objects	Hempel Reichenbach	Russell Van Fraassen

criterion—according to which factual statements are cognitively meaningful only if they are in principle confirmable or disconfirmable on the basis of observational evidence—reflects statement empiricism. Let me hasten to add that it is *not* my intent to argue for or against criteria of cognitive meaning. I have mentioned criteria of the two types only because they serve to draw a useful contrast between the two basic forms of empiricism.

Since my chief concern in this chapter is with logical empiricism, I shall focus primarily on statement empiricism. According to this doctrine, the sole basis upon which we are justified in affirming or denying factual statements is observational evidence. (This doctrine is tantamount, I believe, to the denial that there are synthetic a priori propositions.) Thus, observation itself provides evidence about observed properties of observed entities. Statements about other kinds of properties and entities can be supported or undermined by observational evidence in conjunction with suitable forms of inference. In particular, assuming for the sake of argument (as I am doing and as van Fraassen does) that some rudimentary form of induction is available, we can support statements (including empirical generalizations) about observable properties of observable entities. With the aid of deduction we can support or undermine other statements about observable properties of observable entities. I hope this much will be granted for purposes of the present discussion.

The problem of statements about unobservable entities and/or properties poses greater difficulties. Assuming that unobservables are not—to use Russell's term—logical constructions from observables, it seems evident that observational evidence in conjunction with deductive logic alone cannot support or undermine statements about unobservables. This result follows directly from the nonampliative character of deduction. Therefore, the question for empiricism—the *key question* in my view—is whether inductive reasoning contains the resources to enable us to have observational evidence for or against statements about unobservable entities and/or properties. An affirmative answer to this question would allow us to maintain that empiricism is compatible with scientific realism; a negative answer would lead to the conclusion that they are incompatible. This is the issue to which the present chapter is devoted.

The most surprising thing about the *key question* is how seldom it has been raised explicitly. I should like to present three examples of cases in which the failure to consider it (or some closely related question) has had serious philosophical consequences. The first and most recent occurs in van Fraassen's (1980) development of constructive empiricism. In the opening sentence he says, "The opposition between empiricism and realism is old, and can be introduced by illustrations from many episodes in the history of philosophy" (1). It appears to be a presupposition of his entire discussion that realism and empiricism are incompatible; nothing said later in the book changes that impression.

In clarifying what he means by constructive empiricism, van Fraassen (1980) states, "Empiricism requires theories only to give a true account *of what is observable*, counting further postulated structure as a means to that end. So from an empiricist point of view, to serve the aims of science, the postulates need not be true, except in what they say about what is actual [as opposed to possible] and empirically attestable" (3; van Fraassen's italics). A little later he continues, "On the view that I shall develop, the belief involved in accepting a scientific theory is only that it 'saves the phenomena,' that is, correctly describes what is observable" (4). In summary, he says, "I shall give a momentary name, 'constructive empiricism,' to the specific philosophical position I shall advocate. I use the adjective 'constructive' to indicate my view that scientific activity is one of construction rather than discovery: construction of models that must be adequate to the phenomena, and not discovery of truth concerning the unobservable" (5).

It is, in my opinion, quite easy to read van Fraassen's constructive empiricism as a version of concept empiricism—that is, as holding that only what can be described in observational terms can be established scientifically.[2] He has, however, informed me that he considers this a misinterpretation; he is primarily concerned with the kinds of statements that can be accepted, supported by evidence, or believed to be true. Since he never raises the *key question* explicitly, it is natural to conclude that he has already arrived at an implicit answer. This answer seems to be that induction is merely the inverse of deduction—that is, that the only inductive empirical evidence for a theory consists in its observable consequences. This is, I believe, an inadequate conception of inductive inference.

My second example involves A. J. Ayer's (1946) attempts to formulate a verifiability criterion of cognitive meaningfulness. As is well known, these attempts came to grief upon an elegant refutation by Alonzo Church (1949). This episode played an important role in the general demise of meaning criteria (see Salmon 1966). The story is beautifully recounted by Hempel (1951), whose account may well have constituted the coup de grâce. Without attempting to resuscitate these criteria, I would point out that in none of Ayer's versions of the criterion does he even mention induction. His final, and most carefully formulated version (13), treats the cognitive meaningfulness of a statement entirely in terms of its deductive consequences, not in terms of potential inductive evidence. Again, the view that induction is merely the inverse of deduction seems to be implicit.

My third example comes from Hempel's (1965a) classic survey of arguments and views pertaining to the realism/instrumentalism issue. After careful analysis, he concluded:

> Our argument (5.1), the theoretician's dilemma, took it to be the sole purpose of a theory to establish deductive connections among observation sentences. If this were the case, theoretical terms would indeed be unnecessary. But if it is recognized that a satisfactory theory should provide possibilities also for inductive explanatory and predictive use and that it should achieve systematic economy and heuristic fertility, then it is clear that theoretical formulations cannnot be replaced by expressions in terms of observables only; the theoretician's dilemma, with its conclusion to the contrary, is seen to rest on a false premise. (222).

It strikes me as odd that nowhere in the entire discussion does he address the *key question*—though he comes closer to doing so in his 1963 work (700–701). I am not convinced by Hempel's arguments that theories are *required* for "inductive explanatory and predictive use," though they undoubtedly facilitate such activities. Until we look much more closely at the power of inductive logic and confirmation theory, I do not think we can decide whether they are logically indispensable.

3. Answering the Question

To introduce his discussion of the opposition between empiricism and realism, van Fraassen refers to "the sense of philosophical superiority the participants in the early development of modern science felt toward the Aristotelian tradition"—a tradition that offered explanations of natural phenomena in terms of substantial forms, causal properties, and occult qualities. Such qualities, exemplified by the "dormitive virtue," were satirized by Molière. According to a fairly popular view, modern science succeeded by purging science of occult qualities, thereby rendering it thoroughly empirical. In fact—as van Fraassen clearly realizes—no such thing happened. The actual development is nicely recounted by Keith Hutchison (1982) in "What Happened to Occult Qualities in the Scientific Revolution?" The answer, briefly, is that in the seventeenth century "occult qualities became fully and consciously accepted in natural philosophy..." (223).

To understand what happened, we have to pay careful attention to the meaning of the word "occult." In the medieval tradition, "occult" was the antonym of "manifest," which meant "sensible." Thus, "occult" meant merely "unobservable." At that time there was a strongly held view that God gave humans adequate sensory apparatus to enable them to observe everything they needed to know about, so the occult was also mysterious. Angels or demons were invoked to explain the action of occult qualities—for example, the curative powers of Christian relics. By the time such mechanical philosophers as Robert Boyle condemned occult qualities in the seventeenth century, the term "occult" had shifted its meaning; in that context it meant "incomprehensible" rather than "insensible." This very point is brought out by the passage van Fraassen quotes from Boyle:

> That which I chiefly aim at, is to make it possible to you by experiments, that almost all sorts of qualities, most of which have been by the schools either left unexplicated, or generally referred to I know not what incomprehensible substantial forms, may be produced mechanically; I mean by such corporeal agents as do not appear either to work otherwise than by virtue of the motion, size, figure, and contrivance of their own parts (which attributes I call the mechanical affections of matter). (van Fraassen 1980, 1; quoted from Boyle [1772] 1965–1966, vol. 3, p. 13)

Two remarks are in order. First, the parts to which Boyle refers are atoms; neither they nor their properties are observable. In the medieval meaning of the term, they are totally occult. Second, the properties to which Boyle refers are primary qualities. These are the qualities that account for the fact that these bodies produce secondary qualities for us. In the medieval terminology, secondary qualities are manifest,[3] whereas primary qualities are occult. Late medieval science had a narrowly empiricist character; indeed, as Edward Grant (1962) has persuasively argued, instrumentalism became the dominant view of science after the condemnation of 1277. The mechanical philosophers of the seventeenth century rejected the narrow empiricism of the medieval period and adopted scientific realism. Grant maintains that the major achievement of the Copernican revolution was the switch to the realistic viewpoint. The scientists of the seventeenth century made no effort to exclude *unobservable* entities and properties; they were convinced that they could have scientific knowledge of certain kinds of unobservable entities and properties. They did, of course, attempt to purge their sciences of ad hoc postulations and *incomprehensible* things.

There are two ways in which scientists might go about getting knowledge of unobservables—one direct, the other indirect. The direct method involves the use of instruments as extensions of the normal human senses; the most obvious examples are telescopes and microscopes. It is historically notable that both kinds of instruments were available before the end of the seventeenth century. Around 1610, Galileo used the telescope to discover such phenomena as the phases of Venus and the moons of Jupiter, which are unobservable to normal humans located on the surface of the earth. They are, of course, observable by people using their unaided senses if they are situated at suitable places. For this reason, van Fraassen does not classify telescopically discovered entities like the moons of Jupiter as unobservables.

The case of the microscope is fundamentally different; there is no place to go to make direct observations of microscopic entities. Single magnifying lenses were apparently used in ancient and medieval times, but the first multiple-lens microscope seems to have been made in the sixteenth century. In 1590, Zacharias Jannsen made a simple two-lens microscope. Anton van Leeuwenhoek, who is sometimes credited with the invention of the microscope in the seventeenth century, apparently used only single lenses for magnification. Robert Hooke made an early compound microscope. However, although Galileo's telescopic discoveries in the seventeenth century had cardinal scientific importance, it seems that no serious scientific use of the microscope occurred before the beginning of the nineteenth century (see Hacking 1981a). Nevertheless, I believe, with microscopes of various kinds it is possible to gain knowledge of unobservable entities.

Although it may be acceptable in most contexts to speak of seeing things with a telescope or a microscope, it seems best for the purposes of the present discussion to say that what we see when we look into a telescope or a microscope enables us legitimately to infer something about objects that are not being observed. This direct method of getting knowledge of unobserved—and possibly unobservable—entities and properties does *not* constitute direct observation. Galileo's critics, who refused to look through his telescope, may not have been utterly perverse. They could argue with some plausibility that what Galileo took to be moons of Jupiter might be artifacts of the instrument. One familiar use of lenses at that time was as carnival devices to produce illusions—much as distorting mirrors are used today for the same purpose (see Ronchi 1967). It would not be silly to wonder how a combination of several such tools of deception could disclose realities that were, at that time, contrary to accepted theory and unobservable by any other means. According to Hacking (1981a), the problem of artifacts of the instrument was far more troublesome in the case of the microscope than it was for the telescope, which may explain in part why its scientific use came so late. We need to examine the kind of inductive inference that might be used to support conclusions about entities and properties investigated with the aid of microscopes.

When David Hume ([1739–1740] 1888, III, 1) stated that all of our reasonings about unobserved matters of fact are based upon the relation of cause and effect, he was, I suspect, almost completely correct. One notable exception is induction by simple enumeration (or some other very primitive form of induction). As remarked above, I am assuming for this discussion that some sort of primitive induction is available; I shall not attempt to characterize it or justify it in this context. The type of argument required in connection with microscopic observation is, I think, causal. It is a rather special variety of causal inference that is also analogical. This particular sort of causal/analogical argument is, in my view, quite powerful. To employ it, we must assume that we already have knowledge of cause-effect relations among observables—for example, that hitting one's thumb with a hammer causes pain, that drinking water quenches thirst, and that flipping the switch turns off the light. Such relationships can be established by Mill's methods and controlled experiments. Neither the instrumentalist's nor the constructive empiricist's account of science can get along without admitting knowledge of such relations. Using relations of this sort, we can now schematize what I suspect is the basic argument that enables us to bridge the gap between the observable and the unobservable. It goes something like this:

We observe that

An effect of type E_1 is produced by a cause of type C_1.
An effect of type E_2 is produced by a cause of type C_2.

.

.

.

An effect of type E_k occurred.

We conclude (inductively) that

A cause of type C_k produced this effect of type E_k.

The particular application of this argument that interests us is the case in which $C_1, C_2, \ldots C_k$ are similar in most respects except size. Under these circumstances we conclude that they are similar in causal efficacy.

It should be noted in passing that our everyday experience reveals the fact that inferences from effects to causes are especially strong. From a barking sound we reliably infer the presence of a dog behind a fence. From the charred wood in a pit we infer the existence of a previous fire. From tracks in the soil and other signs we infer that a deer has passed this way. The point is reinforced when we note that effects are often considered records of their causes. Records have great inferential utility, as Reichenbach (1956, chap. IV) explained in detail.

For application to the problem of our main concern, consider a simple instance of this kind of causal/analogical argument. As I open the *Compact Edition of the Oxford English Dictionary* I immediately notice words that I can read with ease, others that I can read only with some difficulty, and others that I cannot read at all without magnification. When I use the magnifying glass that comes with the books, I see that the largest characters appear even larger. Characters that are small and hard to discriminate when viewed without the glass are clear and distinct when viewed through it. Characters that are indiscriminable without the glass can be read with its aid. And a mark of punctuation invisible to the naked eye can be seen at an appropriate place when the glass is used. I conclude that there are black dots and other marks on the pages that are invisible to my naked eye. I have inferred the existence and nature of an unobservable object, and I think that the inference is strong. A similar example involving use of a low-power microscope could easily be constructed.

The foregoing argument regarding the magnifying glass and the microscope can obviously be supplemented by a similar argument involving binoculars and telescopes. With binoculars we can see details of objects (e.g., birds) that cannot be seen with the naked eye, given the relative positions of the bird and the bird watcher. We can, pace Hacking, net such birds and examine them at very close range. Many terrestrial observations can be made with telescopes to reinforce the point even more. Taking a wide variety of such observational evidence, as well as observations of refraction of light rays by glass (Snell's law was established empirically early in the seventeenth century), we can develop geometrical optics, which furnishes a *theory* of the behavior of such instruments of observation. If that theory is correct it explains the phenomena we have been describing and it provides scientific justification for inferences to the existence and properties of objects unobservable to the naked eye. When such a theory is accepted it becomes natural to speak of observations by means of such instruments.

The question that now arises concerns the logical status of arguments from analogy. Instead of attempting to give a completely general answer, let me cite what I consider an extremely important scientific use of such arguments. In my opinion, there is such a thing as scientific confirmation, and Bayes's theorem furnishes the fundamental schema to which it must conform (see chapters 4–6). This view is held not only by so-called Bayesians; both Carnap and Reichenbach also embraced it. It is embedded in Carnap's (1950b, sec. 55) basic definition of "degree of confirmation." Reichenbach ([1935] 1949, 432–433) explicitly advocates this conception of confirmation. One major point on which all of us agree is that prior probabilities are indispensable. The major point of disagreement is the status of the priors. Most theorists who adopt the label "Bayesian" take them to be subjective or personal probabilities. At the most basic level Carnap considered them a priori probabilities. Reichenbach (and I) construe them as objective probabilities whose values may be roughly estimated from empirical evidence. Whatever stand one takes on this question of philosophical interpretation of probability, it seems that analogies often serve to supply information about prior probabilities in the form of plausibility considerations. Let me offer one example each from the physical, biological, and social sciences. (1) The fact that Newton's law of gravitation and Coulomb's law of electrostatic attraction and repulsion both involve inverse square relationships constitutes a fundamental analogy between the two theories. The success of Newton's theory lends plausibility to Coulomb's theory. (2) Studies of the effects of chemicals (e.g., saccharin) on animals (e.g., rats) lend plausibility to the supposition that the same chemical will have similar effects (enhanced propensity for bladder cancer) on humans. (3) Ethnographic analogies are used by archaeologists investigating prehistoric cultures. By finding the use to which a certain artifact is put in an extant primitive culture, one can make plausible hypotheses about its use in a similar primitive prehistoric culture. Analogical arguments are frequently used in the sciences to support plausibility claims—that is, claims pertinent to the estimation of prior probabilities. In a similar fashion, I want to argue, the sort of analogical argument I sketched above can furnish enough of a plausibility consideration to allow the hypothesis that microscopes furnish information about unobservable objects and properties to get off the ground. Physical theories about the behavior of microscopes can then be confirmed by concrete experimental evidence.

The theory of microscopy is discussed in fascinating detail by Ian Hacking (1981) in "Do We See Through a Microscope?" In considering the observation of dense bodies in red blood cells, Hacking mentions a device known as a micro-scopic grid. "Slices of a red blood cell are fixed upon a microscopic grid. This is literally a grid: when seen through a microscope one sees a grid each of whose squares is labeled with a capital letter" (315). He then proceeds to use the grid to address the issue raised by van Fraassen:

> I now venture a philosopher's aside on the topic of scientific realism. Van Fraassen says we can see through a telescope because although we need the telescope to see the moons of Jupiter when we are positioned on earth, we could go out there and look at the moons with the naked eye. Perhaps that fantasy is close to fulfillment, but it is still science fiction. The microscopist avoids fantasy. Instead of flying to Jupiter he shrinks the visible world. Consider the grid that we used for re-identifying dense bodies. The tiny grids are made of metal; they are barely visible to the naked eye. They are made by drawing a very large grid with a pen and ink. Letters are neatly inscribed by a draftsman at the corner of each square on the grid. Then the grid is reduced photographically. Using what are now standard tech-niques, metal is deposited on the resulting micrograph.... The procedures for making such grids are entirely well understood, and as reliable as any other high quality mass production system.
>
> In short, rather than disporting ourselves to Jupiter in an imaginary space ship, we are routinely shrinking a grid. Then we look at the tiny disc through any kind of microscope and see exactly the same shapes and letters as were drawn in the large by the first draftsman. It is impossible seriously to entertain the thought that the minute disc, which I am holding by a pair of tweezers, does not in fact have the structure of a labeled grid. I know that what I see through the microscope is veridical because we *made* the grid to be just that way. (316–317)

Such considerations provide strong confirmation of the theory of the instrument and in consequence to the claim that microscopes inform us about things too small to be observed with unaided human senses.

I distinguished above between two ways of gaining knowledge of the unobservable—direct and indirect—a distinction that is by no means sharp but nevertheless useful. The direct way, it will be recalled, involves the use of in-struments of observation, such as microscopes, as extensions of human sense or-gans. The indirect involves more roundabout theorizing; it can be illustrated historically. Early in the nineteenth century, the botanist Robert Brown observed the random dance of microscopic pollen particles suspended in a fluid, which is now known as Brownian movement. If I am correct in surmising that the microscope was not put to significant scientific use before 1800, then this observation constitutes an early application of first importance. Shortly after the beginning of the twentieth century, Albert Einstein [1905] 1989 published his well-known explanation of Brownian motion, and Jean Perrin initiated his masterful ascertainment of Avogadro's number through the study of Brownian motion. As Mary Jo Nye (1972) recounts in *Molecular Reality*, the culmination of Perrin's work convinced the community of physical scientists (with the notable exception of Ernst Mach) of the reality of atoms and molecules by about 1912. Perrin ([1913] 1916) provided a semipopular account of his work in *Atoms*. In that book he lists thirteen distinct methods of ascertaining Avogadro's number, all of which yield values in close agreement with one another. Perrin, himself, emphasizes the fact that it is the close agreement of results of many independent methods of ascertainment that constitutes the compelling evidence for the existence of atoms and molecules. I believe that Perrin's argument is perfectly sound and that it shows how evidence of observation can be used to make inferences about unobservable entities. In my view, this argument can appropriately be schematized as a type of common cause argument (see chapter 3, "An Empiricist Argument for Realism").

It is noteworthy that Hacking's (1981) example of the microscopic grid furnishes exactly the same sort of argument:

> I know that the process of manufacture [of the grid] is reliable, because we can check the results with any kind of microscope, using any of a dozen unrelated physical processes to produce an image. Can we entertain the possibility that, all the same, this is some kind of gigantic coincidence[?] Is it false that the disc is, in fine, in the shape of a labeled grid? Is it a gigantic conspiracy of 13 totally unrelated physical processes [recall Perrin's table of thirteen ways of ascertaining Avogadro's number] that the large scale grid was shrunk into some non-grid which when viewed using 12 different kinds of microscopes still looks like a grid? (317)

Although Reichenbach (1938) did not say so explicitly, it appears that his main argument for scientific realism against the positivists in *Experience and Prediction* (sec. 14) was a common cause argument. That argument, it will be recalled, involved an observer confined within a cubical world. Birdlike shadows appeared on the translucent ceiling and walls. There was a perfect correlation between the behavior of the shadows on the ceiling and walls. The existence of a common cause was inferred. In this way Reichenbach thought he could make a probabilistic inductive inference to unobservables (120–121). I do not think that Reichenbach's argument stands up well to critical scrutiny; indeed, for many years I was unable to make out what sort of argument he thought he had given. In addition, it does not seem to square with his discussion of the conventional character of "extension rules of language" in the context of quantum mechanics (1946, sec. 5; see also 1951, 55–58). When, however, we go from Reichenbach's oversimplified fictitious example to Perrin's actual case we find a strong inductive argument for the existence of unobservables.

In examining the argument from analogy, it was necessary to face directly the question of its status within inductive logic. The same question arises in connection with common cause arguments. The answer is, in a word, success. One can cite many examples of common cause arguments involving only observables in which the principle of inference leads to correct conclusions. The fact that a man and a woman, who habitually eat lunch in an apparently random way in a variety of restaurants, nevertheless invariably eat in the same restaurant each day, demands explanation in terms of prior arrangement. If microscopically observable entities are allowed, many more examples—such as simultaneous illness among dormitory students who have eaten from the same infected food—can be added. Perhaps the best evidence for the compelling character of common cause arguments is the dismay we feel over the real possibility that they may break down in some quantum mechanical contexts; Bernard d'Espagnat (1979) discusses this point explicitly.

The logical basis for this kind of appeal to the success of common cause arguments is quite straightforward. On the basis of our experience we can make an inductive generalization to the effect that such inferences provide correct conclusions in a large majority of cases. Such inductive justification is obviously not circular, for it appeals to primary induction in support of a secondary or derivative type of inductive inference. The fact that this sort of inference may not work well in quantum mechanics shows merely that the generalizations must be qualified.[4] This is no ground for objection; inductive generalizations often need qualification. All swans are white provided they are not Australian. Appeals to the common cause argument are legitimate, it seems to me, both in the context of Hacking's discussion of the microscope and in Perrin's treatment of molecular reality.

4. Conclusions

The frontispiece of Herbert Feigl's collected essays, *Inquiries and Provocations* (Cohen 1981), is a beautiful photograph of Feigl and Hempel seated side by side on a couch. This heartwarming picture bears the disconcerting caption "the two last empiricists," which is, in all likelihood, inaccurate. Surely the odds heavily favor van Fraassen's surviving both of them [*Editor's note*: at the time this was written, all three were alive], and I trust he is not the only younger empiricist. Perhaps it should read "the two last empiricist/realists," but I fervently hope that that too would be inaccurate. To find that empiricism and realism are genuinely incompatible would be a bitter philosophical pill. It is, nevertheless, a pill we might have to swallow.

The chief aim of this chapter has been to challenge the incompatibility thesis by calling attention to the key question: Does inductive logic contain the resources to enable us to have observational evidence for or against statements about unobservables? An affirmative answer to this question would enable empiricists to be realists. I have tried to argue for the affirmative answer. Even if my arguments are unacceptable, it would be a profound philosophical error to ignore the question. It may turn out that the answer is actually negative—for instance, van Fraassen has maintained (in private correspondence) that no inductive logic of the sort I seem to be discussing even exists[5]—and that we will have to admit that empiricism and realism are incompatible. However, I do not think that we should accept this unpalatable result without even asking the key question. That is the main point of this chapter.

Scientific Realism in the Empiricist Tradition

Contemporary empiricism owes a significant debt to the logical empiricists Carnap, Hempel, and Reichenbach. Their contribution deserves careful consideration. All three were born in Germany, and each attended several major German universities. Although they began their work in philosophy at a time when metaphysics was rampant, they studied with some of the greatest mathematicians and scientists of that time. They had both deep knowledge and profound appreciation of the substantive content of mathematics and empirical science. They helped to lay the foundation and to erect the superstructure of what we can now recognize in retrospect as twentieth-century scientific philosophy. This movement grew chiefly out of the logical positivism of the Vienna Circle (with which Rudolf Carnap was mainly associated) and the logical empiricism of the Berlin Group (with which Hans Reichenbach was connected). Carl Hempel worked closely with both groups. What all had in common was a devotion to the aims and methods of empirical science and an appreciation of the techniques of mathematical logic. By midcentury virtually every important figure in the movement had relinquished some of the more extreme views of early logical positivism, abandoned the designation of "logical positivist," and adopted "logical empiricism" as the name of their movement. After that, logical positivism no longer existed.

It would be folly for anyone to try to say what constitutes the most important intellectual accomplishment of twentieth-century scientific philosophy, for its achievements are many. But let me mention one that appeals strongly to me; it concerns the nature of scientific understanding. At the beginning of the twentieth-century the view was widely held—by philosophers, scientists, and many other people—that science cannot provide an understanding of our world or what occurs within it. Science can describe and predict these phenomena, and even organize our knowledge of them, but it cannot explain them; it cannot say *why* they occur. To answer such why-questions, it was often said, one must appeal to metaphysics or theology. The empirical sciences, in and of themselves, cannot give us "true understanding."

That situation has changed dramatically. Today there is a virtual consensus — at least among philosophers of science, but also among scientists, I think — that science *can* provide bona fide explanations of natural phenomena. Consider a couple of famous historical examples. Brownian movement was discovered early in the nineteenth century, but it took many decades to learn why it occurs. In 1905 Einstein and Smoluchowski provided a theoretical explanation in terms of molecular collisions that still seems fundamentally sound. As a result of their theory, which was brilliantly confirmed experimentally shortly thereafter by Jean Perrin, we now *understand why* microscopic particles suspended in fluids do their random dances. Similarly, Lord Rutherford's nuclear model of the atom explained the large-angle scattering of alpha particles from gold foil, though it left us with a profound problem of explaining the stability of the hydrogen atom.

Neither Carnap nor Reichenbach wrote much on scientific explanation, though it seems clear that both believed in the power of science to provide explanations.[1] Of the three, it was Hempel who devoted substantial efforts to the clarification of that concept. "Studies in the Logic of Explanation" (Hempel and Oppenheim 1948) set forth with great precision what later came to be called "the deductive-nomological model" of scientific explanation. This epoch-making work is a veritable fountain-head from which almost everything published subsequently on the subject is either directly or indirectly derived. It constitutes the dividing line between the history and the prehistory of philosophical discussion of the nature of scientific explanation. Although Karl Popper, R. B. Braithwaite, and Ernest Nagel also made important early contributions, it was to Hempel's work that discussion gravitated. Moreover, Hempel made a number of further contributions, including the articulation of models of statistical explanation, which advanced the topic enormously. His magisterial article "Aspects of Scientific Explanation" (1965a) is the benchmark to which all other work on the subject must be compared.[2]

Scientific explanation is not, however, the chief topic of this chapter. It is really just the point of departure. Both of the above-mentioned examples of explanation, it will be recalled, referred to unobservable entities. The Brownian particles cannot be seen with the naked eye, though they can be viewed with an ordinary optical microscope. But the molecules, whose bombardment of the Brownian particles causes their movement, cannot be seen with any optical microscope. The same can be said of Rutherford's nuclear atom and the alpha particles with which he worked. Without intolerably stretching the meanings of terms, one cannot say that simple molecules, atoms, and subatomic particles are observable. They are entities postulated by theories that seem to have a great deal of explanatory power. These theories appear to be confirmed by an enormous body of empirical evidence.

Although John Dalton's atomic theory was set forth near the beginning of the nineteenth century, it achieved nothing approaching universal assent for approximately 100 years. Around the turn of the twentieth century there were extremely distinguished physical scientists who were skeptical about the existence of such unobservable entities. Ernst Mach was the most famous member of this group, but it also included Wilhelm Ostwald. Ostwald changed his mind in 1908 and accepted the atomic theory, but it seems that Mach never did.[3]

Many members of the Vienna Circle were strongly influenced by Mach, as were Albert Einstein and Bertrand Russell. Early in his career Carnap was greatly impressed by Whitehead and Russell's *Principia Mathematica* (1910–1912) and by what Russell called "the supreme maxim of scientific philosophizing": *Wherever possible, logical constructions are to be substituted for inferred entities.* This motto stood at the beginning of Carnap's (1928a) first major book, *Der Logische Aufbau der Welt*, in which he tried to show how both the world of physics and the world of psychology could be constructed solely on the basis of one individual's immediate experiences. He does not, however, mandate that particular construction; he allows that constructions erected on different bases could be admissible, although the phenomenalistic basis has epistemological priority.

In a later article, "Empiricism, Semantics, and Ontology," Carnap (1950a) discussed the existence of various kinds of entities, mainly abstract entities in logic and mathematics. In such contexts, he says, one adopts a linguistic framework. Once such a framework has been adopted, questions about the reality of numbers, sets, propositions, and so on can be construed in either of two ways—as *internal* or as *external* questions. Internal questions can be answered by reference to the framework. In Peano arithmetic, for example, natural numbers exist, and so do prime numbers. In Zermelo-Fraenkel set theory, sets exist. If, however, one asks whether such entities "really exist"—that is, if one raises the external question of whether the framework itself is justified by the existence of such entities—Carnap rejects it as a pseudoquestion. What matters is the utility and efficiency of the linguistic framework regarded as a tool. The traditional questions of ontology simply do not arise. This is clearly an instrumentalist approach to logic and mathematics, but in a couple of passing remarks he indicates that the same considerations apply to theories of empirical science. Within a Daltonian framework, for example, atoms exist; within a Machian framework they do not. There is no question of which framework is correct or true; the only question is one of utility. Although he made some confusing remarks in a subsequent book, *Philosophical Foundations of Physics* (1966a),[4] it seems clear to me that Carnap never abandoned his 1950 view on the debate between scientific realism and instrumentalism.[5]

Reichenbach's attitude toward scientific realism seems strongly opposed to that of Carnap. Reichenbach's (1938) major epistemological treatise, *Experience and Prediction*, was written and published in English; only recently was it first translated into German for publication in the Vieweg edition of his collected works (Reichenbach 1968–, band 4). He considered it his refutation of logical positivism. It appears to be addressed to Carnap's position in the *Aufbau*. In an earlier review of the *Aufbau* (Reichenbach 1933), he praised Carnap's work highly, but he also criticized it because probability found no place in it. To Reichenbach, probability played *the key role* in the rejection of logical positivism. For example, he formulated the criterion of cognitive meaningfulness in terms of probabilistic verification in place of the absolute verification of the early logical positivists.

In characterizing the relationship between empirical experience and the world of science, Reichenbach (1938) distinguished two types of construction, one of which he called "reductive," the other "projective." A brick wall, for example, is a reductive construction from the bricks; the wall consists of nothing but the bricks (and, I suppose we should add, mortar) placed in a certain spatial arrangement. If the bricks are destroyed, the wall cannot exist; it has no existence independent of the bricks. The reductive construction was the type used by Carnap in the *Aufbau*. It involves, basically, a *definitional* relationship.

A projective construction, according to Reichenbach (1938), involves a *probability* relationship. His chief illustration is a fanciful one, namely, a cubical world with translucent walls. Outside of the walls are a number of birds. They are illuminated by two sets of lights; an arrangement of mirrors makes them cast shadows both on the ceiling and on one of the walls. The inhabitants of this world can observe these shadows, but it is physically impossible for them to penetrate the walls and escape from the cube. They cannot see the birds directly. Through patient observation of the shadows, however, they discover certain correspondences between the shadows in the two sets. They might note that one shadow on the ceiling has a particularly short neck; they can find a shadow on the wall with a similarly short neck. When they observe a shadow on the wall pecking at another and removing some tail feathers, they see similar pecking behavior in a pair of shadows on the

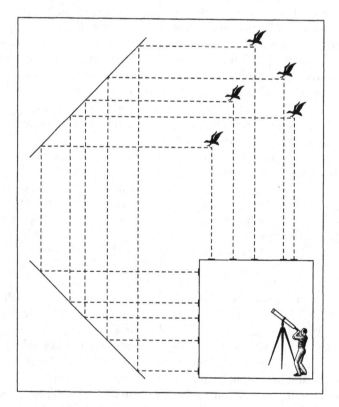

FIGURE 2.1 Reprinted with permission from Hans Rechenbach, *Experience and Prediction*, copyright 1938 by The University of Chicago.

ceiling. Reichenbach refers to the discoverer of these correlations as Copernicus but, rather anachronistically, endows him with a telescope.

Reichenbach (1938) asks what inferences inhabitants of this cubical world would make on the basis of their observations. A positivist in this world, adopting a reductive analysis, would say that the shadows and the observable regularities that they exhibit exhaust physical reality for this fictitious world. Of course, we know better because Reichenbach has revealed to us the external reality consisting of the birds, the mirrors, and the highly contrived lighting arrangements. We know that the positivist answer must be rejected, but what of the people within the cubical world? The physicists, according to Reichenbach, will not accept these correlations as mere coincidences. It can happen by pure chance that, even if the shadows on the wall are physically independent of those on the ceiling, there will occasionally be a case of pecking behavior on the ceiling that corresponds exactly to a simultaneous case of pecking behavior on the wall. When, however, such coincidences occur repeatedly very many times, the physicist will infer *on a probabilistic basis* that there is a physical connection between the two sets of shadows.

At this point it is natural to ask by what argument the physicist comes to this probabilistic conclusion, but when we look in *Experience and Prediction*, we find at best only a few hints. Perhaps the most explicit is this:

> He [our physicist] simply states that, whenever he observed simultaneous changes in dark spots like these [bird shadows], there was a third body different from the spots; the changes happened, then, in the third body and were projected by light rays to the dark spots which he used to call shadow-figures. Freed from all associated representations his inference has this form: Whenever there were corresponding shadow-figures like the spots on the screen, there was in addition, a third body with independent existence; it is therefore highly probable that there is also such a third body in the case in question. It is this probability inference which furnishes a different weight for the projective complex and the reducible complex. (1938, 123)

It is perhaps ironic that in 1935, three years before the publication of *Experience and Prediction*, Einstein, Podolsky, and Rosen published the famous "Can Quantum-Mechanical Description of Physical Reality Be Considered Complete?" It led to profound problems concerning remote correlations. Since, however, these quantum mechanical problems are not raised in *Experience and Prediction*, let us avoid going off on that tangent.[6] Let us return to the context of *Experience and Prediction* and pursue Reichenbach's argument there.

The kind of argument Reichenbach attributes to the physicists in his cubical world can hardly do the job of supporting scientific realism. It may be that physicists in the cubical world often set up screens, lights, and mirrors in various configurations and studied the ways in which shadows were projected by *visible* objects. The question, however, involves an inference to *unobservables*. It is all well and good to make an inference from observed shadows to *unobserved* birds, but that does not answer the question about *unobservables*. One might say that in this case the observables and unobservables are so similar that the probability argument is good enough as it stands. This was *not* Reichenbach's view. It has been quite evident since 1913—the year in which Bohr proposed his famous model of the hydrogen atom—at the very latest that the microcosm is radically different from the macrocosm. *Experience and Prediction* was written long after 1925, when Heisenberg and Schrödinger, in founding quantum theory, gave even stronger support for this radical difference.

Indeed, it is crucial to Reichenbach's (1938) main argument that probabilistic reasoning can take us into *different* realms:

> The probability theory of meaning...allows us to maintain propositions as meaningful which concern facts outside of the domain of the immediately verifiable facts; it allows us to pass beyond the domain of the given facts. *This overreaching character of probability inferences is the basic method of the knowledge of nature.* (127; my emphasis)

Moreover, his critique of logical positivism hinges on this claim:

> It is the neglect of the overreaching character of the probability inferences which leads positivists to their equivalence theory [reductive analysis]...they do not see that the probability inference passes beyond the given observations. *This error about the logical nature of the probability inference is the root of the positivistic doctrine of existence.* (130; my emphasis)

We see, then, that Reichenbach places great stress on what he calls the "overreaching character" of probabilistic reasoning, but he tells us very little about how this is accomplished. In fact, from Reichenbach's general characterization of his theory of probability and induction, as given in the final chapter of *Experience and Prediction*, it is by no means obvious how probabilistic reasoning can achieve his goal. The two fundamental features of his theory are, first, that the limiting frequency interpretation is the only legitimate interpretation of the calculus of probability and, second, that his *rule of induction* (induction by enumeration) is the only rule of nondemonstrative inference. The rule of induction permits us to make posits, for example, of the limiting frequency with which a die will show side six if tossed in a certain way, on the basis of the observed frequency with which that side has come up in an observed initial section of a sequence of such tosses. It is extremely difficult to see how such reasoning can extend our knowledge from the macrocosm to the microcosm. Only in a footnote headed "Remark for the mathematician" does he give us *any* hint of the details (1938, 124). And all he says there is that the reasoning of the physicist is based on Bayes's theorem. He does not tell us how. In *Experience and Prediction* he does not even write out Bayes's theorem, let alone explain how the symbols that occur within it are to be interpreted. He insists, nevertheless, that the invocation of probability is essential to our understanding of nature.

Experience and Prediction contains many references to Reichenbach's (1935) *Wahrscheinlichkeitslehre*, but even in the second English edition, *The Theory of Probability* ([1935]1949)—to which he added discussions of the probability of theories not present in the first German edition—it is hard to see how an appeal to Bayes's theorem will do the trick, especially in view of his commitment to the limiting frequency interpretation of the probability calculus. Although he assures us that as science progresses we will have as reliable statistics on success rates of various types of scientific hypotheses—as good as an insurance company's statistics on death rates for given classes of people—the last forty years of scientific experience does not seem to bear out that prediction.

At the time of his death in 1953, Reichenbach was working on a major book, *The Direction of Time*, which was published posthumously in 1956. There he articulated a fundamental principle, which he called "the principle of the common cause": "if an improbable coincidence has occurred, there must exist a common cause" (157). He recognizes that this statement is too strong, so he proceeds to modify it: "Chance coincidences, of course, are not impossible. . . . The existence of a common cause is therefore . . . not absolutely certain, but only probable. This probability is greatly increased if coincidences occur repeatedly" (157–158). Reichenbach explicates the structure involved in the common cause situation in terms of a set of statistical relations that he called a *conjunctive fork* (159).

Reichenbach offers a number of commonsense examples. Both of the lights in a room go off simultaneously, even though they have no common switch. It is possible that the bulbs just happened to burn out simultaneously, but we do not accept such an improbable explanation. We attribute the occurrence to a blown fuse or some other interruption of the common power supply. Several members of a theatrical company become violently ill on the same evening. We attribute this coincidence to the consumption of tainted food in a common meal. Scientific cases are also easy to find; the recent discovery of the gravitational lens is a good example. The spectra of two quasar images resembled one another so strikingly that astronomers inferred that they were two images of one and the same quasar; the dual image was created by the gravitational bending of light passing close to an extremely massive galaxy (or, as was later discovered, a cluster of galaxies).

One remarkable feature of the common cause principle is the fact that, although it has been widely used since time immemorial, it had never (to the best of my knowledge) been previously articulated with any degree of clarity or precision. Indeed, it seems clear to me that this was precisely the kind of reasoning Reichenbach (1938) attributed to the physicists in his cubical world. His comments in *Experience and Prediction* on the correlations between the shadows on the ceiling and those on the wall, and the remarks about the improbability of sheer chance coincidences, strongly suggest the common cause principle. Moreover, the explanation offered by the physicists is obviously a common cause explanation. The task that remains, then, is to relate the discussion of unobservables in *Experience and Prediction* to the common cause principle in *The Direction of Time* (1956). The answer, it seems to me, lies in connecting the common cause principle and Bayes's theorem. We can perhaps say that theories embodying common causes have much higher prior probabilities than theories involving improbable chance coincidences. This argument needs, of course, to be worked out in detail, but I am optimistic about the possibility of doing so successfully.[7]

Carl G. Hempel's main contribution to the realism/instrumentalism controversy can be found in his famous article "The Theoretician's Dilemma," first published in 1958 and reprinted in his *Aspects of Scientific Explanation and Other Essays in the Philosophy of Science* (1965b). The point of departure for his discussion is this: "Scientific research in its various branches seeks not merely to record particular occurrences in the world of our experience: it tries to discover regularities in the flux of events and thus to establish general laws which may be used for prediction, postdiction, and explanation" (173). According to Hempel, the three types of inference—prediction, postdiction, and explanation—all have a common logical schema, namely, a set of premises containing general laws and particular initial conditions and a conclusion asserting that a particular fact obtains.[8] The differences among the three types are entirely pragmatic, relating to the context in which the argument is given. The comprehensive term for these three scientific activities is *systematization*. Hempel carefully notes that, in some cases, at least some of the laws are statistical generalizations and the arguments are inductive. We must, therefore, distinguish two types of systematization—deductive and inductive.

Given that the purpose of scientific theories is to provide systematization of the empirical world, Hempel (1965c) asks whether such theories must necessarily involve references to *unobservable* entities. The instrumentalist argument is, roughly, that such references are not needed if the theory is successful. If connections among observables can be established by invoking so-called theoretical entities, these very connections *among observables only* can be used instead of the theory for purposes of systematizing the world of observables. Hempel shows that, in principle, it can be done, at least with respect to deductive systematization. As a result, he poses the dilemma mentioned in the title of the article:

> The Theoretician's Dilemma: If the terms and principles of a theory serve their purpose they are unnecessary, as just pointed out; and if they do not serve their purpose they are surely unnecessary. But given any theory, its terms and principles either serve their purpose or they do not. Hence, the terms and principles of any theory are unnecessary. (186)

Hempel identifies theories as axiomatic systems; if these theories are to have any relevance to empirical science they must be at least partially interpreted—that is, the nonlogical terms that occur in them must somehow be connected with the physical world. After carefully canvasing the possibilities for explicit definition of all theoretical terms in our observational vocabulary, he remains skeptical of the possibility of carrying out that program. Thus, in effect, he abandons Russell's *supreme maxim of scientific philosophizing* with which Carnap had prefaced his *Aufbau*. If the theoretical terms are to have any empirical meaning, it must be by way of *partial* interpretation. Hempel (1965) asks whether Carnap's *reduction sentences*, which serve as partial definitions, can provide the partial interpretations required, and again he is skeptical. Moreover, he finds no reason to be so restrictive. "In order to obtain a general concept of partial interpretation," he writes, "we will now admit as interpretative statements any sentences, of whatever logical form, which contain theoretical and observational terms" (208). This degree of latitude is required to do justice to the kind of partial interpretation actually found, especially in the advanced sciences. We can now think of an *interpreted theory T'* as consisting of a formal uninterpreted theory T and a finite set J of interpretative statements, where the uninterpreted theory contains terms of a theoretical vocabulary V_T and J contains terms both of the theoretical vocabulary and an antecedently understood basic vocabulary V_B.

The question now becomes whether it is possible to reformulate the interpreted theory T' in such a manner as to remove all theoretical terms and nevertheless to retain all of the systematic power of the interpreted theory. The answer is yes if we consider only *deductive* systematization. From a heuristic standpoint the resulting theory has many undesirable traits, but logically speaking, the systematization remains (Hempel 1965c, 211). If, however, we consider *inductive* systematization as well, the answer is no (215). Given that science seeks and provides inductive, as well as deductive, systematization, the theoretician's dilemma is defeated. The theoretical concepts of science are not dispensable.

Can we then regard our theory as consisting of meaningful discourse? Hempel (1965c) argues that we can, for such theories are amenable to empirical test by examining their observational consequences. If they are highly confirmed in this way, we can consider them true, and we can consider the terms that occur in them meaningful. If those terms designate unobservable entities, the theory provides good evidence of their existence. Hempel thus concludes with a positive argument regarding scientific realism.[9]

In his 1963 article, "Implications of Carnap's Work for the Philosophy of Science," which he contributed to the Schilpp volume on Carnap, Hempel again takes up several of the issues he discussed in "The Theoretician's Dilemma."[10] In particular, he argues again, and even more persuasively, that scientific theories must achieve inductive, as well as deductive, systematization. He offers the following example to show what he has in mind. Let the theoretical statement be "The parts obtained by breaking a rod-shaped magnet in two are again magnets." The term "magnet" is construed as a dispositional predicate, and consequently, it is not part of the observational vocabulary. In addition, we have an interpretative statement: "If x is a magnet, then, if iron-filings are brought into the vicinity of x they will cling to x" (700). On the basis of these two statements we can say that if b and c were obtained by breaking a in two, and a was a magnet, then iron-filings brought into the vicinity of b will cling to b. In order to establish the observational conclusion that iron-filings will cling to b, we must first establish that a was a magnet, and this can only be done by induction. Hempel reiterates the point that theoretical terms like "magnetic" are needed to achieve the inductive systematization we expect scientific theories to accomplish.

Carnap's (1963) response to this part of Hempel's article is remarkable:

> Hempel discusses . . . the following methodological question. Since the purpose of scientific theories is to establish predictive connections between data of experience, is it not possible to avoid the theoretical language and work with the observation language alone? In a detailed discussion Hempel gives convincing reasons for the thesis that this is not possible, in other words, that theoretical terms are indispensable for the purposes of science. [Carnap's bibliographical footnote is omitted here.] His main argument is based on the point that a scientific theory has the task of establishing not only deductive relations but also inductive relations among observational data. I believe that Hempel was the first to emphasize clearly this important point. (960)

Carnap is, I take it, expressing complete agreement with Hempel's thesis about the indispensability of the theoretical vocabulary for purposes of inductive systematization.

One's first reaction to Carnap's response might be to suppose that Hempel had persuaded Carnap to embrace some form of scientific realism—at any rate, that was my immediate reaction.[11] I now think, however, that this was a misinterpretation. Carnap does not say anything about the existence of unobservable entities; he comments only on the scientific vocabulary. Referring to the views expressed in his 1950a article, "Empiricism, Semantics, and Ontology," we can understand Carnap as saying that Hempel has addressed an *external* question, not in a metaphysical way but rather in a pragmatic sense. He has shown that a linguistic framework incorporating the theoretical vocabulary is better suited practically than is a language lacking such vocabulary for purposes of doing science. No ontological commitment is involved. Carnap's (1963) remarks in another section of his "Replies and Systematic Expositions"—namely, section 4A, "The realism controversy" (868–871)—reinforce this point. In answer to Adolf Grünbaum he writes, "If 'realism' is understood as a preference for the reistic [physical thing] language over the phenomenal language, then I am also a realist. However, if 'realism' is understood, in the customary sense, as an ontological thesis, then the arguments against it were given in my monograph [*Scheinprobleme in der Philosophie*; *Pseudoproblems in Philosophy* (1928b)]" (870).[12] So, although he made a valiant effort, Hempel does not seem to have converted Carnap to a realistic position. It appears to me that Carnap did not retreat from the position he held in 1950.

Let us consider, however, what Hempel *did* accomplish. In "Empiricism, Semantics, and Ontology," as well as in other writings, Carnap refers to a whole range of languages among which we may choose. However, as John Earman has remarked in conversation, it is by no means clear that any such choice exists. It has been widely accepted by now that phenomenal languages—including the type advocated by Carnap in the *Aufbau*—are simply not adequate to deal with the ordinary middle-sized material objects we meet in everyday life. As Carnap and Hempel later agreed, that alternative is not available. Now Hempel has argued that the physical thing language, containing a vocabulary suitable for dealing with observable material objects but lacking the theoretical vocabulary of science, is inadequate to incorporate the scientific systematizations it is the aim of the advanced sciences to provide. That seems to leave us with languages of only one type, and within such languages we can, with the help of suitable empirical investigation, answer *internal* questions about the existence of such things as atoms, molecules, and subatomic particles.

When I began discussing Reichenbach's views on the realism/instrumentalism issue above, I stressed the contrast between him and Carnap. The later Carnap of 1950 and 1963 did not find such opposition. In the paragraph immediately following his reply to Grünbaum he writes:

> Later, Reichenbach gave to the thesis of realism an interpretation in scientific terms, asserting the possibility of induction and prediction; a similar interpretation was proposed by Feigl. On the basis of these interpretations, the thesis is, of course,

meaningful; in this version it is a synthetic, empirical statement about a certain structural property of the world. I am doubtful, however, whether it is advisable to give to old theses and controversies a meaning by reinterpretation. . . . (1963, 870)

Thus, he seems to be saying, Reichenbach can be construed as dealing with the *internal* problem of existence of unobservables. He is doing so, we might add, in the kind of linguistic framework Hempel has shown to be required for the task of scientific systematization. Moreover, Reichenbach has shown us the type of argument that is able to give us, in that context, knowledge of unobservables. It rests upon the principle of the common cause.

At the beginning of his famous *Lectures on Physics*, the late physicist Richard Feynman said:

> If, in some cataclysm, all of scientific knowledge were to be destroyed, and only one sentence passed on to the next generation of creatures, what statement would contain the most information in the fewest words? I believe it is the *atomic hypothesis* (or the atomic *fact*, or whatever you wish to call it) that *all things are made of atoms—little particles that move around in perpetual motion, attracting each other when they are a little distance apart, but repelling upon being squeezed into one another.* In that one sentence, you will see, there is an *enormous* amount of information about the world, if just a little imagination and thinking are applied. (Feynman et al. 1963, 1–2; emphasis in original)

Is the atomic hypothesis true? Of course it is, as Carnap, Hempel, and Reichenbach all agree.[13] It is supported by overwhelming quantities of scientific evidence. Do atoms exist? Of course they do. Jean Perrin and others—using what may be described basically as a common cause argument—provided compelling scientific evidence for their existence in the first decade of this century (see Nye 1972). There is as much reason to believe that there are atoms as there is to believe that there are mountains on the other side of the moon. I am as sure that my body contains atoms as I am that it contains a brain—indeed, at times, I am not as sure about the brain. If, fully cognizant of the scientific facts, someone were to ask, "But do atoms *really* exist?" the only suitable response, I think, would be "What kind of question is *that?*" Early in his philosophical career Carnap gave the resounding answer: it is a pseudoproblem. I think Reichenbach and Hempel would agree.

I began this chapter with some remarks about scientific explanation, and I should like to conclude on the same theme. In the passage just quoted, Feynman mentions the amount of information about the world contained in the atomic hypothesis. We can properly add that this hypothesis has enormous explanatory value; for example, it explains Brownian motion. Combined with what has subsequently been learned about the nature of atoms, it explains myriad facts about chemical combination and about the structure of ordinary material objects. If atoms were some kind of fiction—merely hypothetical entities—they would provide no genuine explanations at all.

We find, then, that science can provide bona fide explanations of the phenomena that occur in our world. There are, obviously, many phenomena that present-day science cannot explain, so the work of explanation is not finished. But despite any metaphysical yearnings we may have, we find that it is *science*, not metaphysics or theology, that provides genuine understanding based on secure objective foundations.

Addendum

I have recently read, with great pleasure, the popular book *Seven Clues to the Origin of Life* (Cairns-Smith, 1985), in which the author advances in a tentative spirit a scientific hypothesis to the effect that all life on earth has literally evolved from clay. The physicochemical mechanisms through which this might have come about are discussed in detail. If this hypothesis is false it does not provide a genuine explanation of the origin of life; if it is true it does supply a bona fide explanation. One cannot fail to be struck by its superficial similarity to the biblical account in *Genesis*, according to which God created life by breathing on clay. Whatever the emotive value of the religious account may be, there can hardly be any doubt that the scientific account provides far greater intellectual understanding. This is a telling example for anyone who might still maintain that explanations are not to be found in science but only in theology or metaphysics.

An Empiricist Argument for Realism

As I sit in my office and glance about, I see many ordinary things—some chairs, my briefcase and umbrella, pencils, books, and various other objects. With a modicum of effort I can touch them. Like Descartes, I know that I could be dreaming or hallucinating and that in the past I have sometimes been mistaken about such matters. Unlike Descartes, I am not epistemologically disturbed by this fact. Certainty about contingent matters of fact is beyond our reach, but I am convinced, "with a probability bordering on certainty,"[1] that such middle-sized material objects exist and that we are aware of many of their properties.

There is no temptation to retreat to any version of phenomenalism. In the first place, it appears that the efforts of such philosophers as A. J. Ayer, Rudolf Carnap, C. I. Lewis, Ernst Mach, and Bertrand Russell to show how ordinary material objects could be constructed out of the data of immediate experience have simply been unsuccessful, and unsuccessful in such a grand way as to arouse considerable skepticism about the possibility in principle of any such program.[2] We can, of course, discriminate qualities of the objects of our experience, but if there are such things as sense data or immediate impressions they are constructs of our folk-psychological theories, not the fundamental data for all of our knowledge of the physical world. Moreover, we now realize, thanks primarily to Richard Jeffrey (1968), that the maxim of C. I. Lewis (1946, 186)—if anything is to be probable something must be certain—is simply false. Thus, as I undertake a discussion of the reality of such unobservables as atoms, I shall assume that common sense and everyday experience provide us with a great deal of reasonably reliable knowledge of the kinds of things we normally consider observable.

For purposes of the present discussion, I shall also assume that we have some degree of scientific knowledge of the world; the only problem is how to interpret it. Instrumentalism, positivism, constructive empiricism, and scientific realism are the leading options. Instrumentalism denies the existence of unobservable objects, positivism holds that sentences purporting to refer to unobservables are cognitively meaningless, constructive empiricism remains agnostic regarding unobservables, and scientific realism affirms their existence. Moreover, referring to arguments offered in chapter 1, I shall assume that such instruments of observation as optical microscopes and telescopes provide us with reliable information about objects and their properties that we cannot obtain through use of our normal, unaided human senses. One example, crucial to this chapter, is our ability to observe the motions of microscopic Brownian particles suspended in fluids.

Finally, for purposes of the present discussion, I shall assume that our scientific investigations have revealed various regularities that have held within our past experience and that we can extrapolate to the past, to the future, and to the elsewhere. Without some primitive form of induction, even the most elementary parts of our science could not get off the ground. The question upon which we shall focus is whether some more advanced form of inductive reasoning enables us to extend our knowledge to the various realms of unobservables.

The basic question comes down to this: Do we have empirical evidence that gives us just about as much reason to believe in the existence of such entities as molecules, atoms, ions, and subatomic particles as we have for our belief in middle-sized material objects? When we refer to middle-sized material objects we should keep in mind that this category includes objects considerably more problematic than my office furnishings—for example, Charon (Pluto's moon), eggs of the echidna (a mammal), and one's own pineal gland. According to the realist thesis I shall attempt to defend, we have no sound basis for making an ontological distinction among middle-sized, very small (microscopic), and extremely small (submicroscopic) entities. When I talk about the reality of these things, it should be understood that I am not pounding the table, raising my voice, invoking Kantian noumenal objects, or referring to REALITY. I am thinking about existence in a straightforward, down-to-earth sense. As nearly as I can tell, this corresponds to what Arthur Fine (1984) has called the "natural ontological attitude." I call it *scientific realism*. According to this form of realism, assertions about reality that go beyond this ordinary type of existence have no explanatory import in the sciences.

Although, as I said in the preceding chapter, I find Reichenbach's approach to the problem of scientific realism the most promising, it seems to me that he did not articulate it clearly or work it out in sufficient detail. As a result of my general dissatisfaction with the arguments of philosophers regarding the existence of unobservables, I decided that it might be fruitful to take a look at the arguments that convinced the overwhelming majority of competent physical scientists of the existence of such entities as atoms and molecules. That is where, I believe, we can find a satisfactory answer to *the key question*.

1. The Atomic Theory: The First Half Century

When John Dalton advanced his atomic theory early in the nineteenth century, he obviously was not the first person to suppose that ordinary middle-sized objects are composed of much smaller particles known as atoms. In antiquity Democritus and Lucretius had proposed this hypothesis; in the seventeenth century such mechanical philosophers as Robert Boyle and Robert Hooke had embraced atomism.

Dalton's hypothesis was importantly different because he made certain very specific claims about these atoms. As Clark Glymour (1980) has put it:

> Atomic theories were common enough at the beginning of the nineteenth century; what distinguished Dalton's theory was his systematic explanation of simple (but not uncontroversial) chemical regularities in terms of physical properties ascribed to atoms. In Dalton's theory, the laws of definite and multiple proportions do not result just because all things are made of atoms. They result because every compound substance is made of elementary substances; because every pure sample of an elementary substance is constituted of some definite number of atoms; because all atoms of the same elementary kind have the same mass; because every pure sample of a compound substance is constituted of some definite number of molecules; and, finally, because every molecule of a compound substance contains the same number of atoms of the respective kinds that compose it, as does every other molecule of that compound substance. (226–227)

As Glymour points out, these claims refer to such properties of atoms as mass and number.

It is no simple matter to determine the values of these atomic quantities with genuine accuracy; it can reasonably be said that it took precisely 100 years—from the publication of Dalton's (1808) theory to Jean Perrin's (1908) first publication of his ascertainment of Avogadro's number N. This century is split exactly in half by Stanislao Cannizzaro's (1858) publication of an unambiguous list of *relative* atomic and molecular weights, for example, the ratio of the mass of an oxygen atom to that of a hydrogen atom. This was the problem of atoms and molecules that occupied chemists for the first half of the century—indeed, Dalton designated this as the principal aim of his work.

In attacking this problem Dalton introduced some rules based on rather ad hoc simplicity considerations. If, for example, a compound contains two elements, and no other compound of the same two elements is known, it is to be assumed that the molecule of that compound contains just one atom of each element. According to this rule the chemical formula for water would be HO, and the ratio of the masses of the two atoms would be 8:1; in fact it is about 16:1. If there is a second compound of the same two elements, then one of the molecules would contain three atoms, one atom of one element and two atoms of the other. For example, we have two compounds containing nitrogen and oxygen, nitric oxide, and nitrous oxide. Today we recognize nitric oxide as the diatomic molecule NO and nitrous oxide as N_2O, conforming to Dalton's rule. But Dalton's rule gives us no basis for preferring N_2O to NO_2; indeed, it does not even tell us which substance has a diatomic and which a triatomic molecule.

Dalton's atomic theory did not meet with anything like universal acceptance. When Humphrey Davy presented a Royal Medal to Dalton in 1826, he said that it was for the law of combining proportions, "usually called the Atomic Theory," but he made it clear that he construed atomism in a thoroughly instrumentalist manner. Dalton, in contrast, interpreted his theory realistically (Gantner 1979, 4–5).

There were various reasons for reservations about Dalton's theory. In addition to the ambiguity and ad hoc character of his rules, there were empirical problems.

Whereas in 1808 Dalton had articulated a law of combining proportions based on weight, in 1809 Joseph Louis Gay-Lussac established a law of combining proportions of gases based on volumes. There were difficulties in reconciling these regularities. They were overcome by Amedeo Avogadro who in 1811 postulated that at equal pressures and temperatures, equal volumes of gas contain the same number of molecules, but the reconciliation required the assertion that molecules of some elemental gases are diatomic. Dalton's principles of simplicity would seem to imply that an elemental gas, such as hydrogen, nitrogen, or oxygen, should consist of monatomic rather than diatomic molecules. It was difficult to understand what kind of bond could unite two identical atoms to form such a molecule. It was even harder to understand how the molecules of different elements in their gaseous states could contain different numbers of atoms: mercury vapor, one; hydrogen, two; phosphorus vapor, four; sulfur vapor, six (Nye 1984, xviii).

Between 1811 and 1858 the atomic theory suffered many vicissitudes. These are well recounted by Glymour (1980) and other authors. As mentioned above, 1858 stands out as the year in which Stanislao Cannizzaro published a systematic table of relative atomic weights that were consistent with the empirical data involving combining weights, combining volumes, specific heats, and Avogadro's law of equal numbers. Does this achievement establish the atomic hypothesis with reasonable scientific certainty? For two reasons the answer is no.

First, the most that can be said is that Cannizzaro showed that the atomic hypothesis furnished a consistent model for the chemical facts that were available at the time. That was a remarkable achievement, and I do not intend in the least to minimize its magnitude or value. For example, it laid the groundwork for the development of the periodic table of elements by Dmitri Mendeléev in 1869. It did not, however, sufficiently undermine the instrumentalistic interpretation of the atomic hypothesis. Physical scientists could reasonably say that Cannizzaro succeeded in producing an effective instrument—one that was not plagued by ambiguities and inconsistencies—but just an instrument nevertheless. In fact, historically, that is the attitude taken by a significant portion of the scientific community for almost fifty years thereafter.

Second, it should be recalled that the development of the atomic hypothesis from Dalton to Cannizzaro did not yield a definite atomic magnitude. At that point it was impossible to establish the mass of any individual atom or molecule, and it was impossible to determine the number of atoms or molecules in a given sample of a chemical element or compound. The atomic hypothesis still lacked a handle on the fundamental atomic quantities explicitly mentioned in Dalton's theory.[3]

2. The Atomic Theory: The Second Half Century

Mary Jo Nye's (1984) historical collection of scientific papers begins with the Karlsruhe Congress of 1860, which is closely associated in time and theme with Cannizzaro's work. Cannizzaro was, in fact, a member of the steering committee. As Nye explains in her introduction, the second half century involved controversies among three physical hypotheses—atomism, electromagnetism, and energeticism—

and a philosophical ideology, namely, positivism.[4] The spirit of positivism was succinctly expressed in Marcelin Berthelot's 1877 rhetorical question, "Who has ever seen, I repeat, a gaseous molecule or an atom?" (quoted by Nye 1984, xix), This sort of skepticism regarding unobservables had a strong influence on physical scientists beyond the turn of the twentieth century. At that time the question of the existence of atoms and molecules was regarded by many as purely metaphysical. Such skeptics could, of course, adopt an instrumentalist position and—doing as Ernst Mach did quite late in life—accept the atomic theory *but only as a useful tool* (Wolters 1989). That did not evade the question of the real existence of atoms.

Two fields of physics, thermodynamics and electrodynamics, flourished greatly in the nineteenth century. Thermodynamics was pursued in at least two different ways. One of these was mechanical, and it led to the development of the molecular-kinetic theory of gases, which presupposes the atomic-molecular theory of the constitution of matter, at least in the gaseous state. This was an assumption that could not, at that time, be empirically verified in any direct way. The other approach was energeticism; it rejected the construction of mechanical models and dealt with thermodynamic phenomena directly. As Leon Cooper (1968) describes it,

> What is perhaps most elegant about the concepts of heat, temperature, and entropy is that they are not necessarily wedded to mechanical notions: force, mass, acceleration. They do not deal with hypothetical particles or assume very much about the nature of matter. That the concepts of temperature, entropy, and heat can be made independent of mechanics seemed to some a virtue. They were directly observable. Temperature, for example, can be defined as the level of mercury in a thermometer without any assumptions about hypothetical particles that make up matter. In this sense, what was called energetics seemed an alternative to the mechanics of Newton—an alternative to the proposed corpuscular nature of matter. (332)

Positivistic scruples did not constitute the only objection of energeticists to the molecular-kinetic theory. The fundamental laws upon which this latter theory is based are those of Newtonian mechanics, which were clearly recognized as temporally reversible. The second law of thermodynamics—the law of increase of entropy—is manifestly time-asymmetric and provides an account of many observable irreversible phenomena in the physical world.[5] How Newton's time-symmetric laws could furnish the basis for irreversible phenomena was a profound mystery.

The first law of thermodynamics—the law of conservation of energy—was also a source of inspiration to energeticists. It tells us that energy, which can take on many diverse forms, is an invariant[6]—a conserved quantity. Invariants are good indicators of physical reality, and to such energeticists as Wilhelm Ostwald, energy—not particulate matter—was the basic constituent of the physical world. Moreover, the fundamental phenomena of thermodynamics are described by differential equations, which seem to preclude the discontinuity involved in atomism. In energeticism not only did energy replace matter but also continuity replaced discontinuity. At the turn of the century, the energeticists and the atomists could be designated, respectively, as the "partisans of the plenum" and the "partisans of the void" (Nye 1972, 83).

From almost the very beginning Dalton's theory faced problems concerning the nature of the chemical bond. If atoms are hard lumps of matter devoid of internal structure and incapable of internal movement or modification, and if all atoms of a given element are identical to one another, it is difficult to see how two or more atoms of the same element could stick together. Obviously the gravitational attraction is too weak by many orders of magnitude, and even if gravitation were responsible, it would be difficult to see why two atoms of hydrogen would stick together to form a molecule without attracting many more. Phenomena such as the spectra of excited gases and the heat capacity of various substances strongly suggested internal structure and movement. It was tempting to turn to electricity and magnetism for solutions of some of these puzzles; indeed, it was tempting to say that atoms are composed of particles of electricity.

Some physicists—for example, William Thomson (Lord Kelvin)—attempted to construct atoms on the basis of the electromagnetic aether and electrically charged particles, thus creating a kind of compromise between Newtonian corpuscles and Cartesian vortices (Nye 1984, xxi). For the most part, however, the electrodynamic approach was conjoined to atomism in an effort to explain the nature and behavior of atoms. As we now know, such an approach cannot succeed; on strictly classical principles the hydrogen atom would be unstable, suffering collapse in a time on the order of nanoseconds.[7] Suffice it to say that problems of atomic structure and stability, of specific heats, and of emission spectra turned out to be intractable within classical physics—Newtonian mechanics, Maxwellian electrodynamics, and thermodynamics—and could only be treated satisfactorily within twentieth-century quantum mechanics.

At the end of the nineteenth century and the beginning of the twentieth there were two approaches to thermodynamics, the energeticist hypothesis that denies the existence of atoms and molecules and the molecular-kinetic hypothesis that affirms their existence. There were physical, as well as philosophical, reasons for rejecting atomism and embracing energeticism; the two theories are not empirically equivalent. These reasons persisted through the end of the nineteenth century and into the beginning of the twentieth century. Gantner (1979) and Nye (1972) provide excellent accounts of this situation. A number of the most important scientific papers bearing on these issues are collected in Nye (1984).

Two basic developments in the decisive resolution of the question of the atom occurred around 1905. One was the Einstein-Smoluchowski theory of Brownian motion; as presented by Einstein ([1905] 1989) it offers the possibility of a crucial experiment to decide between energeticism and the molecular-kinetic theory. The other was the initiation by Jean Perrin (1909) of his famous experimental studies of Brownian motion, which provided, among other things, strong empirical support for the Einstein-Smoluchowski theory. These two events occurred independently; Perrin was unaware of the theoretical work of Einstein and Smoluchowski, and they were unaware of his. The first publication of Perrin's results occurred in 1908, marking the end of the second half century.[8]

3. Einstein and Perrin on Brownian Motion

Early in his scientific career—before the beginning of the twentieth century, when many scientists were saying that the question of the existence of atoms and molecules is a completely metaphysical one—Perrin set himself the goal of providing *empirical scientific evidence* that would definitively resolve the question. The experimental work on Brownian motion that he began in 1905 had culminated in complete success by the time he published *Les Atomes* in 1913. Einstein's 1905 work also deals with observational evidence concerning phenomena that occur on the molecular level. From the molecular-kinetic theory he derives results concerning the rate of diffusion of Brownian particles; if the phenomena turn out as predicted, they show that the second law of thermodynamics is not absolutely valid. Violations should appear at the microscopically observable level. He concludes the work with the "hope that a researcher will soon succeed in solving the problem posed here, which is of such importance in the theory of heat" ([1905] 1989, 134). The results that Perrin published in 1908 fulfill Einstein's wish magnificently.

If one accepts the general idea of the molecular-kinetic theory—that fluids (liquids and gases) are composed of molecules in constant motion—and the hypothesis that in a fluid in thermal equilibrium the average kinetic energy of the particles of each type is the same,[9] then it appears superficially that there is a fairly direct way of determining the mass of a molecule of a given substance—for example, water—by observation of Brownian motion. Early in the nineteenth century the botanist Robert Brown had observed the apparently random agitation of microscopic particles of pollen suspended in liquid. He was at first tempted to regard this movement as some manifestation of life, but further observations of particles of inorganic substances convinced him that his first hunch was wrong. Subsequent investigators sought to explain the motions of Brownian particles on the basis of evaporation, convection, electric charge, and other factors, but none of these efforts proved satisfactory. As we now realize, the correct explanation is that these particles, although much larger than molecules, are constituents of the fluid and share its thermal agitation through collisions with molecules. Given this fact, one can set the average kinetic energy of the Brownian particles equal to the average kinetic energy of the molecules of the suspending fluid,

$$1/2 m_m v_m^2 = 1/2 m_b v_b^2 \qquad (1)$$

where the subscripts m and b refer to molecules and Brownian particles, respectively. It would seem possible to establish values for three of the four terms in this equation more or less directly, allowing one to solve for the fourth. Since the Brownian particles are microscopically observable, it would seem reasonable to suppose that values for both terms on the right could be established in that way. The velocities of molecules can be ascertained rather straightforwardly by experiment,[10] or they can be calculated from other observable quantities via Herapath's formula

$$P = 1/3 \rho v^2 \qquad (2)$$

where P stands for the pressure of the gas in a container and ρ for its density, both of which are macroscopically measurable.[11] This approach proved impractical, however, because the velocities of Brownian particles cannot be observed microscopically. Their direction of motion changes so rapidly as a result of molecular collisions that attempts to determine their velocities by microscopic observation give values that are much too low.

Perrin (1908) realized, however, that there were other approaches that would attain the same end. If, for example, tiny particles denser than water are placed in a container of water they have a tendency to fall to the bottom under the influence of gravity. At the same time, molecular collisions from below exert an upward force so that the particles do not all settle on the bottom. The result, at equilibrium, is an exponential vertical distribution of particles, denser toward the bottom, less dense toward the top. In a remarkable set of extremely difficult experiments Perrin ascertained the distribution empirically, and from that result he calculated Avogadro's number N, the number of molecules in a mole of the substance in which the particles are suspended (see Nye 1972, 102–110). Thus, given N and the *relative* molecular weight of water one can derive the mass of a single molecule.

Not satisfied with this achievement, Perrin performed another series of experiments on the rate of horizontal diffusion of particles introduced at one side of a container of fluid. Given their molecular bombardment, the Brownian particles will eventually arrive at a uniform horizontal distribution. From the rate at which this occurs Perrin was again able to calculate N. The result of this empirical determination of N agreed with the value resulting from the first set of experiments. The results of these experiments were published in 1908. Mayo (1996) provides a careful analysis of Perrin's methodology.

As it turned out, the horizontal diffusion of Brownian particles is precisely the phenomenon with which Einstein ([1905] 1989) was concerned.[12] It opens with the following statement:

> It will be shown in this paper that, according to the molecular-kinetic theory of heat, bodies of microscopically visible size suspended in liquids must, as a result of thermal molecular motions, perform motions of such magnitude that these motions can easily be detected by a microscope. It is possible that the motions to be discussed here are identical with the so-called "Brownian molecular motion"; however, the data available to me on the latter are so imprecise that I could not form a definite opinion on this matter.
>
> If it is really possible to observe the motion to be discussed here, along with the laws it is expected to obey, then classical thermodynamics can no longer be viewed as strictly valid even for microscopically distinguishable spaces, and an exact determination of the real size of atoms becomes possible. Conversely, if the prediction of this motion were to be proved wrong, this fact would provide a weighty argument against the molecular-kinetic theory of heat. (123)[13]

Here Einstein describes a crucial experiment to decide between strict energeticism and the molecular-kinetic theory. He was delighted subsequently to learn that Perrin had performed the experiment and that the outcome favored the molecular-kinetic theory.

Let us briefly review the history we have been considering. In 1808 Dalton advanced his atomic theory, which referred explicitly to masses and numbers of atoms. By 1858 Cannizzaro produced a consistent table of *relative* molecular weights, but there was still no access to the properties of individual atoms and molecules to which Dalton referred. Dalton's theory was revised and augmented as a result of the molecular-kinetic theory of gases. Almost a hundred years after Dalton's enunciation of the atomic hypothesis and almost eighty years after the discovery of Brownian motion, Einstein (and independently Smoluchowski) provided a theoretical explanation of Brownian motion, given the molecular-kinetic theory. Perrin's experiments verified Einstein's prediction and provided numerical parameters that yielded values for Avogadro's number and masses of individual atoms and molecules. This is the story of an outstanding hypothetico-deductive confirmation of a scientific theory.

Perrin did not, however, stop there. He subsequently conducted a third set of experiments, dealing with the rotational motion of Brownian particles. According to the molecular-kinetic theory, a portion of the thermal energy of the system made up of the Brownian particles suspended in a fluid consists of rotational kinetic energy. By ascertaining the velocities of rotation of the particles, Perrin (1908) obtained a third type of empirical basis for ascertaining N, and the results of this method were in excellent agreement with those of the preceding two methods. What more could anyone ask for?

4. The Completion of the Argument

In several of the chapters in part II of this book I argue that the hypothetico-deductive schema is *not* an adequate characterization of scientific confirmation. In the context of this chapter, let me suggest the sort of response that would have been appropriate on the part of antiatomists in the early years of the twentieth century.

From the very beginning, Dalton's atomic theory received one sort of response (among others), namely, that it provided a useful heuristic model for chemists but that it did not necessarily furnish an accurate description of reality. With the work of Cannizzaro and Mendeléev the utility of the model was greatly improved. The molecular-kinetic theory increased the value of the model by showing how it could deal with thermodynamic phenomena, which are, of course, closely connected to chemistry. Then Einstein and Perrin showed how the model could, in addition, treat Brownian motion, which constituted a further demonstration of its utility. Nevertheless, it remained a model that one could accept without making any commitment about the nature of physical reality.

One of the major antiatomists of the early twentieth century was Wilhelm Ostwald, who had stated categorically in 1906 that "atoms are only hypothetical things" (quoted by Nye 1972, 151). A thoroughgoing energeticist, he was committed to the view that energy constituted physical reality even on the microlevel. This thesis was vulnerable to the critique of the universal validity of the second law of thermodynamics offered by Perrin and Einstein. In the fourth edition of his *Grundiss der physikalischen Chemie* (1908) he revised his opinion drastically:

> I have satisfied myself that we arrived a short time ago at the possession of experimental proof for the discrete or particulate nature of matter—proof which the atomic hypothesis has vainly sought for a hundred years, even a thousand years.... [The results] entitle even the cautious scientist to speak of an experimental proof for the atomistic constitution of space-filled matter. (Quoted by Nye 1972, 151)

In this passage Ostwald is referring not only to Perrin's work on Brownian motion but also to J. J. Thomson's work on the kinetic theory of gases. Given his previous commitment to the universal truth of the laws of thermodynamics, the new evidence was, for him, decisive.

Considering the situation more generally, however, we must recall that scientists of the nineteenth century, especially English scientists, engaged in a great deal of model building that was not intended as a description of physical reality. As is well known, James Clerk Maxwell constructed elaborate models of the electromagnetic field, only to discard them later in favor of a set of differential equations. One possible reaction to this approach was to maintain that

> mathematical analogies—employing invisible entities such as vortices, hidden masses or atoms—could be used to formulate differential equations and their validity experimentally verified according to their ability to predict physical results. A variety of different models might easily result in identical physical measurements, in which circumstance no judgment could really be made upon the relationship of any one model to physical reality. Such a method was followed by the English school of physicists almost in entirety, and it was no happenstance that Boltzmann's kinetic theory of gases was most avidly accepted by those scientists who were most enthusiastic about the application of mechanical analogies. (Nye 1972, 15–16)

Consequently, even if Boltzmann's theory were regarded as a splendid model, that was no reason to think that it provided a true description of reality.

Moreover, there were those who were not enchanted with the making of models. Duhem (1954), who considered all questions about the physical reality of unobservables to be entirely metaphysical, heaped scorn on English models in a famous passage:

> Here is a book [O. Lodge, 1890] intended to expound the modern theories of electricity and to expound a new theory. In it there are nothing but strings which move around pulleys, which roll around drums, which go through pearl beads, which carry weights; and tubes which pump water while others swirl and contract; toothed wheels which are geared to one another and engage hooks. We thought we were entering the tranquil and neatly ordered abode of reason, but we find ourselves in a factory. (70–71; quoted in part by Nye 1984, xx)

What is required to support strongly the conclusion that atoms and molecules are more than merely constituents of an extremely useful model? How can it be shown that atoms are physically real and not "only a construct of his [the physicist's] imagination" (Pearson 1911, xii)? What would, in Perrin's words, give "the real existence of the molecule...a probability bordering on certainty" ([1913] 1916, 207)? Perrin was clearly aware of this question, and he offered a straightforward answer—one that turned out to be convincing to the vast majority of knowledgeable physical scientists of that period.

One simple way to formulate the answer is to say that the argument rests on the variety of evidence that was brought to bear on this question. Beginning in 1908, Perrin took pains to cite, not only the results of his own research, but also the results of other completely diverse methods. In his earliest set of papers he mentioned not only his own experiments on Brownian motion but also such other ascertainments of Avogadro's number as those based on Planck's work on black-body radiation and Rutherford's studies of the alpha radiation of radium. From a strictly phenomenological point of view these experiments have nothing to do with one another—Rutherford counting tiny scintillations on screens in a darkened room, Planck doing spectral analysis of radiation emitted from the mouth of a blast furnace, and Perrin counting microscopic particles of gamboge suspended in water. In retrospect we can easily see what they have in common, namely, atomicity or quantization in one guise or another. Our justification for looking at these experiments in this manner is our acceptance of Perrin's conclusion. At the end of his comprehensive 1908 article Perrin offers a table of fifteen distinct ways to ascertain the value of N, although seven of them provide only upper limits, lower limits, or very wide ranges of possible values (Nye 1984, 598). Immediately thereafter he comments, "I think it impossible that a mind, free from all pre-conception, can reflect upon the extreme diversity of the phenomena which thus converge to the same result, without experiencing a very strong impression, and I think it will henceforth be difficult to defend by rational arguments a hostile attitude to molecular hypotheses which, one after another, carry conviction ..." (Quoted by Nye 1984, 599).

At the end of his 1913 book he provides a different table of thirteen distinct ways to determine the value of N; in this table each value is a definite number, even though some were admittedly subject to considerable error. The set of values ranges from 6×10^{23} to 7.5×10^{23}. The following comment comes right after this table:

> Our wonder is aroused at the very remarkable agreement found between values derived from the consideration of such widely different phenomena. Seeing that not only is the same magnitude obtained by each method when the conditions under which it is applied are varied as much as possible, but that the numbers thus established also agree among themselves, without discrepancy, for all the methods employed, the real existence of the molecule is given a probability bordering on certainty. ([1913] 1916, 206–207)

Perrin goes on to point out that by a purely formal maneuver one can eliminate N from thirteen equations, thereby obtaining twelve equations that make no reference to molecules: "These equations express fundamental connections between the phenomena, at first sight entirely independent, of gaseous viscosity, the Brownian movement, the blueness of the sky, black body spectra, and radioactivity" (207). With this approach

> an expression is obtained that enables us to predict the rate of diffusion of spherules 1 micron in diameter in water at ordinary temperatures, if the intensity of the yellow light in the radiation issuing from the mouth of a furnace containing molten iron has been measured. Consequently a physicist who carries out

observations on furnace temperatures will be in a position to check an error in the observation of the microscopic dots in emulsions! And this without the necessity of referring to molecules. (207)

Henri Poincaré was one of the most distinguished opponents of atomism in the beginning years of the twentieth century, but he was converted. At the conclusion of a conference in 1912, reflecting on the papers presented that day, he said,

> A first reflection is sure to strike all the listeners; the long-standing mechanistic and atomistic hypotheses have recently taken on enough consistency to cease almost appearing to us as hypotheses; atoms are no longer a useful fiction; things seem to us in favour of saying that we see them since we know how to count them.... The brilliant determinations of the number of atoms made by M. Perrin have completed this triumph of atomism.... In the procedures deriving from Brownian movement or in those where the law of radiation is invoked... in the blue of the sky... when radium is dealt with.... The atom of the chemist is now a reality. (Quoted by Nye 1972, 157)

Perrin stated repeatedly, emphatically, and unambiguously that the variety of experimental evidence plays a crucial role in his argument for the reality of molecules. As we have just seen, Poincaré reiterated the same point. Nye, in her extended and authoritative analysis of Perrin's work, emphasizes the same aspect of the situation.

Miller (1987) denies Perrin's claim that the variety of methods for ascertaining Avogadro's number plays a significant part in the establishment of molecular reality:

> An appeal to the coincidence of differently developed numerical estimates cannot, remotely, account for the distinctive outcome of the twentieth-century argument, as compared with the nineteenth-century ones. For Cannizzaro could also point to a variety of coincidences in important magnitudes, when they are estimated using independent parts of molecular theory. A variety of different routes warranted similar atomic weights. Indeed, molecular weights are, if anything, more important to molecular theory than Avogadro's number. (479)[14]

As I noted above, Cannizzaro's ascertainment of a consistent set of *relative* atomic and molecular weights had supreme importance for the development of chemistry in the latter part of the nineteenth century. But one can hardly compare the breadth of Cannizzaro's evidence—consisting mainly of combining weights, combining volumes, and specific heats—with the range involved in Perrin's argument. Indeed, at the time of the Karlsruhe Congress, there remained a serious question in the minds of some scientists whether the atoms and molecules of chemistry were the same or different from the atoms and molecules of physics. Nye's (1984) *The Question of the Atom* traces the history, by means of a collection of primary sources, from the Karlsruhe Congress (1860) to the First Solvay Congress (1911), almost exactly the period from Cannizzaro's major publication to the dates of the statements by Perrin and Poincaré quoted above. As we have seen, Cannizzaro's evidence can reasonably be classified as narrowly chemical. The work of Perrin on Brownian motion along with the work of others that he cited in making his argument from the "remarkable agreement" ranges over a wide variety of phenomena from diverse branches of physical science. It should be added that Cannizzaro's results had only a small amount of redundancy in terms of cross-checking of particular values by multiple methods of testing. Perrin's ascertainment of Avogadro's number offered thirteen distinct ways, each of which, by itself, provided an independent determination of its value.[15] It should also be recalled that Cannizzaro was unable to establish any such properties as mass or size of atoms and molecules. With the aid of Avogadro's number, many such microquantities can be calculated from macroscopically measurable magnitudes. As Eyvind H. Wichmann (1967, 20) has put it, "Avogadro's number . . . is the link which connects microphysics with macrophysics."

5. The Common Cause Principle

As I suggested in the preceding chapter, Reichenbach seemed to be appealing to some sort of common cause principle on which to found his defense of scientific realism against the antirealism of some of the logical positivists. In *Scientific Explanation and the Causal Structure of the World* (1984b, chap. 8) I attempt to spell out the argument of Perrin in terms of common causes. I still believe, first, that the argument corresponds closely to Reichenbach's basic conceptions as expounded in his 1956 book, even though, to the best of my knowledge, he never discussed this particular episode in the history of science, and second, that it is fundamentally a correct argument. Since various features of my preceding treatment of this issue have received significant criticisms, I want to try to spell out the argument anew in ways that will take them into account.

Both everyday life and science are full of examples of inference to a common cause. The standard example is the simultaneous illness of a number of people who have shared a common meal. It is used in Reichenbach (1956, 157) and in Salmon (1984b, 158). Lest anyone think this example is merely an invention of philosophers, I can report that during 1992 I was called for jury duty in civil court. The case to which I was assigned involved the simultaneous consumption of frozen fish sticks by three members of a family, each of whom experienced severe gastrointestinal distress in the middle of the night following the meal.[16] Although there is a small probability that any given member of that family might suffer such illness during any given night, the chance that the three would become ill during the same night in the absence of a common cause is extremely small. It is more reasonable to seek a common cause than to assume a fortuitous coincidence. A number of other examples, real or fictitious, from common sense and science, can be found in Reichenbach (1956) and Salmon (1984b).

In *Scientific Explanation and the Causal Structure of the World* (1984b), I discussed the "twin quasars," the first discovered case of a gravitational lens, as a particularly fine scientific example of use of the common cause principle:

> The twin quasars 0957 + 561 A and B are separated by an angular width of 5.7 seconds of arc. Two quasars in such apparent proximity would be a rather improbable occurrence given simply the observed distribution of quasars [circa

1980]. Examination of their spectra indicates equal red shifts, and hence, equal distances. Thus these objects are close together in space, as well as appearing close together as seen from earth [a much greater coincidence]. Moreover, close examination of their spectra reveals a striking similarity—indeed, they are indistinguishable. This situation is in sharp contrast to the relations between the spectra of any two quasars picked at random. Astronomers immediately recognized the need to explain this astonishing coincidence in terms of some sort of common cause. One hypothesis that was entertained quite early was that the twin quasars had somehow (no one had the slightest idea how this could happen in reality) developed from a common ancestor. Another hypothesis was the gravitational lens effect—that is, there are not in fact two distinct quasars, but the two images were produced from a single body by the gravitational bending of the light by an intervening massive object. This result might be produced by a black hole, it was theorized, or by a very large elliptical galaxy. Further observation, under fortuitously excellent viewing conditions, has subsequently revealed the presence of a galaxy that would be adequate to produce the gravitational splitting of the image. This explanation is now, to the best of my knowledge, accepted by virtually all of the experts.[17] (159)

Since then a number of additional cases of gravitational lensing have been discovered. However, inasmuch as Ian Hacking (1989) has expressed skepticism regarding gravitational lensing, I shall return to this example in the appendix to this chapter.

Other examples can obviously be found in evolutionary biology. As Elliott Sober (1988, 211–212) has written, "Evolutionists attempt to infer the phylogenetic relationships of species from facts about their sameness and difference. We observe that chimps and human beings are similar in a great many respects; we infer from this that there exists a common ancestor of them both, presumably one from whom these shared traits were inherited as homologies." I have quoted this example from Sober because he uses it to introduce a penetrating critique in which he raises serious questions about limitations on the applicability of the common cause principle in precisely this context. I shall return to his argument below.

Reichenbach (1956, 157) stated his *principle of the common cause* as follows: "*If an improbable coincidence has occurred, there must exist a common cause.*" This statement is obviously too strong, so he immediately added a qualification: "Chance coincidences, of course, are not impossible.... The existence of a common cause is therefore ... not absolutely certain, but only probable" (157–158). In order to give a precise analysis to the foregoing principle, Reichenbach offered the following statistical characterization of the *conjunctive fork* (159):[18]

$$P(A \cdot B|C) = P(A|C)\, P(B|C) \tag{3}$$

$$P(A \cdot B|{\sim}C) = P(A|{\sim}C)\, P(B|{\sim}C) \tag{4}$$

$$P(A|C) > P(A|{\sim}C) \tag{5}$$

$$P(B|C) > P(B|{\sim}C) \tag{6}$$

From these four relations, assuming $0 < P(C) < 1$, Reichenbach deduces

$$P(A \cdot B) > P(A)\, P(B) \tag{7}$$

In these formulas A, B, and C stand for classes of particulars.[19] Since the definition of statistical independence is the same as (7) except for equality instead of the inequality, formula (7) says that instances of types A and B do not occur in a statistically independent fashion. Formula (3) says, however, that instances of types A and B occur independently, given that C obtains; and (4) says that instances of types A and B occur independently, given that C does not obtain. Formulas (5) and (6) say that C is positively relevant to the occurrence of A and to the occurrence of B.

Nowhere does Reichenbach (1956, 159) say that every conjunctive fork represents a common cause; indeed, he implicitly denies it.[20] Moreover, a clever example by Ellis Crasnow shows how an antecedent of type C, which is not a common cause, can fulfill formulas (3)–(6) (see Salmon 1984b, 167–168). Reichenbach's principle must be construed to mean that whenever there is an improbable coincidence there most probably will exist *some common cause or other* that forms a conjunctive fork, but the existence of types A, B, and C satisfying those conditions does not mean that type C, rather than some other type D, is a common cause. In order for the C in (3)–(6) to qualify as a common cause there must be suitable causal processes connecting the C's with the A's and with the B's, but none connecting the A's with the B's in such a way as to make the A's causes of the B's or vice versa.[21] In such cases let us say that we have a *conjunctive common cause*.

Not all common causes are conjunctive. In Salmon (1984b, 168–174) I offer examples, such as collisions of balls in games of pool (an example that has been vigorously disputed) and collisions of photons and electrons in Compton scattering, which are more properly called *interactive common causes*. They differ from conjunctive forks in important ways, as can easily be seen by a trivial reformulation of (3). According to the multiplication rule of the probability calculus

$$P(A \cdot B | C) = P(B | C)\, P(A | B \cdot C) \tag{8}$$

Since the left-hand side of (3) is the same as that of (8) we can equate their right-hand sides, yielding

$$P(B | C)\, P(A | B \cdot C) = P(A | C)\, P(B | C) \tag{9}$$

By virtue of (6), $P(B | C)$ must be greater than zero, so we may divide by it, with the result

$$P(A | B \cdot C) = P(A | C) \tag{10}$$

Equation (10) is the definition of *screening off*; given C, B is irrelevant to A. Since irrelevance is a symmetrical relation, C screens off B from A and it screens off A from B; in the presence of a common cause each of the effects is statistically irrelevant to the other.

Consider, for example, a photon with a fairly high-energy E that collides with an electron whose kinetic energy is small enough to be neglected. There is a certain probability ($0 < P < 1$) that the electron will be ejected with kinetic energy E_1 ($< E$) and that a scattered photon with energy E_2 ($< E$) will emerge. Since energy is conserved in Compton scattering, $E_1 + E_2 = E$. Hence, given the collision C, there is a probability considerably less than 1 that the scattered electron will have energy E_1, but given the collision and the energy E_2 of the scattered photon, the probability that the scattered electron will have energy E_1 equals 1. The collision is an interactive common cause, and it does not screen off the effects from each other. Consequently, not all common causes satisfy formulas (3)–(6).

There is a fundamental *physical* difference between interactive and conjunctive common causes. In the case of Compton scattering the world lines of the photon and of the electron actually intersect, and each has a direct effect on the other. In the case of food poisoning there is no actual intersection of the alimentary canals of the separate individuals; indeed, the individuals may not share any particular event. The common cause is a background condition—fish sticks consumed from the same package. In a strict sense, it seems fair to say that the conjunctive common cause—in sharp contrast with the interactive common cause—is not an event. This point has fundamental importance for my treatment of atomism. Since I construe Perrin's argument as an appeal to a *conjunctive* common cause, failure to find an *event* that serves as a common cause is irrelevant. We are looking for a *background* condition to fulfill the role of common cause, and it can readily be identified as the atomic composition of matter.

In Salmon (1984b, 217–219)—"emulating the venerable tradition established by St. Thomas"—I discuss, in a bit of detail, five ways of ascertaining Avogadro's number that were available early in the twentieth century. Three of them appear in Perrin's table; the other two do not. Let me begin by mentioning two of these methods, which were cited by Perrin in his first publications on the ascertainment of N and will be used to show in detail how the conjunctive fork can be applied to Perrin's argument.

1. Alpha radiation. Radium emits alpha particles, each of which is a helium nucleus. Soon after emission this nucleus picks up two electrons from the environment and becomes a neutral helium atom. These atoms can be collected and the collection can be weighed. Knowing the rate at which radium decays, one can easily calculate the number of atoms in the collection. The value of N follows directly. This result can be checked by determining the diminution of radium in the source of alpha radiation. The determination of Avogadro's number by this method, carried out by Ernest Rutherford and Frederick Soddy, was cited by Perrin in 1908.
2. Brownian motion. We have already discussed Perrin's work in this area in some detail.

    In the application of the definition of the conjunctive fork below, A and B will be associated with alpha radiation and Brownian motion, respectively.

3. Blackbody radiation. From the law of blackbody radiation, which he established in December 1900 (the final month of the nineteenth century), Max Planck derived the equation

$$\lambda_{max} \times kT/c = 0.2014 \times h \qquad (11)$$

where k is Boltzmann's constant, T is the temperature of the black body, c is the speed of light, h is Planck's constant, and λ_{max} is the wavelength at which the maximum amount of energy is radiated; T, c, and λ_{max} are macroscopically measurable. Einstein's theory of the photoelectric effect provides a basis for the empirical determination of Planck's constant. From equation (11) we can obviously calculate the value of k. The ideal gas law

$$PV = nRT \qquad (12)$$

furnishes the universal gas constant R; this is the gas constant per mole. Boltzmann's constant k is the gas constant per molecule. Consequently,

$$R/k = N \qquad (13)$$

Perrin also cited the value of N that results from this approach.

4. Electrochemistry. The faraday (F), which amounts to 96,500 coulombs, is the charge required to deposit one mole of a monovalent metal, such as silver, on an electrode. Since it takes the charge e of one electron to deposit each ion,

$$F/e = N \qquad (14)$$

The value of e was determined by Robert A. Millikan in 1911 by means of his famous oil-drop experiment.

5. If a crystal is bombarded by X rays a diffraction pattern results. If the wavelength of the X rays is known, the geometrical relations among the atoms in the crystal can be ascertained. This information makes it possible to ascertain how many atoms a given crystal contains. From the outcome of this type of experiment, first conducted in 1912, Avogadro's number can be determined immediately.

It is my contention, in agreement with Perrin and Poincaré, that the enormous superficial diversity of such experiments is a crucial part of the argument for the reality of atoms and molecules. In one experiment we have observers in a darkened room counting flashes on scintillation screens and subsequently weighing a sample of gas that has been collected. In the next we have scientists looking through microscopes at the distribution of tiny spheres of gamboge suspended in water. In another we have spectroscopists checking the wavelengths of light issuing from the mouth of a blast furnace. In another the observer is checking on the amount of electricity required to accomplish the electroplating of silver. Finally, we have measurements of diffraction patterns made on film by X rays bouncing off of a crystal. If these experiments were not telling us something about the fundamental structure of matter, how could they possibly yield substantially equal values for Avogadro's number N? If they are not counting something real, the agreement among these values would be an incredibly improbable coincidence. According to the principle of the common cause, an apparent coincidence this improbable requires an explanation in terms of a common cause.

6. The Logical Status of the Common Cause Principle

Certainly no one would propose that Reichenbach's version of the principle of the common cause, articulated solely in terms of conjunctive forks, holds with logical necessity in all possible worlds. Indeed, it seems clear that it does not hold universally in this world. Even if the principle is expanded to include interactive common causes, as well as conjunctive common causes, it seems to be violated in the domain of quantum mechanics, at least if "spooky action at a distance"[22] is precluded. (The outlandish character of action at a distance was exemplified by a remark of Woody Allen to the effect that "I take a bath in New York and my brother in Australia gets clean.")[23] In experiments on remote correlations, inspired by the classic article by Einstein, Podolsky, and Rosen (1935) and by Bell's theorem, we have clear cases in which what looks like a common cause does not screen off the effects from one another.[24] Moreover, introduction of interactive forks does not solve the problem. As David Mermin (1985) clearly explains, the causal processes involved in the interaction cannot possibly carry the information or causal influence required to explain these perplexing remote correlations. The characteristic that distinguishes the cases in which the common cause principle breaks down in the quantum domain is a mysterious 'process' known as "the collapse of the wave function" or "the reduction of the state vector." In contexts in which this phenomenon plays a significant role we cannot rely upon the principle of the common cause.[25]

Suppose, then, that we steer clear of situations plagued by quantum weirdness. What is the status of the principle of the common cause in other areas of empirical science? Sober (1988) maintains that even in the area of evolutionary biology, where one might suppose that this principle would have considerable applicability, it has to be handled with great care. Recall Sober's example of chimps and humans. No evolutionary biologist seriously doubts the existence of a common ancestor in this case, or, for that matter, the existence of a common ancestor of all primates. Indeed, it is plausible to suppose a common origin for all forms of life on earth (see Cairns-Smith, 1985). Common causes abound. The question for evolutionary biologists is how far back one has to go to find the common ancestor. For example, some serious authors believe it likely that humans and orangutans are more closely related than are humans and chimps (Schwartz 1987); the question is, therefore, whether the common ancestor of chimps and humans is earlier or later than the common ancestor of orangs and humans. As Sober points out, if current cosmological theory is correct, then everything has a common cause—the big bang. Important as that question is for cosmologists, it is irrelevant to the work of evolutionary biologists. It is also irrelevant to our evaluation of Perrin's argument concerning molecular reality.

Sober (1988, 212) also discusses Reichenbach's famous example of the theatrical company, supplying some plausible numerical values of the probabilities involved. Suppose that each of two actors—say, the leading lady and the leading man—suffer severe gastrointestinal distress on the average of once in 100 days, but their illnesses always occur on the same night. This would be a remarkable coincidence. Consider a long sequence of days, $x_1, x_2, x_3, \ldots$, let A be the class of days on which the leading lady suffers severe gastrointestinal distress, and let B be the class of days on which the leading man is similarly afflicted. We are assuming that 1% of these x's fall into A and that they are randomly distributed, and that 1% of the x's fall into B and that they are randomly distributed. If the illnesses were statistically independent, 0.01% ($= 0.0001$) of the x's would belong to both classes. This means that, on average, a day that falls into both A and B would occur once in about twenty-eight years. If this happens three times in a single year, we have strong grounds for believing that the illnesses of the two players are not independent events. To bring out the point even more sharply, suppose that the simultaneous illness were to happen three times in a single year, with ten, rather than two, actors involved. This is not an outlandish size for a theatrical troupe. In that case, if the illnesses were independent, the probability that all ten would be ill on the same night is 10^{-20}; that is, on average it would happen *once* in about $2.73/10^{18}$ years. According to current estimates, the age of the universe is of the order of 10^{10} years.

In this example Sober (1988, 220) agrees that we would look for a common cause, such as tainted food in shared meals. But, he asks, what if there is extremely careful control of the food preparation, so that there is only one chance in a billion that tainted food would be served? Even if we rule out tainted food as a common cause, we would be strongly impelled, I should think, to seek for a different common cause rather than attributing this coincidence to chance. The principle of the common cause does not, in itself, give us any clue to the nature of the cause. Perhaps it is an infection coming from a different source. I am strongly reminded of a remark attributed to the fictional secret agent James Bond by David Armstrong (1989): "Once is happenstance, twice is coincidence, thrice is enemy action." If accidental food poisoning and other plausible medical explanations were ruled out it would not be paranoid to suspect a devious plot. Is there, for example, a competing troupe that includes an extremely clever magician?

A basic moral that Sober draws from his discussion is that in particular cases we need to compare common cause explanations with other available explanatory theories. For example, the common cause explanation of the similarities between chimps (or orangs) and humans seems correct. A common cause explanation of wings of birds and wings of insects is less strongly founded; in this case, a separate cause explanation is more acceptable.[26] Sober (1988) puts his general conclusion as follows:

> So what I'm suggesting is that the principle of the common cause is not a *first* principle. It is not an ultimate and irreducible component of our methodology of

nondeductive inference. What validity it has in the examples that have grown up around this problem can be described pretty well within a Bayesian framework. And of equal importance is the fact that a Bayesian framework allows us to show why the preference for common cause explanations can sometimes be misplaced. (224)

It seems to me that Sober is exactly correct in offering a Bayesian interpretation of the principle of the common cause, as well as in his emphasis upon the importance of comparing different possible theoretical explanations of various phenomena, including ones that appear superficially to exhibit highly improbable coincidences. In the initial chapters in part II of this volume I explore the Bayesian approach; in chapter 6, I discuss the problem of theory comparison in Bayesian terms in considerable detail. My analysis of theory preference fits harmoniously with Sober's Bayesian treatment of the principle of the common cause.

7. Empiricism: The Key Answer

I have been arguing that the agreement in results of many diverse methods for ascertaining Avogadro's number would constitute an incredibly improbable coincidence if there were no common cause. Consider, for example, Rutherford's work on alpha radiation, discussed in Cooper (1968). It is discovered that alpha radiation produces helium. Now, without appealing to any other experimental ascertainment of Avogadro's number, we ask ourselves how many observable scintillations are required to produce 4 grams of helium. If we do not already assume the atomic-molecular theory, it seems to me, we are not even entitled to assume that each of the scintillations is associated with an equal amount of stuff. (How many pieces of coal does it take to make a ton?) The range of possible values for the number of bits of alpha radiation is pretty wide open. There does not appear to be any way even to pin down the order of magnitude (10^{23}).

Consider now the electrochemical experiment. It takes 96,500 coulombs of electric charge to deposit approximately 108 grams of silver. Rutherford's experiment at least gave rather direct evidence of discreteness; superficially viewed, this experiment does not even suggest discreteness. J. J. Thomson's work on cathode rays did provide evidence of the corpuscular nature of their constituents—electrons—and did yield a value of the ratio m/e of the mass to the charge. The value was small, but as Thomson says himself, "The smallness of m/e may be due to the smallness of m or the largeness of e, or to a combination of these two" (quoted by Cooper 1968, 442). It would not have been possible to predict with any degree of accuracy what the ratio of the faraday to the electron charge would turn out to be. And why should the number of scintillations associated with the production of 4 grams of helium agree almost precisely with the number of electron charges required to deposit approximately 108 grams of silver?

Avogadro's number N also appears in Einstein's diffusion formula. Apart from the molecular-kinetic theory there is no reason to expect that number to bear any resemblance to the values found in the alpha-radiation and electrolysis experiments. This value of N is, of course, the value discovered by Perrin (1908) in his experiment on horizontal diffusion of Brownian particles.

I could continue elaborating the lack of any basis for supposing—in the absence of the atomic/molecular theory—that the numerical results from all of the various experiments would even be of the same order of magnitude, but the point is clear enough. In fact, all thirteen entries in Perrin's 1913 table fall into an extremely narrow range, namely, 6×10^{23} to 7.5×10^{23}. Each of the thirteen values represents the results of numerous measurements, each of which has a statistical distribution. The agreement among the thirteen different methods signifies that the mean values of the different distributions are very close together. This agreement cannot be assumed to be due to chance.[27] It cannot be a result of artifacts of the experiment; given the diversity of the experiments, it would be an incredible coincidence if all of them produced the same artifactual result. Energeticism offers no explanation whatever of the agreement. The only hypothesis that remains in the running is the atomic-molecular theory.

Let us now look at this situation in terms of Reichenbach's (1956) conjunctive fork—formulas (3)–(6)—recalling that (7) follows logically from them. Considerable care must be used in interpreting these formulas because they are written in an abbreviated notation. According to Reichenbach, a probability expression of the general form "$P(B|A)$" implicitly refers to two ordered sequences, $\{x_i\}$ and $\{y_i\}$, and it stands for the frequency with which an $x_i \in A$ corresponds to a $y_i \in B$; that is, $P(y_i \in B | x_i \in A)$.[28]

For present purposes let us take as our universe of discourse W the class of all experiments intended to ascertain Avogadro's number by any method. (For simplicity we shall, however, discuss only two types, those involving alpha radiation and helium production and those involving Brownian motion.) When one sets up any reasonably sophisticated scientific experiment it is important to make sure that the supposed initial conditions are satisfied. For example, it required tremendous labor on Perrin's part to create spheres of gamboge of uniform size and density that would be suitable for his experiments on Brownian motion. Many other experimental problems had to be overcome. Rutherford's work on alpha radiation depended crucially on the ability of his observers to detect minute scintillations in a completely darkened room; the observers had to spend long periods of time adapting their eyes to the dark in order to make accurate counts. Clearly, if the initial conditions are not fulfilled, we cannot expect successful results, except perhaps occasionally by chance.

With this point in mind we next partition W into two classes: S, the class of experiments in which the initial conditions are satisfied within tolerable limits; $\sim$S, the class of experiments in which they are not thus satisfied. We assume that neither of these subclasses is empty. Within W we select the class X of experiments employing alpha radiation and order its members $x_1, x_2, x_3, \ldots$ in some manner. We select the class Y of experiments involving Brownian motion and order its members $y_1, y_2, y_3, \ldots$. The order might be chronological. The subscripts furnish a correspondence between the members of the two sequences even though the members of a given pair will, generally speaking, be performed at different times. We define a new sequence Z, consisting of such ordered pairs; that is, $z_i = \langle x_i, y_i \rangle$. A given pair $\langle x_i, y_i \rangle$ belongs to C just in case both members belong to S; it belongs to $\sim$C if either experiment fails to satisfy initial conditions.[29] Let A be the class of experiments on alpha radiation that yield results falling within a range to be specified; let B be the class of experiments on Brownian motion that yield results falling within that range. In the end we will think of experiments falling within that range as successful. What is meant by a successful result? The values in Perrin's 1913 table run from 6.0×10^{23} to 7.5×10^{23}. Since these experiments are exceedingly difficult, let us allow a more generous range, say, from 4.5×10^{23} to 9.0×10^{23}, to count as successful. Because of the delicacy of these experiments we cannot expect successful results in every case, even when the initial conditions are fulfilled. We can, however, expect successful results more frequently when the initial conditions are fulfilled than when they are not.

In order to interpret the formulas in Reichenbach's definition of the conjunctive fork we must consider the meanings of the probability expressions. To that end, we now write out the formulas in the expanded notation.[30]

$$P(x_i \in A \cdot y_i \in B | z_i \in C) = P(x_i \in A | z_i \in C) \; P(y_i \in B | z_i \in C) \qquad (3^*)$$

$$P(x_i \in A \cdot y_i \in B | {\sim} z_i \in C) = P(x_i \in A | {\sim} z_i \in C) \; P(y_i \in B | {\sim} z_i \in C) \qquad (4^*)$$

$$P(x_i \in A | z_i \in C) > P(x_i \in A | {\sim} z_i \in C) \qquad (5^*)$$

$$P(y_i \in B | z_i \in C) > P(y_i \in B | {\sim} z_i \in C) \qquad (6^*)$$

In the expanded notation, the logical consequence becomes

$$P(x_i \in A \cdot y_i \in B) > P(x_i \in A) \; P(y_i \in B) \qquad (7^*)$$

Having exhibited the expanded notation, we can now revert to the contracted forms. Since we can expect successful outcomes more frequently when the initial conditions are fulfilled than when they are not, the two inequalities in the definition of the conjunctive fork,

$$P(A|C) > P(A|{\sim}C) \qquad (5)$$

$$P(B|C) > P(B|{\sim}C) \qquad (6)$$

will be fulfilled. The first equation,

$$P(A \cdot B|C) = P(A|C) \; P(B|C) \qquad (3)$$

is satisfied because, given the satisfaction of initial conditions in both types of experiments, the success rate of one type of experiment is independent of the success rate of the other type. The second equation,

$$P(A \cdot B|{\sim}C) = P(A|{\sim}C) \; P(B|{\sim}C) \qquad (4)$$

is satisfied for essentially the same reason. The accidental success of an experiment on alpha radiation conducted when the initial conditions of the experiment are not fulfilled has no bearing on the frequency of success of experiments on Brownian motion whether or not their initial conditions are fulfilled.

In the foregoing analysis we have considered satisfaction or violation of initial conditions in experiments without picking out any particular type of violation. To make the analysis more directly relevant to the problem of molecular reality, one particular initial condition is crucial, namely, the atomic/molecular constitution of matter. Since the matter involved in all of these experiments has this character, formulas (4)–(6) present a problem, for $\sim C$ becomes an empty reference class.[31] The appropriate way to address this problem is with a little imagination; that is, we need to introduce hypothetical probabilities that would obtain if not all matter had an atomic/molecular structure. Since we have a reasonably well-formulated alternative, namely, energeticism, it is not unreasonable to see what would happen if that theory were true:

> Consider Brownian motion. Perhaps water is a continuous medium that is subject to internal microscopic vibrations related to temperature. Suspended particles might behave in just the way Brown and others observed. Moreover, these vibrations might bear some relationship to the chemical properties of different substances that we (mistakenly) identify with molecular weight. Conceivably, such vibratory behavior might lead to precisely the results of experiments on Brownian motion that Perrin found, leading us (mistakenly) to think that we had successfully ascertained the number of molecules in a mole of the suspending fluid. It is for this reason that no *single* experimental method of determining Avogadro's number, no matter how ingeniously and beautifully executed, could serve to establish decisively the existence of molecules. The opponent could plausibly respond that, with respect to Brownian motion, fluids behave merely *as if* they were composed of molecules, and, indeed, as if they contained a specific number of molecules.
>
> When we turn to helium production by alpha radiation, we must find another alternative explanation. Assuming that radium is *not* composed of atoms, we may still suppose that tiny globs of stuff are ejected from a mass of radium at a statistically stable rate. We may further claim that, for some reason, all of these globs are the same size. Moreover, we may assert that the globs are bits of helium that can be collected and weighed. We find, consequently, that it takes a given number of little globs of helium to make up four grams. Even given all of these assumptions, it is extremely implausible to suppose that the number of globs in four grams of helium should equal the number that Perrin (mistakenly) inferred to be the number of molecules in 18 grams of water. It is difficult—to say the least—to see any connection between the vibratory motions of water that account for Brownian motion of the noncorpuscular hypothesis and the size of the globs of helium that pop out of chunks of radium. And even if—in some fashion that is not clear to me—it were possible to provide a noncorpuscular account of the substances involved in experiments of these two types, we would still face the problem of extending this noncorpuscular theory to deal, not just qualitatively, but also quantitatively with the results of the many other types of experiments to which we have made reference. To scientists like Ostwald, the task must have seemed insuperable. (Salmon 1984b, 225–226)

Under the assumption that atoms and molecules do not exist, it is evident that the probability of an experiment on Brownian motion yielding a value for N in the above specified range from 4.5×10^{23} to 9.0×10^{23} is independent of the probability of an experiment on alpha radiation doing so. The problematic equation

$$P(A \cdot B | {\sim} C) = P(A | {\sim} C)\ P(B | {\sim} C) \qquad (4)$$

is thus fulfilled. Similarly, if we expand our universe W to include the hypothetical experiments (supposing energeticism to be true), as well as the actual experiments (supposing the atomic/molecular theory to be true), the inequalities

$$P(A | C) > P(A | {\sim} C) \qquad (5)$$

$$P(B | C) > P(B | {\sim} C) \qquad (6)$$

are clearly satisfied. These results can be asserted whether one thinks of these probabilities as counterfactual physical probabilities or reasonable degrees of belief.

The preceding quotation brings out one of Sober's basic points, namely, that evaluation of common cause arguments requires comparison with alternative hypotheses that might serve to explain the same coincidence. The chief *scientific* alternatives at the turn of the century were energeticism and atomism; energeticism, even according to its most distinguished proponents, was unequal to the task. No novel hypothesis was forthcoming. Thus, if one wanted a scientific theory to explain the great coincidence involved in the results of thirteen or more distinct ways of ascertaining Avogadro's number, the atomic-molecular theory was it.

One could, of course, hew to a *philosophical* line and insist that we cannot know anything about such unobservables as atoms and molecules. As mentioned above, toward the end of his life Mach *accepted* the atomic theory, *but only as an instrument*. A similar position, it seems to me, is advocated by van Fraassen (1980, 3)—namely, that we accept the atomic theory, not as providing a veridical picture of reality, but only as empirically adequate. As van Fraassen has correctly noted, given any body of evidence, the statement that a theory is empirically adequate is at least as probable as (and, in general, more probable than) the statement that it is true. High probability is not, however, our only value; it is easily had by refusing to endorse anything beyond truisms. Information is also important. Recall Feynman's highly relevant comment: "If, in some cataclysm, all of scientific knowledge were to be destroyed, and only one sentence passed on to the next generation of creatures, what statement would contain the most information in the fewest words? I believe it is the *atomic hypothesis* . . ." (Feynman et al. 1963, 1–2).[32] In my opinion we can afford to sacrifice a small amount of probability for a great amount of information, especially when, in Perrin's ([1913] 1816) words, the theory has "a probability bordering on certainty."

One could also return to the closely associated view that science is concerned only with models, as van Fraassen (1980, 5) has put it: "I shall give a momentary name, 'constructive empiricism', to the philosophical position I shall advocate. . . . I use the adjective 'constructive' to indicate my view that scientific activity is one of construction rather than discovery: construction of models that must be adequate to the phenomena, and not discovery of truth concerning the unobservable." When we look at Perrin's argument for the atomic hypothesis, however, we must keep in mind that we are not dealing with one model but with many. We have a model for chemical combination, one for electrolysis, one for radioactive decay, one for the behavior of gases, one for blackbody radiation, one for thermodynamics, and so on. How do these models relate to physical reality? Can we draw any conclusions concerning the correspondence (or lack of it) to what exists in the world?

The answer to this question, it seems to me, hinges greatly on the question of how the models fit together. If we have a large number of models to cover a variety of types of physical situations, and the models fit together so harmoniously as to give us the same value for Avogadro's number, it seems to me that by a common cause argument we can conclude that they mirror reality—not perfectly, of course, but in certain crucial respects. We still do not know, to this day, everything there is to know about atoms and molecules, but we do know that they exist and that they have various important properties that were well known by 1913. The existence of such unobservables as atoms and molecules is surely better founded than the (in principle observable) occurrence of an impact of a large meteor or comet with the earth approximately 65 million years ago, in which many scientists are currently prepared to believe.

So what is the answer to the key question, "whether inductive reasoning contains the resources to enable us to have observational evidence for or against statements about unobservable entities and/or properties"? I have addressed the question in two stages. In chapter 1, I considered realism with respect to microscopic objects. In this chapter I have extended the argument to the submicroscopic domain. The answer is affirmative in both cases. (Going from small to large, I argue for "extragalactic realism" in the appendix below.) The realism I have supported is not a new or unique view; it is held by many philosophers on the basis of a variety of arguments. However, the argument on which I base my realistic claims has not previously, to the best of my knowledge, been spelled out in detail by any other philosopher. It rests solidly on developments in the history of science and fits comfortably into a Bayesian framework. The Bayesian approach will be examined in part II.

Appendix: Extragalactic Reality

Ian Hacking, whose work on microscopic observation I enthusiastically endorsed and used in chapter 1, has subsequently raised questions concerning scientific realism on a cosmic scale (Hacking 1989). He directs his attention mainly to gravitational lenses; in section 5 of this chapter I cited the first alleged discovery of one (quasars O957 + 561 A, B) as a prime example of application of the principle of the common cause. A good deal of Hacking's discussion involves technical details regarding the interpretation of the evidence. For example, he raises the well-known question concerning the number of images. At first only two images were seen, but theoretical considerations led to the conclusion that there should be three. Subsequent observation involving better resolution of the images convinced some of the investigators that there were actually three images, two of which were very close together. Others were not convinced. The observational and theoretical difficulties involved in this case are enormous. All of this is good old scientific controversy over theory and the interpretation of the data.

I happen to believe that the supporters of the gravitational lens hypothesis are correct, but that is beside the point as far as scientific realism is concerned. Suppose that the skeptics are right, and that O957 + 561 A, B is not a case of gravitational lensing. Go even further; suppose that no actual gravitational lens has been discovered. Scientific existence claims are always open to question: N rays, cold fusion, gravitational lenses, and so on. The scientific realist is not committed to belief in every kind of entity or process hypothesized at any particular stage in the history of science. Hacking (1989), however, is making a much more sweeping claim—one that goes far beyond the denial of gravitational lenses—leading him to identify himself as "an anti-realist about [extragalactic] astrophysics" (576). He offers two primary reasons for taking this position.

First, in his discussions of scientific realism, Hacking has often emphasized the importance of experiment, intervention, and manipulation. Astronomy is, however, an old and honored science in which, until quite recently, it was impossible to manipulate its objects of interest and, even now, is severely limited in that respect. Still, Hacking seems willing to admit that we can and do have astronomical knowledge of many entities in our own galaxy. We have reasonable scientific confidence in the existence of nearby stars, and we know some of their properties. He would admit, I presume, that the Milky Way contains innumerable individual stars, even though a telescope is required to discern them. "I did not say in my (1983) that an experimental argument is the only viable argument for scientific realism about unobserved entities," he remarks. "I said only that it is the most compelling one, and perhaps the only *compelling* one" (1989, 560–561; emphasis in original).

Some astrophysicists believe that there is a black hole at the center of our galaxy; I am not convinced of this particular existence claim, and Hacking is skeptical far more generally about black holes. Some existence claims about types of objects in our own galaxy are highly problematic. Since, however, he admits that telescopic observation, even in the absence of intervention, allows us to establish existence claims about objects in our galaxy, I see no reason why he should not allow similar telescopic observation to furnish evidence for the existence of other galaxies beyond our own. It is not immediately evident, of course, that the observed galaxies (which used to be called "nebulae" and were thought to reside within our galaxy) are actually located outside of our galaxy. It seems to me, however, that a strong enough case can now be made for their extragalactic location. Even on Hacking's principles, I think, wholesale extragalactic antirealism cannot be sustained. We have to pick and choose.

Existence claims about black holes are on weak ground, gravitational waves not much stronger, gravitational lenses somewhat stronger, quasars quite a bit stronger, and other galaxies very strong indeed. I am an extragalactic realist; I hold that we have compelling evidence at least for items in last mentioned category. I am also a fallibilist. In 1968, when Joseph Weber announced that he had detected gravitational waves on the basis of correlations between vibrations in two widely separated detectors, I accepted his claim as a superb example of a common cause argument. The conclusion of that argument was wrong. Perhaps there was a significant correlation. If so, there was probably a common cause. Almost certainly the passage of gravitational radiation was not the common cause.

Hacking's second reason for extragalactic antirealism concerns models. Citing his earlier work (1983, 216) he offers "a simple reason for not regarding models as true-of-the-world, namely, that it is absolutely standard practice to use a number of different models in solving the same class of problems, even though the models are formally inconsistent with each other, if taken to be making literal assertions about the world. In short one must be anti-realist about models . . ." (1989, 573). He goes on to express his suspicion that "there is no branch of natural science in which modeling is more endemic than astrophysics, nor one in which at every level modeling is more central" (573–574). According to Hacking, ancient astronomy was a science devoted to 'saving the phenomena' in the sense of making models that would be consistent with what is observed. A good deal of current astrophysics, especially extragalactic astrophysics, follows exactly the same method. In this regard it differs radically from the kind of experimental natural science that was born in the scientific revolution. For this reason we are entitled to be skeptical about extragalactic existence assertions (sec. 19, 576–578).

The issue of modeling brings us back to the main theme of the present chapter, namely, the existence of atoms and molecules. I hold a realistic position in that domain, and so does Hacking. I have tried to spell out my reasons. We should recall, in this connection, that models played a crucial role in the whole development from Dalton to the present day. Clearly we must distinguish at least two radically different types of models. The first is the mechanical model that played an important role in nineteenth-century physics—the type consistent with what is observed. We now consider models of that type utterly dispensable in the presence of a well-developed theory such as Newtonian mechanics[33] or Maxwellian electrodynamics. Models of the other type arose in response to partial knowledge and partial ignorance. A great deal could be learned about gases in the nineteenth century by treating them as collections of tiny spheres in constant motion that collide elastically with one another and with the walls of the container of the gas. This model is literally "false" because the molecules are not impenetrable spheres; but gases do consist of small particles—that is, the molecules exist and have certain properties such as size. Obviously we want to learn more about them.

Molecules are composed of atoms, but the structure of the atom was not well understood. J. J. Thomson found that he could strip electrons from atoms, and he offered his 'plum-pudding model' with electrons embedded throughout. Antoine Henri Becquerel discovered that atoms are subject to spontaneous radioactive decay. Rutherford studied the behavior of alpha radiation, and concluded that Thomson's model was untenable; he proposed his own planetary model, which was highly unstable according to classical electrodynamics. Niels Bohr provided a quantized model, which, though highly problematic, dealt with the hydrogen spectrum. Many of these developments occurred during the period in which Perrin was attempting to demonstrate the real existence of atoms and molecules. The fact that a succession of different models was employed to deal with different problems about the structure of matter does not call into question the value of Perrin's work or the validity of his conclusion about molecular reality. In fact, it was impossible to understand atomic or molecular structure in the absence of quantum mechanics, in particular, Pauli's exclusion principle. The fact that many different patently inadequate and conflicting models were devised to deal with various phenomena does nothing to undermine the claim that ordinary matter in its gaseous, liquid, and solid states is composed of minute particles, some of whose properties—such as mass, size, velocity, and number—could be established.

The conclusion to be drawn, in my opinion, is that the practice of modeling does not, in itself, have ontological implications. The question of molecular reality was settled by the convergence of a variety of types of evidence. Questions about extragalactic entities must be settled on the basis of the particular kinds of evidence that become available to us. Gravitational lenses are regarded by some astrophysicists as a powerful research tool. Even if we cannot manipulate them, their utility as an instrument of observation may carry conviction about their actual existence.

I must add one final remark about Hacking's (1989, 568) characterization of my position. He attributes to me an argument of the type known as "inference to the best explanation." This is inaccurate. It seems to me that the phrase "inference to the best explanation" is nothing but a slogan; it does not designate a clearly articulated mode of reasoning in science or about science. In the absence of some consensus on the nature of scientific explanation—and nothing even approaching one exists at present (see Salmon [1989] 1990)—the slogan itself has no precise meaning. As I have said consistently, my argument appeals to the principle of the common cause. Common causes often have great explanatory value, but not all good explanations are of that type. I therefore strongly resist the use of that expression in connection with my argument for scientific realism.

PART II

RATIONALITY

Introduction

Thomas Kuhn's (1962) *The Structure of Scientific Revolutions* had a profound effect on philosophy of science in the second half of the twentieth century. Reactions were mixed. My main impression was that Kuhn based much of his argument upon a defective conception of scientific confirmation—something akin to the traditional *hypothetico-deductive* (H-D) method. His claim that the choice among scientific theories goes beyond empirical evidence and logic, even when logic is taken to include both deduction and inductive confirmation, resulted from overlooking certain important aspects of confirmation that are revealed by a Bayesian analysis. His insistence that *judgment* and *persuasion* play indispensable roles in theory choice suggests that *plausibility considerations* are involved. His rejection of the distinction between the *context of discovery* and the *context of justification* hints strongly that his conception of justification (which surely involves confirmation) is inadequate. Despite these views, Kuhn unequivocally rejects the notion that science is irrational. This denial cries out for a clarification of the concept of *scientific rationality* itself.

The first three chapters in part II deal with these issues. "Plausibility Arguments in Science" (chapter 4) exhibits the shortcomings of the H-D method and introduces the Bayesian approach. It argues that plausibility considerations can be identified with assessment of the prior probabilities of hypotheses. Since prior probabilities are required for application of Bayes's theorem, plausibility considerations are not only admissible for scientific confirmation but also indispensable. "Discovery and Justification" (chapter 5) maintains that a proper analysis of plausibility arguments enables us to retain a philosophically useful distinction between these two contexts. It further claims that the Bayesian analysis forges a strong link between history of science and philosophy of science—one that has been largely ignored. "Rationality and Objectivity in Science" (chapter 6) shows how the rationality of science can be upheld within the Bayesian framework.

It offers a Bayesian account of choice between competing theories, as opposed to the evaluation of individual theories viewed in isolation from one another. This comparative approach, which echoes a major theme in Kuhn's work, avoids the aggravating Bayesian problem of the so-called *catchall hypothesis*.

"Revisions of Scientific Convictions" (chapter 7) pursues the Kuhnian topic of scientific change, showing how a Bayesian approach accounts for revisions of scientific opinion. It deals with Bayesian conditionalization in the light of new evidence and with other types of revisions that are made under other circumstances. It includes a resolution of the notorious Bayesian *problem of old evidence*.

Bayes's theorem itself is an uncontroversial proposition in the mathematical calculus of probability. Controversy arises when we try to interpret the probability concept in order to apply the abstract mathematical schema in concrete situations, especially in connection with scientific confirmation. Most so-called Bayesians adopt a subjective or personalist interpretation; they excuse the subjectivity by appealing to a kind of convergence of opinion that results from using Bayes's theorem. However, although intersubjective agreement can often be achieved, we are left with profound questions about the objectivity of science. Kuhn's approach raises a similar problem. One wonders whether a scientific paradigm reflects more than some sort of communal consensus. "Dynamic Rationality" (chapter 8) deals with the relationships between subjective and objective concepts of probability, showing how objectivity can be attained within the Bayesian framework.

"Hume's Arguments on Cosmology and Design" (chapter 9) is a dividend. A long tradition maintains that we can have empirical or scientific proof for the existence of God on the basis of the *design argument*. David Hume's ([1779] 1970) philosophical masterpiece, *Dialogues Concerning Natural Religion*, provides a profound analysis of that argument. If, as I claim, Bayes's theorem constitutes the correct schema for scientific confirmation, then it should be not only possible but also fruitful to apply that schema to Hume's analysis. This shows how the Bayesian approach provides insight into a fundamental philosophical problem in a field outside of the realm of pure confirmation theory. The conclusion is that, if the existence of God is decidable on the basis of empirical evidence, then in all probability God does not exist.

The aforementioned six chapters in part II are fundamentally concerned with Bayesian themes. The remaining three chapters deal, in one way or another, with analogies that have been drawn explicitly or implicitly between deduction and induction. They show unequivocally that extreme caution must be used in drawing such analogies and that casual intuitions are extraordinarily poor guides.

"The 'Almost-Deduction' Theory" (chapter 10) takes the theorem on total probability as its point of departure. Although it has not often been explicitly recognized as such, this theorem, which was mentioned along with Bayes's theorem in several of the foregoing chapters, is the transitivity rule in the calculus of probability. Recognition of this fact enables us to sort out several valid and invalid inductive transitivity relations. Next, this chapter sets down conditions under which a probabilistic analogue of deductive contraposition is admissible and when it is not. The closely related "probabilistic *modus tollens*"—an extremely important form of argument—is handled by the same analysis. We see that it is legitimate under certain specifiable conditions; otherwise it is fallacious. To my mind, this chapter conclusively refutes what might be called "the almost-deduction theory of inductive logic."

Many philosophers have been attracted to the idea that inductive logic can be based on a relation of partial entailment in much the same way that deductive logic rests on a relation of full entailment. This notion is explored in "The 'Partial-Entailment' Theory" (chapter 11). Analysis shows that partial entailment cannot do the job many philosophers had hoped for. It is easy to demonstrate, as a corollary, that the notorious Popper-Miller "proof" of the impossibility of inductive probabilistic support is fallacious.

This work on partial entailment was the point of departure for the doctoral dissertation, "Content and Confirmation," of Kenneth Bruce Gemes (1991). Gemes rejects the Bar Hillel and Carnap (1953–1954) definition of semantic content, which I adopt in chapter 4, and develops an alternative concept that blocks intuitively irrelevant statements from entering into the content of a statement. He thereby avoids some extremely counterintuitive consequences of the Bar Hillel and Carnap concept. Gemes also introduces a valuable distinction between genuine confirmation and mere "content-cutting," where the latter process is typified by the absence of learning from experience. It was precisely this distinction that led Carnap to reject the Wittgensteinian confirmation function $c^{\dagger}$ and introduce his function c^{*}.

As "The 'Partial-Entailment' Theory" brings out, the concept of confirmation is radically ambiguous; "Confirmation and Relevance" (chapter 12) explores this ambiguity. In some cases when we say that a hypothesis is confirmed we mean that it has a high degree of confirmation; I call this *absolute confirmation*. In many other cases when we say that a hypothesis is confirmed we mean that its probability or degree of confirmation has increased to some degree; I call this *incremental confirmation*. Carnap (1950b, chap. VI) studied the characteristics of incremental confirmation in great detail, showing that this relation has strange and counterintuitive properties. These results were widely ignored or misunderstood; in his preface to the second edition (Carnap 1962b), Carnap drew the distinction even more carefully and emphatically. My chapter reiterates Carnap's results, extends them to a certain degree, and applies them in philosophical contexts. I try to show that much of the counterintuitive character of incremental confirmation stems from a strong tendency to think about inductive reasoning as analogous to two-valued truth-functional logic.

The mathematical theory of probability provides the foundation for all of the chapters in part II. Yet, the mathematical technicalities are minimal. The point is to show how consideration of extremely simple formulas enables us to answer serious philosophical questions. All of these chapters are self-contained; they do not presuppose any prior knowledge of mathematical probability. Although it is not required for an understanding of the chapters in part II, the following appendix provides easy proofs of the theorem on total probability and the theorem of Bayes.

Appendix

Let $P(B|A)$ be the probability of B, given A, where these variables may stand for statements or for sets. We begin with the derivation of the theorem on total probability. Trivially

$$P(C|A) = P(C|[B \vee {\sim}B]A) = P(B{\cdot}C \vee {\sim}B{\cdot}C|A) \tag{1}$$

Since $B{\cdot}C$ and ${\sim}B{\cdot}C$ are mutually exclusive

$$P(B{\cdot}C \vee {\sim} B{\cdot}C|A) = P(B{\cdot}C|A) + P({\sim}B{\cdot}C|A) \tag{2}$$

By the multiplication rule

$$P(B{\cdot}C|A) = P(B|A)\,P(C|A{\cdot}B) \tag{3}$$
$$P({\sim}B{\cdot}C|A) = P({\sim}B|A)\,P(C|{\sim}B{\cdot}A) \tag{4}$$

Substituting in (1) using (3) and (4) we have

$$P(C|A) = P(B|A)\,P(C|A{\cdot}B) + P({\sim}B|A)\,P(C|{\sim}B{\cdot}A) \tag{5}$$

which is the theorem on total probability. To derive Bayes's theorem we write the multiplication rule twice:

$$P(B{\cdot}C|A) = P(B|A)\,P(C|A{\cdot}B) \tag{6}$$
$$P(C{\cdot}B|A) = P(C|A)\,P(B|A{\cdot}C) \tag{7}$$

Since $B{\cdot}C$ is logically equivalent to $C{\cdot}B$ we can equate the right-hand sides of (6) and (7) with the result

$$P(C|A)\,P(B|A{\cdot}C) = P(B|A)\,P(C|A{\cdot}B) \tag{8}$$

Assuming $P(C|A) \neq 0$ we divide both sides by it:

$$P(B|A{\cdot}C) = \frac{P(B|A)\,P(C|A{\cdot}B)}{P(C|A)} \tag{9}$$

which is one form of Bayes's theorem. Using the theorem on total probability to replace the denominator on the right-hand side of (9) yields another useful form:

$$P(B|A{\cdot}C) = \frac{P(B|A)\,P(C|A{\cdot}B)}{P(B|A)\,P(C|A{\cdot}B) + P({\sim}B|A)\,P(C|{\sim}B{\cdot}A)} \tag{10}$$

Plausibility Arguments
in Science

In 1950, L. Ron Hubbard published *Dianetics: The Modern Science of Mental Healing*, which purported to provide a comprehensive explanation of human behavior and which recommended a therapy for the treatment of all psychological ills. According to Hubbard's theory, psychological difficulties stem from "engrams," or brain traces, that are the results of experiences the individual has undergone while unconscious as a result of sleep, anesthesia, a blow to the head, or any other cause. Of particular importance are experiences that occur before birth. Hubbard gives strikingly vivid accounts of life in the womb, and it is far from idyllic. There is jostling, sloshing, noise, and a variety of rude shocks. Any unpleasant behavior of the father can have serious lasting effects upon the child. On a Saturday night, for example, the father comes home drunk and in an ugly mood; he beats the mother and with each blow he shouts, "Take that, take that!" The child grows up and becomes a kleptomaniac. Martin Gardner (1957) provides many additional, and often amusing, details.

It is perhaps worth remarking that the author of this work had no training whatsoever in psychology or psychiatry. The basic ideas were first published in an article in *Astounding Science Fiction*. In spite of its origins, this book was widely read, the theory was taken seriously by many people, and the therapy it recommended was practiced extensively. A psychologist friend of mine remarked at the time, "I can't condemn this theory before it is carefully tested, but afterwards I will."

In the same year—it seems to have been a vintage year for things of this sort—Immanuel Velikovsky (1950) published *Worlds in Collision*, a book that attempted to account for a number of the miracles alleged in the *Old Testament*, such as the flood and the sun's standing still. This latter miracle, it was explained, resulted from a sudden stop in the earth's rotation about its axis, which was brought about, along with the various other cataclysms, by the very close approach to the earth of a giant comet that later became the planet Venus (Gardner 1957, 28–29). One of the chief difficulties encountered by Velikovsky's explanation is that, on currently accepted scientific theory, the rotation of the earth simply would not stop as a result of the postulated close approach of another large body. In order to make good his explanation, Velikovsky must introduce a whole body of physical theory that is quite incompatible with what is generally accepted today and for which he can summon no independent evidence. The probability that Velikovsky's explanation is correct is, therefore, no greater than the probability that virtually every currently accepted physical theory is false. Again, Gardner provides many more interesting details.

Before the publication of his book, parts of Velikovsky's (1950) theory were published serially in *Harper's Magazine*. When the astounding new theory did not elicit serious consideration from the scientific community, the editors of *Harper's* expressed outrage at the lack of scientific objectivity exhibited by the scientists.[1] They complained, in effect, of a scientific establishment with its scientific orthodoxy, which manifests such overwhelming prejudice against heterodox opinions that anyone like Velikovsky, with radically new scientific ideas, cannot even get a serious hearing. They were not complaining that the scientific community rejected Velikovsky's views, but rather that it dismissed them without any serious attempt to test them.

1. The Status of Plausibility Judgments

The foregoing are only two examples of scientific prejudgment of a theory; many other fascinating cases can be found in *Fads and Fallacies in the Name of Science* (Gardener 1957). Yet, there is a disquieting aspect of this situation. We have been told on countless occasions that the methods of science depend upon the objective observational and experimental testing of hypotheses; science does not, to be sure, prove or disprove its results absolutely conclusively, but it does demand objective evidence to confirm or disconfirm them. This is the scientific ideal. Yet scientists in practice do certainly make judgments of plausibility or implausibility about newly suggested theories, and in cases like those of Hubbard and Velikovsky, they judge the new hypothesis too implausible to deserve further serious consideration. Can it be that the editors of *Harper's* had a point, that there is a large discrepancy between the ideal of scientific objectivity and the actual practice of prejudgment on the basis of plausibility considerations alone? One could maintain, of course, that this is merely an example of the necessary compromise we make between the abstract ideal and the practical exigencies. Given unlimited time, talent, money, and material, perhaps we should test every hypothesis that comes along; in fact, we have none of these commodities in unlimited supply, so we have to make practical decisions concerning the use of our scientific resources. We have to decide which hypotheses are promising and which are not. We have to decide which experiments to run and what equipment to buy. These are all practical decisions that have to be made, and in making them, the scientist (or administrator) is deciding which hypotheses will be subjected to serious testing and which will be ignored. If *every* hypothesis that comes along had to be tested, I shudder to think how scientists would be occupied with antigravity devices and refutations of Einstein.

Granted that we do, and perhaps must, make use of these plausibility considerations, the natural question concerns their status. Three general sorts of answers suggest themselves at the outset. First, they might be no more than expressions of the attitudes and prejudices of individual scientists or groups of scientists. The editors of *Harper's* might be right in claiming that they arise simply from the personal attitudes of individual scientists. Second, they might be thought to have a purely practical function. Perhaps they constitute a necessary but undesirable compromise with the ideal of scientific objectivity for the sake of getting on with the practical work of science. Or maybe these plausibility considerations have a heuristic value in helping scientists discover new and promising lines of research, but their function is solely in relation to the discovery of hypotheses, not to their justification. Third, it might be held that somehow plausibility arguments constitute a proper and indispensable part of the very logic of the justification of scientific hypotheses (see the following chapter). This is the view I shall attempt to elaborate and defend. I shall argue that plausibility arguments are objective in character and that they must be taken into account in the evaluation of scientific hypotheses on the basis of evidence.[2]

2. The Hypothetico-Deductive Method

The issue being raised is a logical one. We are asking what ingredients enter into the evaluation of scientific hypotheses in the light of evidence. In order to answer such questions, it is necessary to look at the logical schema that represents that logical relation between evidence and hypotheses in scientific inference. Many scientific textbooks, especially the introductory ones, attempt to give a brief characterization of the process of confirming and disconfirming hypotheses. The usual account is generally known as the *hypothetico-deductive (H-D) method*. As it is frequently described, the method consists in deducing consequences from the hypothesis in question and checking by observation to determine whether these consequences actually occur. If they do, that counts as confirming evidence for the hypothesis; if they do not, the hypothesis is disconfirmed.

There are two immediate difficulties with the foregoing characterization of the hypothetico-deductive method. First, from a general hypothesis alone it is impossible to deduce any observational consequences. Consider, for example, Kepler's first two laws of planetary motion: the first states that the orbits of the planets are elliptical, and the second describes the way the speed of the planet varies as it moves through the ellipse. With this general knowledge of the motion of Mars, for instance, it is impossible to deduce its location at midnight tonight, and so to check by observation to see whether it fulfills the prediction or not. But with the addition of some further observational knowledge, it may be possible to make such deductions—for instance, if we know its position and velocity at midnight last night. This additional observational evidence is often referred to as the *initial conditions*. Second, in many cases the observational prediction does not follow from the hypothesis and initial conditions alone, but so-called *auxiliary hypotheses* are also required. For instance, if an astronomical observation is involved, optical theories concerning the behavior of telescopes may be implicitly invoked. From the hypothesis together with statements about initial conditions and the auxiliary hypotheses it is possible to deduce a concrete prediction that can be checked by observation. With these additions, the hypothetico-deductive method can be represented by the following simple schema:

H-D schema: H   (hypothesis being tested)
               A   (auxiliary hypotheses)
               I   (initial conditions)
               _____
               O   (observational prediction)

For purposes of the present discussion, let us make two simplifying assumptions. First, although it is always possible for errors of observation or measurement to occur, and consequently for us to be mistaken about the initial conditions, I shall assume that we have correctly ascertained the initial conditions. This is one useful simplifying assumption. Second, a false prediction can, in principle, be the occasion to call the auxiliary hypotheses into question, so that the most that can be concluded is that *either* an auxiliary hypothesis *or* the hypothesis up for testing is false, but we cannot say which. However, I shall assume that the truth of the auxiliary hypotheses is not in question, so the hypothesis we are trying to test is still the only premise of the argument whose truth is open to question. Such simplifying assumptions are admittedly unrealistic, but things are difficult enough with them, and relinquishing them does not help with the problem we are discussing.

Under the foregoing simplifying assumptions a false prediction provides a decisive result: If the prediction is false the hypothesis is falsified, for a valid deduction with a false conclusion *must* have at least one false premise and the hypothesis being tested is the only premise about which we are admitting any question. However, if the prediction turns out to be true, we certainly cannot conclude that the hypothesis is true, for to infer the truth of the premises from the truth of the conclusion is an elementary logical fallacy. And this fallacy is not mitigated in the least by rejecting the simplifying assumptions and admitting that other premises might be false. The fallacy, called *affirming the consequent*, is illustrated by the following example: If one has chicken pox, one will run a fever; this patient is running a fever; therefore, this patient has chicken pox. The difficulty is very fundamental and very general. Even though a hypothesis gives rise to a true prediction, there are always other, different hypotheses that would provide the same prediction. This is the *problem of the alternative hypotheses.* It is especially apparent in any case in which one wishes to explain data that can be represented by points on a graph in terms of a mathematical function that can be represented by a curve. There are always many different curves that fit the data equally well; in fact, for any finite number of data, there are infinitely many such curves. Additional observations will serve to disqualify some of these (in fact, infinitely many), but infinitely many alternatives will still remain.

Since we obviously cannot claim that the observation of a true consequence establishes the truth of our hypothesis, the usual claim is that such observations tend to support or confirm the hypothesis, or to lend it probability.[3] Thus, it is often said, the inference from the hypothesis, initial conditions, and auxiliary hypotheses (if any) to the prediction is deductive, but the inference in the opposite direction, from the truth of the prediction to the hypothesis, is inductive. Inductive inferences do not pretend to establish their results with certainty; instead, they confirm them or make them probable. The whole trouble with looking at the matter in this way is that it appears to constitute an automatic transformation of deductive fallacies into correct inductive arguments. When we discover, to our dismay, that our favorite deductive argument is invalid, we simply rescue it by saying that we never intended it to be deductive in the first place, but that it is a valid induction. Surely inductive logic, if it plays a central role in scientific method, must have better credentials than this.

When questions about deductive validity arise, they can usually be resolved in a formal manner by reference to an appropriate logical system. It has not always been so. Modern mathematical logic dates from the nineteenth century, and it has undergone extraordinary development, largely in response to problems that arose in the foundations of mathematics. One such problem concerned the foundations of geometry, and it assumed critical importance with the discovery of non-Euclidean geometries. Another problem concerned the status of the infinitesimal in the calculus, a concept that was the center of utter confusion for two centuries after the discovery of the 'infinitesimal' calculus.[4] Thanks to extensive and fruitful investigations of the foundations of mathematics, we now have far clearer and more profound understanding of many fundamental mathematical concepts, as well as an extremely well-developed and intrinsically interesting discipline of formal deductive logic. The early investigators in this field could never have conceived in their wildest imaginings the kinds of results that have emerged.[5]

It is an unfortunate fact that far less attention has been paid to the foundational questions that arise in connection with the empirical sciences and their logic. When questions of inductive validity arise, there is no well-established formal discipline to which they can be referred for definitive solution. A number of systems of inductive logic have been proposed, some in greater and some in lesser detail, but none is more than rudimentary, and none is widely accepted as basically correct. Questions of inductive correctness are more often referred to scientific or philosophical intuitions, and these are notoriously unreliable guides.

3. The Theorem of Bayes

We do have one resource, which although not overlooked entirely, is not exploited as fully as it could be. I refer to the mathematical calculus of probability. The probability calculus will not, by itself, solve all of our foundational problems concerning scientific inference, but it will provide us with a logical schema for scientific inference that is far more adequate than the H-D schema. And insofar as the probability calculus fails to provide the answers to foundational questions, it will at least help us to pose those problems in intelligible and, hopefully, more manageable form.

In order to show how the probability calculus can illuminate the kinds of questions I have been raising, I should like to introduce a very simple problem involving quality control. A certain factory, which manufactures automobiles, has two assembly lines. One of these is new and efficient; it is equipped with the latest automated devices. On the average it creates three-fourths of all cars produced in this factory, but 1% of its products are defective. The other assembly line is somewhat antiquated and inefficient; it produces one-fourth of the cars, and 2% of them are defective. There is a simple formula for calculating the probability that a car produced in this factory is defective. Letting P(B|A) stand for the probability from A to B (that is, the probability of B, given A), and letting A stand for the set of all cars produced in this factory, B for cars produced on the new assembly line, and C for defective cars, the following formula, which is a special case of the *theorem on total probability*, yields the desired computation:

$$P(C|A) = P(B|A) \ P(C|A \cdot B) + P(\sim B|A) \ P(C|A \cdot \sim B) \tag{1}$$

The dot means "and," and the tilde before the symbol negates it. Accordingly, the probabilities appearing in the formula are

P(C|A)—probability that a car from this factory is defective
P(B|A)—probability that a car from this factory was produced on the new assembly line ($=0.75$)
P($\sim$B|A)—probability that a car from this factory was produced on the old assembly line ($=0.25$)
P(C|A·B)—probability that a car produced on the new assembly line is defective ($=0.01$)
P(C|A·$\sim$B)—probability that a car produced on the old assembly line is defective ($=0.02$)

The theorem on total probability yields the result

$$P(C|A) = 0.75 \times 0.01 + 0.25 \times 0.02 = 0.0125$$

Suppose, now, that an inspector selects a car at random from those produced in this factory and finds it to be defective. Supposing (perhaps unrealistically) that this car bears no mark identifying the assembly line from which it came, we ask for the probability that it came off of the new assembly line. Again, the probability calculus provides a simple formula to compute the desired probability. This time it is a special form of Bayes's theorem, and it can be written in either of two completely equivalent ways:

$$P(B|A \cdot C) = \frac{P(B|A) \ P(C|A \cdot B)}{P(C|A)} \tag{2}$$

$$= \frac{P(B|A) \ P(C|A \cdot B)}{P(B|A) \ P(C|A \cdot B) + P(\sim B|A) \ P(C|A \cdot \sim B)} \tag{3}$$

The theorem on total probability (1) assures us that the denominators of the two fractions are equal; we must, of course, impose the condition that P(C|A) $\neq$ 0 in order to avoid an indeterminate fraction; this is not, however, a serious restriction because $P(C|A) = 0$ would mean that the evidence C has no chance whatever of occurring. The expression on the left evidently represents the probability that the defective car was produced on the new assembly line. Substituting known values in equation (2) yields

$$P(B|A \cdot C) = [0.75 \times 0.01]/[0.0125] = 0.6$$

There is nothing controversial about either of the foregoing theorems or their applications to simple examples of the type just described. A simple derivation of the theorem on total probability and Bayes's theorem is given in the appendix to the introduction to part II.

4. The Application of Bayes's Theorem

In order to get at our logical questions about the nature of scientific inference, let me redescribe the example and what we learned about it, although in so doing I may admittedly be stretching some meanings a bit. It is nevertheless illuminating. We can think of the defective car as an effect that can be produced in either of two ways, by the old assembly line or by the new. When we asked for the probability that a defective car was produced by the new assembly line, we were asking for the probability that the first of the two possible causes, rather than the second, was operative in bringing about this effect. In fact, there are two causal hypotheses, and we were calculating the probability that was to be assigned to one of them, namely, the hypothesis that the car came from the new assembly line. Notice that the probability that the defective car came from the old assembly line is considerably less than one-half; the odds are much better than 50:50 that it was produced by the new assembly line, even though the old assembly line has a propensity twice as great as that of the new to produce defective cars. The reason, obviously, is that many more cars are made on the new assembly line. This point has fundamental philosophical importance.

Continuing with our unusual use of terms, let us look at the probabilities used to carry out the computation via Bayes's theorem. $P(B|A)$ and $P(\sim B|A)$ are known as *prior probabilities*; they are the probabilities, respectively, that the particular cause is operative or not, regardless of whether the car is defective. They are linked in a simple manner,

$$P(\sim B|A) = 1 - P(B|A)$$

so that knowledge of one of them suffices to determine the other. $P(C|A \cdot B)$ and $P(C|A \cdot \sim B)$ are usually known as *likelihoods*. $P(C|A \cdot B)$ is the probability that the car is defective, given that it came from the new assembly line (given the truth of the causal hypothesis), whereas $P(C|A \cdot \sim B)$ is the probability that the car is defective if it came from the old assembly line (given the negation of our hypothesis, i.e., the truth of the alternative). Note, however, that *the likelihood of a hypothesis is not a probability of that hypothesis*; it is, instead, the probability of a result, given that the hypothesis holds. Note, also, that the two likelihoods need not add up to one; they are logically independent of one another and both need to be known—knowledge of one only does not suffice. These are the probabilities that appear on the right-hand side of the second form of Bayes's theorem (3).

In the first form of Bayes's theorem (2) we do not need the second likelihood, $P(C|A \cdot {\sim} B)$, but we require $P(C|A)$ instead. This probability is known as the *expectedness*; it is the probability that the effect in question occurs regardless of which cause is operative. Expectedness is the opposite of surprisingness; a low value of $P(C|A)$ means that C is an unexpected or surprising occurrence. Since the expectedness occurs in the denominator, the smaller its value the greater the posterior probability; the occurrence of surprising evidence enhances the probability of the hypothesis. But whichever form of the theorem is used, we need three logically distinct probabilities in order to carry out the calculation. $P(B|A \cdot C)$, the probability we endeavor to establish, is known as the *posterior probability* of the hypothesis. When we entertain the two causal hypotheses about the randomly selected car, we may take the fact that it is defective as observational evidence relevant to the causal hypotheses. (We already know that the probability that a car, defective or not, came from the new assembly line is 0.75.) Thus, we may think of our posterior probability, $P(B|A \cdot C)$, as the probability of a hypothesis in the light of observational evidence. This is precisely the kind of question that arose in connection with the hypothetico-deductive method and in connection with our attempt to understand how evidence confirms or disconfirms scientific hypotheses. Bayes's theorem therefore constitutes a logical schema in the mathematical calculus of probability that shows some promise of incorporating the main logical features of the kind of inference the H-D schema is intended to describe.

The striking difference between Bayes's theorem and the H-D schema is the relative complexity of the former compared with the latter. In some special cases the H-D schema provides just one of the probabilities required in Bayes's theorem, but never does it yield either of the other two required. Thus, the H-D schema is inadequate as an account of scientific inference because it is a gross oversimplification that omits reference to essential logical features of the inference. Bayes's theorem fills these gaps. The H-D schema describes a situation in which an observable result is deducible from a hypothesis (in conjunction with initial conditions, and possibly auxiliary hypotheses, all of which we are assuming to be true); thus, if the hypothesis is correct, the result must occur and cannot fail to occur. In this special case, $P(C|A \cdot B) = 1$, but without two other probabilities, say, $P(B|A)$ and $P(C|A \cdot {\sim} B)$, no conclusion at all can be drawn about the posterior probability. Inspection of Bayes's theorem makes it evident that $P(C|A \cdot B) = 1$ is compatible with $P(B|A \cdot C) = 0$—that is, if $P(B|A) = 0$.[6] At best, the H-D schema yields the likelihood of the hypothesis for that given evidence, but we need a prior probability and the likelihood of an alternative hypothesis on the same evidence—or the expectedness, $P(C|A)$, of the evidence.

That these other probabilities are indispensable, as well as the manner in which they function in scientific reasoning, can be indicated by examples. Consider *Dianetics* once more (Hubbard 1950). As remarked above, this book contains not only a theory to explain behavior; it also contains recommendations for a therapy to be practiced for the treatment of psychological disturbances. The therapeutic procedure bears strong resemblances to psychoanalysis; it consists of the elimination of those 'engrams' that are causing trouble by bringing to consciousness, through a process of free association, the unconscious experiences that produced the engrams in the first place. The theory, presumably, enables us to deduce that practice of the recommended therapy will produce cures of psychological illness.

At the time the theory was in vogue, this therapy was practiced extensively, and there is every reason to believe that "cures" did occur. There were unquestionably cases in which people with various neurotic symptoms were treated, and they experienced a remission of their symptoms. Such instances would seem to count, according to the hypothetico-deductive method, as confirming instances. That they cannot actually be so regarded is due to the fact that there is a far better explanation of these "cures." We know that there is a phenomenon of *faith healing* that consists in the efficacy of any treatment the patient sincerely believes to be effective. Many neurotic symptoms are amenable to such treatment, so anyone with such symptoms who believed in the soundness of the dianetic approach could be "cured" regardless of the truth or falsity of the theory upon which it is based. The reason, in terms of Bayes's theorem, is that the second likelihood—the probability $P(C|A \cdot \sim B)$ that the same phenomenon would occur even if the hypothesis were false—is very high. Since this term occurs in the denominator, the value of the whole fraction tends to be small when the term is large.

A somewhat similar problem arises in connection with psychotherapy based upon more serious theoretical foundations. The effectiveness of any therapeutic procedure has to be compared with the so-called *spontaneous remission rate*. Any therapy will produce a certain number of cases in which there is a remission of symptoms; nevertheless, in a group of people with similar problems who undergo no therapy of any kind, there will also be a certain percentage who experience remission of symptoms. For a therapy to be judged effective, it has to improve upon the spontaneous remission rate; it is not sufficient that there be some remission among those who undergo the treatment. In terms of Bayes's theorem, this means that we must look at both likelihoods, $P(C|A \cdot B)$ and $P(C|A \cdot \sim B)$, not just the one we have been given in the standard H-D schema. This is just what experimental controls are all about. For instance, vitamin C has been highly touted as a cold remedy, and many cases have been cited of people recovering quickly from colds after taking massive doses. But in a *controlled* experiment in which two groups of people of comparable age, sex, state of general health, and severity of colds are compared, where one group is given vitamin C and the other is not, no difference in duration of severity of colds is detected.[7] This gives us a way of comparing the two likelihoods.

Let me mention, finally, an example of a strikingly successful confirmation, showing how the comparative likelihoods affect this sort of situation. At the beginning of the nineteenth century, two different theories of light were vying for supremacy: the wave theory and the corpuscular theory. Each had its strong advocates, and the evidence up to that point was not decisive. One of the supporters of the corpuscular theory was the mathematician Poisson, who deduced from the mathematical formulation of the wave theory that, if that theory were true, there should be a bright spot in the center of the shadow of a disk. Poisson declared that this absurd result showed that the wave theory is untenable, but when the experiment was actually performed the bright spot was there. Such a result was unthinkable in the corpuscular theory, so this turned into a triumph for the wave theory because the probability on any other theory then available was negligible (Born and Wolf, 1964). It was not until about a century later that the need for a combined wave-particle theory was realized. Arithmetically, the force of this dramatic confirmation is easily seen by noting that if $P(C|A \cdot {\sim} B) = 0$ in (3), the posterior probability $P(B|A \cdot C)$ automatically becomes 1.

In addition to the two likelihoods, Bayes's theorem requires us to have a prior probability $P(B|A)$ or $P({\sim}B|A)$ in order to ascertain the posterior probability. These prior probabilities are probabilities of hypotheses without regard to the observational evidence provided by the particular test we are considering. In the quality-control example described above, the prior probability was the probability that a car came from one particular assembly line regardless of whether it was defective or not. In the more serious cases of the attempt to evaluate scientific hypotheses, the probability of a hypothesis without regard to the outcome of a test being contemplated is precisely the sort of plausibility consideration that was discussed at the outset. How plausible is a given hypothesis? What is its chance of being a successful one? This is the type of consideration that is demanded by Bayes's theorem in the form of a prior probability. The traditional stumbling block to the use of Bayes's theorem as an account of the logic of scientific inference is the great difficulty of giving a description of what sorts of things these prior probabilities could be.

5. Plausibility Arguments

In spite of this problem, it seems possible to give many examples of plausibility arguments, and even to classify them into very general types. Such arguments may then be regarded as criteria that are used to evaluate prior probabilities—criteria that indicate whether a hypothesis is plausible or implausible, whether its prior probability is to be rated high or low. I shall mention three general types of criteria, and give some instances of each.

1. Let us call criteria of the first general type *formal criteria*, for they involve formal logical relations between the hypothesis under consideration and other accepted parts of science. This idea was illustrated at the outset by Velikovsky's (1950) theory, which contradicts virtually all of modern physics. Because of this formal relationship we can say that Velikovsky's theory must have a very low prior probability since it is incompatible with so much we accept as correct. Another example of the same type can be found in those versions of the theory of telepathy that postulate the *instantaneous* transference of thought from one person to another, regardless of the distance that separates them, because the special theory of relativity stipulates that information cannot be transmitted at a speed greater than the speed of light, and so it would preclude instantaneous thought transmission. It would be even worse for precognition, the alleged process of direct perception of

future occurrences, for this would involve messages being transmitted backward in time. Such parapsychological hypotheses must be given extremely low prior probabilities because of their logical incompatibility with well-established portions of physical science. A hypothesis could, of course, achieve a high prior probability on formal grounds by being the logical consequence of a well-established theory. Kepler's laws, for example, are extremely probable (as approximations) because of their relation to Newtonian gravitational theory.

2. I shall call criteria of the second type *pragmatic criteria*. Such criteria have to do with the evaluation of hypotheses in terms of the circumstances of their origins—for example, the qualifications of the author. This sort of consideration has already been amply illustrated by the example of *Dianetics* (Hubbard 1950). Whenever a hypothesis is dismissed as being a "crank" hypothesis, pragmatic criteria are being brought to bear. In Martin Gardner's (1957, 11–14) fascinating book, he offers some general characteristics by which cranks can be identified.[8]

One might be tempted to object to the use of pragmatic criteria on the grounds, as we have all been taught, that it is a serious fallacy to confuse the *origin* of a theory with its *justification*. Having been told the old story about how Newton was led to think of universal gravitation by seeing an apple fall, we are reminded that that incident has nothing to do with the truth or justification of Newton's gravitational theory. That issue must be decided on the evidence.[9] Quite so. But there are factors in the origin of a hypothesis, such as the qualifications of the author, which have an *objective* probability relationship to the hypothesis and its truth. On the one hand, crank hypotheses seldom, if ever, turn out to be sound; they are based upon various misunderstandings, prejudices, or sheer ignorance. It is not fallacious to conclude that they have low prior probabilities. On the other hand, as Abner Shimony (1970) has pointed out, hypotheses seriously proposed by competent scientists have a nonnegligible chance of being right. Modern science has enjoyed a good deal of success during the past four centuries.

3. Criteria of the third type are by far the most interesting and important; let us call them *material criteria*. They make reference, in one way or another, to what the hypothesis actually says, rather than to its formal relation to other theories or to the circumstances surrounding its origins. These criteria do, however, depend upon comparisons of various theories or hypotheses; they make reference to analogies or similarities among different ones. Again, a few examples may be helpful.

Perhaps the most frequently cited criterion by which to judge the plausibility of hypotheses is the property of simplicity. Curve drawing illustrates this point aptly. Given data that can be represented graphically, we generally take the smoothest curve—the one with the simplest mathematical expression—that comes sufficiently near the data points as representing the best explanatory hypothesis for those data. This factor was uppermost with Kepler, who kept searching for the simplest orbits to account for planetary motion and finally settled upon the ellipse as filling the bill. Yet, we do not *always* insist upon the simplest explanation. We do not take seriously the 'hypothesis' that television is solely responsible for the breakdown of contemporary morals, assuming that there is such a breakdown, for it is an obvious oversimplification. It may be that simplicity is more to be prized in the physical than in the social sciences, or in the advanced than in the younger sciences. But it does seem that we need to exercise reasonable judgment as to just what degree of simplicity is called for in any given situation.

Another consideration that may be used in plausibility arguments concerns causal mechanisms. There was a time when much scientific explanation was teleological in character; even the motion of inanimate objects was explained in terms of the endeavor to achieve their natural places. The physics of Galileo and Newton removed all reference to purpose from these realms. Although there have been a few attempts to read purpose into such laws as least action ("The Absolute is lazy"), it is for the most part fully conceded that physical explanation is nonpurposive.[10]

The great success of Newtonian physics provided a strong plausibility argument for Darwin's account of the development of the biological species. The major difference between Darwin's evolutionary theory and its alternative contenders is the thoroughgoing rejection of teleological explanation by Darwin. Although teleological-sounding language may sometimes creep in when we talk about natural selection, the concept is entirely nonpurposive. We ask, "Why is the polar bear white?" We answer, "Because that color provides a natural camouflage." It sometimes sounds a bit as if we are saying that the bear thinks the situation over and decides before it is born that white would be the best color, and thus chooses that color. But, of course, we mean no such thing. We are aware that, literally, no choice or planning is involved. There are chance mutations, some favorable to escaping from enemies and finding food. Those animals that have the favorable characteristics tend to survive and reproduce their kind, whereas those with unfavorable characteristics tend to die out without reproducing. The cause-effect relations in the evolutionary account are just as mechanical and without purpose as are those in Newtonian physics. This nonteleological theory is in sharp contrast to the theory of special creation, according to which God created the various species because it somehow fit a divine plan.[11]

The nonteleological character of Newton's theory surely must lend plausibility to a nonteleological biological theory such as Darwin's. If physics, which was far better developed and more advanced than any other science, got that way by abandoning teleological explanations for efficient causation, then it seems plausible for those sciences that are far less developed to try the same approach. When this approach paid off handsomely in the success of evolutionary theory, it becomes much more plausible for other branches of science to follow the same line. Thus, for theories in psychology and sociology, for example, higher plausibility and higher prior probability would now attach more to those hypotheses that are free from teleological components than to those that retain teleological explanation. Thus, we can regard as implausible theories according to which human institutions are necessarily developing toward an ideal goal or those that characterize human nature is approaching a preestablished ideal form. And when a biological hypothesis comes along that regresses to the pre-Darwinian teleology, such as Lecomte du Noüy's (1947), it must be assigned a low prior probability.[12]

Let me give one final example of material criteria. Our investigations of the nature of physical space, extending over many centuries, have led to some rather sophisticated conceptions. To early thinkers, nothing could have been more implausible than to suppose that space is homogeneous and isotropic. Everyday experience seems clearly to demonstrate that there is a preferred direction—down. This view was expressed poetically by Lucretius (1951) in *The Nature of the Universe*, in which he describes the primordial state of affairs in which all the atoms are falling downward in space at a uniform speed.[13] On this view, not only was the downward direction preferred, but also it was possible to distinguish absolute motion from absolute rest. By Newton's time it seemed clear that space had no preferred direction; rather, it was isotropic—possessed of the same structure in every direction. This consideration lent plausibility to Newton's inverse square law, for if space is Euclidean and it has no preferred directions, then we should expect any force, such as gravitation, to spread out uniformly in all directions. In Euclidean geometry, the surface of a sphere varies with the square of the radius, so if the gravitational force spreads out uniformly in the surrounding space, it should diminish with the square of the distance.

Newton's theory, though it regards space as isotropic, still makes provision for absolute motion and rest. Einstein, reflecting on the homogeneity of space, enunciated a principle of relativity that precludes distinguishing physically between rest and uniform motion. In the beginning, if we believe Einstein's (1949) own autobiographical account, this principle recommended itself entirely on the grounds of its very great plausibility. The matter does not rest there, of course, for it had to be incorporated into a physical theory that could be subjected to experimental test. His special theory of relativity has been tested and confirmed in a wide variety of ways, and it is now a well-established part of physics, but prior to the tests and its success in meeting them, it could be certified as highly plausible on the basis of very general characteristics of space.

6. The Meaning of "Probability"

Up to this point I have been attempting to establish two facts about prior probabilities: (1) Bayes's theorem shows that they are needed, and (2) scientific practice shows that they are used. But their status has been left very vague indeed. There is a fundamental reason. In spite of the fact that the probability calculus was established early in the seventeenth century, not much serious attention was given to the analysis of the meaning of the concept of probability until the latter part of the nineteenth century.[14] There is nothing especially unusual about this situation. Questions about the meanings of fundamental concepts are foundational questions, and foundational investigations usually follow far behind the development of a discipline. Even today there is no real consensus on this question; there are, instead, three distinct interpretations of the probability concept, each with its strong adherents. A fortiori, there is no widely accepted answer to the question of the nature of the prior probabilities, for they seem to be especially problematic in character.

Among the three leading probability theories, the *logical theory* regards probability as an *a priori measure* that can be assigned to propositions or states of affairs, the *personalistic theory* regards probability as a *subjective measure* of degrees of belief, and the *frequency theory* regards probability as a *physical characteristic* of types of events.[15]

The logical theory is the direct descendent of the famous classical theory of Laplace. According to the classical theory, probability is the ratio of favorable to equally possible cases. The equipossibility of cases, which is nothing other than the equal probability of these cases, is determined a priori on the basis of a *principle of indifference*; namely, two cases are equally likely if there is no reason to prefer one to the other. This principle gets into deep logical difficulty. Consider, for example, a car that makes a trip around a 1-mile track in a time somewhere between 1 and 2 minutes, but we know no more about it. It seems reasonable to say that the time could have been in the interval from 1 to 1 1/2 minutes, or it could have been in the interval of 1 1/2 to 2 minutes; we don't know which. Since these intervals are equal, we have no reason to prefer one to the other, and we assign a probability of 1/2 to each of them. Our information about this car can be put in other terms. We know that the car made its trip at an *average* speed somewhere in the range of 60 to 30 miles per hour. Again, it seems reasonable to say that the speed could have been in the range 60–45 miles per hour, or it could have been in the range 45–30 miles per hour; we don't know which. Since the two intervals are equal, we have no reason to prefer one to the other, and we assign a probability of 1/2 to each. But we have just contradicted our former result, for a time of 1 1/2 minutes corresponds with an average speed of 40, not 45, miles per hour (see Salmon 1967b, 65–68).

This contradiction, known as the Bertrand paradox, brings out the fundamental difficulty with any method of assigning probabilities a priori. Such a priori decisions have an unavoidable arbitrary component, and in this case the arbitrary component gives rise to two equally reasonable, but incompatible, ways of assigning the probabilities. Although the logical interpretation, in its current form, escapes this particular form of paradox, it is still subject to philosophical criticism because of the same general kind of aprioristic arbitrariness.

The personalistic interpretation is the twentieth-century successor of an older and more naive subjective concept. According to the crude subjective view, a probability is no more or less than a subjective degree of belief; it is a measure of our ignorance. If I assign the probability value of 1/2 to an outcome of heads on a toss of the coin, this means that I expect heads just as much as I expect tails, and my uncertainty is equally divided between the two outcomes. If I expect twice as strongly as not that an American will be the first human to set foot on the moon, then that event has a probability of 2/3. (Obviously I won this bet.)

The major difficulty with the old subjective interpretation arises because subjective states do not always come in sizes that will fit the mathematical calculus of probability. It is quite possible, for example, to find a person who believes to the degree 1/6 that a 6 will turn up on any toss of a given die, and who also believes that the tosses are independent of one another (the degree of belief in an outcome of 6 on a given toss is unaffected by the outcome of the previous tosses). This same individual may also believe to the degree 1/2 in getting at least one 6 in three tosses of that die. There is, of course, something wrong here. If the probability of 6 on a given toss is 1/6, and if the tosses are independent, this probability is considerably less than 1/2 (it is approximately 0.42). For four tosses, the probability of at least one 6 is well over 1/2. This is a trivial kind of error that has been recognized as such for hundreds of years, but it is related to a significant error that led to the discovery of the mathematical calculus of probability. In the seventeenth century, the view was held that in twenty-four tosses of a pair of dice, there should be at least a 50-50 chance of tossing at least one double 6. In fact, the probability is just under 1/2 in twenty-four tosses; in twenty-five it is just over 1/2. The point of these examples is very simple. If probabilities are just subjective degrees of belief, the mathematical calculus of probability is mistaken because it specifies certain relations among probabilities that do not obtain among degrees of belief.

Modern personalists do not interpret probabilities merely as subjective degrees of belief but rather as *coherent* degrees of belief. To say that degrees of belief are coherent means that they are related in such manner as to satisfy the conditions imposed by the mathematical calculus of probability. The personalists have seen that degrees of belief that violate the mathematical calculus involve some sort of error or blunder that is analogous to a logical inconsistency. Hence, when a combination of degrees of belief is incoherent, some adjustment or revision is called for in order to bring these degrees into conformity with the mathematical calculus.[16] The chief objection to the personalist view is that it is not objective; we shall have to see whether and to what extent the lack of objectivity is actually noxious.

The frequency interpretation goes back to Aristotle, who characterized the probable as that which happens often. More exactly, he regards a probability as a relative frequency of occurrence in a large sequence of events. For instance, a probability of 1/2 for heads on tosses of a coin would mean that in the long run the ratio of the number of heads to the number of tosses approaches and remains close to 1/2. To say that the probability of getting a head on a particular toss is 1/2 means that this toss is a member of an appropriately selected large class of tosses within which the overall relative frequency of heads is 1/2. It seems evident that there are many contexts in which we deal with large aggregates of phenomena, and in these contexts the frequency concept of probability seems well suited to the use of statistical techniques—for example, in quantum mechanics, kinetic theory, sociology, and the games of chance, to mention just a few. But it is much more dubious that the frequency interpretation is at all applicable to such matters as the probability of a scientific hypothesis in the light of empirical evidence. In this case where are we to find the large classes and long sequences to which to refer our probabilities of hypotheses? This difficulty has seemed insuperable to most authors who have dealt with the problem. The general conclusion has been that the frequency interpretation is fine in certain contexts, but we need a radically different probability concept if we are to deal with the probability of hypotheses (see, however, Shimony 1970).

Returning to our main topic of concern, we easily see that each of the foregoing three probability theories provides an answer to the question of the nature of plausibility considerations and prior probabilities. According to the logical interpretation, hypotheses are plausible or not on the basis of certain a priori considerations; on this view, reason dictates which hypotheses are to be taken seriously and which not. According to the personalistic interpretation, prior probabilities represent the prior opinion or attitude of the investigator toward the hypothesis before its testing begins. Different investigators may, of course, have different views of the same hypothesis, so prior probabilities may vary from individual to individual. According to the frequency interpretation, prior probabilities arise from experience with scientific hypotheses, and they reflect this experience in an objective way. To say that a hypothesis is plausible, or has a high prior probability, means that it is of a type that has proved successful in the past. We have found by experience that hypotheses of this general type have often worked well in science.

From the outset, the personalistic interpretation enjoys a major advantage over the other two. It is very difficult to see how we are to find nonarbitrary a priori principles to use as a basis for establishing prior probabilities of the a priori type for the logical interpretation, and it is difficult to see how we are reasonably to define classes of hypotheses and count frequencies of success for the frequency interpretation. But personal probabilities are available quite unproblematically. Each individual has a degree of belief in the hypothesis, and that's all there is to it. Coherence demands that degrees of belief conform to the mathematical calculus, and Bayes's theorem is one of the important relations to be found in the calculus. Bayes's theorem tells us how, if we are to avoid incoherence, we must modify our degrees of belief in the light of new evidence.[17] The personalists, who constitute an extremely influential school of contemporary statisticians, are indeed so closely wedded to Bayes's theorem that they have even taken its name and are generally known as "Bayesians."

The chief objection to the personalist approach is that it injects a purely subjective element into the testing and evaluation of scientific hypotheses; we feel that science should have a more objective foundation. The Bayesians have a very persuasive answer. Even though two people may begin with radically different attitudes toward a hypothesis, accumulating evidence will force a convergence of opinion. This is a basic mathematical fact about Bayes's theorem; it is easily seen by an example. Suppose a coin that we cannot examine is being flipped, but we are told the results of the tosses. We know that it is either a fair coin or a two-headed coin (we don't know which), and we have very different prior opinions on the matter. Suppose your prior probability for a two-headed coin is 1/100 and mine is 1/2. Then as we learn that various numbers of heads have been tossed (without any tails, of course), our opinions come closer and closer together as follows:

	Prior probability that coin has two heads	
	1/100	1/2
Number of heads		
	Posterior probability on given evidence	
1	2/101	2/3
2	4/103	4/5
10	1024/1123 × 91	1024/1025 × 99

After only ten tosses, we both find it overwhelmingly probable that the coin that produced this sequence of results is a two-headed one. This phenomenon is sometimes called "the washing out of the priors" or "the swamping of the priors," for their influence on the posterior probabilities becomes smaller and smaller as evidence accumulates. The only qualification is that we must begin with somewhat open minds. If we begin with the certainty that the coin is two-headed or with the certainty that it is not, that is, with prior probability of zero or 1, evidence will not change that opinion. But if we begin with prior probabilities differing ever so little from those extremes, convergence will sooner or later occur. As L. J. Savage (1954) remarked, it is not necessary to have an open mind; it is sufficient to have one that is slightly ajar.

The same consideration about the swamping of prior probabilities also enables the frequentist to overcome the chief objection to his approach. If it were necessary to have clearly defined classes of hypotheses, within which exact values of frequencies of success had to be ascertained, the situation would be pretty hopeless, but because of the swamping phenomenon, it is sufficient to have only the roughest approximation. All that is really needed is a reasonable guess as to whether the value is significantly different from zero. In the artificial coin-tossing example, where there are only two hypotheses, it is possible to be perfectly open-minded and give each alternative a nonnegligible prior probability; but in the serious cases of evaluation of scientific hypotheses, there are infinitely many alternative hypotheses, all in conflict with one another, and they cannot all have nonnegligible prior probabilities. This is the problem of the alternative hypotheses again. For this reason, it is impossible to be completely open-minded, so we must find some basis for assigning negligible prior probabilities to some possible hypotheses. This is tantamount to judging some hypotheses to be too implausible to deserve further testing and consideration. It is my conviction that this is done on the basis of experience; it is not done by means of purely a priori considerations, nor is it a purely subjective affair. As I tried to suggest by means of the examples of plausibility arguments, scientific experience with the testing, acceptance, and rejection of hypotheses provides an objective basis for deciding which hypotheses deserve serious testing and which do not.

I am not suggesting that we proceed on the basis of plausibility considerations to summary dismissal of almost every hypothesis that comes along; on the contrary, the recommendation would be for a high degree of open-mindedness. However, we need not and cannot be completely open-minded to any and every hypothesis of whatever description that happens to be proposed by anyone. This approach shows how we can be reasonably open-minded in science without being stupid about it. It provides an answer to the kind of charge made by the editors of *Harper's*: Science is objective, but its objectivity embraces two aspects, objective testing and objective evaluation of prior probabilities. Plausibility arguments are used in science, and their use is justified by Bayes's theorem. In fact, Bayes's theorem shows that they are indispensable. The frequency interpretation of probability enables us to view them as empirical and objective.

It would be an unfair distortion of the situation for me to conclude without remarking that the view I have been advocating is very definitely a minority view among inductive logicians and probability theorists. There is no agreed-upon majority view. One of the most challenging aspects of this sort of investigation lies in the large number of open questions and the amount that remains to be done. Whether my view is correct is not the main issue. Of far greater importance is the fact that there are many fundamental problems that deserve extensive consideration, and we cannot help but learn a great deal about the foundations of science by pursuing them.

Discovery and Justification

I n his splendid introduction to *Minnesota Studies in the Philosophy of Science,*
Volume V (Steuwer 1970), Herbert Feigl rightly stresses the central importance
of the distinction between the context of discovery and the context of justification.[1]
These terms were introduced by Hans Reichenbach to distinguish the social and
psychological facts surrounding the discovery of a scientific hypothesis from the
evidential considerations relevant to its justification.[2] The folklore of science is full
of dramatic examples of the distinction, for example, Kepler's mystical sense of
celestial harmony (Pannekoek 1961, 235) versus the confrontation of the postulated
orbits with the observations of Tycho Brahe; Kekule's drowsing vision of dancing
snakes (Partington 1964, 553ff.) versus the laboratory confirmation of the hexagonal
structure of the benzine ring; Ramanujan's visitations in sleep by the Goddess of
Namakkal (Hardy et al. 1927, xii) versus his waking demonstrations of the mathematical theorems. Each of these examples offers a fascinating insight into the
personality of a working scientist, and each provides a vivid contrast between those
psychological factors and the questions of evidence that must be taken into account in order to assess the truth or probability of the result. Moreover, as we all
learned in our freshman logic courses, to confuse the source of a proposition with
the evidence for it is to commit the genetic fallacy.

If one accepts the distinction between discovery and justification as viable,
there is a strong temptation to maintain that this distinction marks boundaries
between history of science and philosophy of science. History is concerned with
the *facts* surrounding the growth and development of science; philosophy is concerned with the *logical structure* of science, especially with the evidential relations
between data and hypotheses or theories. As a matter of fact, Reichenbach (1938)
described the transition from the context of discovery to the context of justification
in terms of a *rational reconstruction*. On the one hand, the scientific innovator
engages in thought processes that may be quite irrational or nonrational, manifesting no apparent logical structure: this is the road to discovery. On the other

hand, when the scientist wants to present the results to the community for judgment, he or she provides a reformulation in which the hypotheses and theories are shown in logical relation to the evidence that is offered in support of them: this is the rational reconstruction. The items in the context of discovery are *psychologically* relevant to the scientific conclusion; those in the context of justification are *logically* relevant to it. Since philosophers of science are concerned with logical relations, not psychological ones, they are concerned with the rationally reconstructed theory, not with the actual process by which it came into being.

Views of the foregoing sort regarding the relationships between the context of discovery and the context of justification have led to a conception of philosophy of science that might aptly be characterized as a "rational reconstructionist" or "logical reconstructionist" approach. This approach has been closely associated with logical empiricism or logical positivism, though by no means confined to them. Critics of the reconstructionist view have suggested that it leaves the study of vital, growing science to the historian, while relegating philosophy of science to the dissection of scientific corpses—not the bodies of scientists but of theories that have grown to the point of stagnation and ossification. According to such critics, the study of completed science is not the study of science at all. One cannot understand science unless one sees how it grows; to comprehend the logical structure of science, it is necessary to take account of scientific change and scientific revolution. Certain philosophers have claimed, consequently, that philosophy of science must deal with the logic of discovery, as well as the logic of justification (see Hanson 1961). Philosophy of science, it has been said, cannot proceed apart from study of the history of science. Such arguments have led to a challenge of the very distinction between discovery and justification (Kuhn 1962, 9). Application of this distinction, it is claimed, has led to the reconstructionist approach, which separates philosophy of science from real science and makes philosophy of science into an unrealistic and uninteresting form of empty symbol manipulation.

The foregoing remarks make it clear, I hope, that the distinction between the context of discovery and the context of justification is a major focal point for any fundamental discussion of the relations between history of science and philosophy of science. As the dispute seems to shape up, the reconstructionists rely heavily upon the viability of a sharp distinction, and they apparently conclude that there is no very significant relationship between the two disciplines. Such marriages as occur between them—for example, the International Union of History and Philosophy of Science, the National Science Foundation Panel for History and Philosophy of Science, and the Departments of History and Philosophy of Science at Melbourne, Indiana, and Pittsburgh—are all marriages of convenience. The antireconstructionists, who find a basic organic unity between the two disciplines, seem to regard a rejection of the distinction between discovery and justification as a cornerstone of their view. Whatever approach one takes, it appears that the distinction between the context of discovery and the context of justification is the first order of business.

I must confess at this point, if it is not already apparent, that I am an unreconstructed reconstructionist, and I believe that the distinction between the context of discovery and the context of justification is viable, significant, and fundamental to the philosophy of science. I do not believe, however, that this view commits me to an intellectual divorce from my historical colleagues; in the balance of this chapter I should like to explain why. I shall not be concerned to argue in favor of the distinction but shall instead try (1) to clarify the distinction and repudiate certain common misconceptions of it, (2) to show that a clear analysis of the nature of scientific confirmation is essential to an understanding of the distinction and that a failure to deal adequately with the logic of confirmation can lead to serious historical misinterpretations, and (3) to argue that an adequate conception of the logic of confirmation leads to basic, and largely unnoticed, logical functions of historical information. In other words, I shall be attempting to show how certain aspects of the relations between history and philosophy of science can be explicated within the reconstructionist framework. Some of my conclusions may appear idiosyncratic, but I shall take some pains along the way to argue that many of these views are widely shared.

1. The Distinction between Discovery and Justification

When one presents a distinction between the two sorts of things, it is natural to emphasize their differences and to make the distinction appear more dichotomous than is actually intended. In the present instance, some commentators have apparently construed the distinction to imply that, first, a creative scientist goes through a great succession of irrational (or nonrational) processes, such as dreaming, being hit on the head, pacing the floor, or having dyspepsia, until a full-blown hypothesis is born. Only after these processes have terminated does the scientist go through the logical process of mustering and presenting the evidence to justify the hypothesis. Such a conception would, of course, be factually absurd; discovery and justification simply do not occur in that way. A more realistic account might go somewhat as follows. A scientist, searching for a hypothesis to explain some phenomenon, hits upon an idea but soon casts it aside because it is seen to be inconsistent with some other accepted theory or because it does not really explain the phenomenon in question. This phase undoubtedly involves considerable logical inference; it might, for instance, involve a mathematical calculation that shows that the explanation in question would not account for a result of the correct order of magnitude. After more searching the scientist—who in the meantime has perhaps attended a cocktail party and spent a restless night—hits upon another idea, which also proves to be inadequate but can be improved by some modification or other. Again, by logical inference, it is determined that this new hypothesis bears certain relations to the available evidence. Our scientist further realizes, however, that although the present hypothesis squares with the known facts, further modification would make it simpler and give it a wider explanatory range. Perhaps additional tests are devised and executed to check the applicability of the latest revision in new domains. And so it goes. What I am trying to suggest, by such science fiction, is that the processes of discovery and justification are intimately intertwined, with steps of one type alternating with steps of the other. There is no reason to conclude, from a distinction between the context of discovery and the context of justification, that the entire process of discovery must be completed before the process of justification can begin and that the rational reconstruction can be undertaken only after the creative work has ended. Such conclusions are by no means warranted by the reconstructionist approach.

There is, moreover, no reason to suppose that the two contexts must be mutually exclusive. Not only may elements of the context of justification be temporally intercalated between elements of the context of discovery, but also the two contexts may have items in common. The supposition that this cannot happen is perhaps the most widespread misunderstanding of the distinction between the two contexts. The most obvious example is the case in which a person or a machine discovers the answer to a problem by applying an algorithm, such as doing a sum, differentiating a polynomial, or finding a greatest common divisor. Empirical science also contains routine methods for finding answers to problems—which is to say, for discovering correct hypotheses. These are often the kinds of procedures that can be delegated to a technician or a machine, for example, chemical analyses, ballistic testing, or determination of physical constants of a new compound. In such cases, the process of discovery and the process of justification may be nearly identical, though the fact that the machine blew a fuse or the technician took a coffee break could hardly qualify for inclusion in the latter context. Even though the two contexts are not mutually exclusive, the distinction does not vanish. The context of discovery consists of a number of items related to one another by psychological relevance, whereas the context of justification contains a number of items related to one another by (inductive and deductive) logical relevance. There is no reason at all why one and the same item cannot be both psychologically and logically relevant to some given hypothesis. Each context is a complex of entities that are interrelated in particular ways. The contexts are contrasted with one another, not on the ground that they can have no members in common, but rather on the basis of differences in the types of relationships they incorporate. The fact that the two contexts can have items in common does not mean that the distinction is useless or perverse, for there *are* differences between logical and psychological relevance relationships that are important for the understanding of science.

The problem of scientific discovery does not end with the thinking up of a hypothesis. One has also to discover evidence and the logical connections between the evidence and the hypothesis. The process of discovery is, therefore, involved in the very construction of the rational reconstruction. When one puts forth a scientific hypothesis as acceptable, confirmed, or corroborated, along with the evidence and arguments upon which the claim is based, one is offering *one's own* rational reconstruction (that discovered by one's self) and presumably claiming that it is logically sound. This is a fact about the scientist, namely, the acceptance of this evidence and these arguments as satisfactory. A critic—scientist or philosopher— might, of course, show that the scientist has committed a logical or methodological error and, consequently, that the rational reconstruction is unsound. Such an occurrence would belong to the context of justification.[3] However, even if the argument seems compelling to an entire scientific community, it may still be logically faulty. The convincing character of an argument is quite distinct from its validity; the former is a *psychological* characteristic, the latter, *logical*. Once more, even though there may be extensive overlap between the contexts of discovery and justification, it is important not to confuse them.

Considerations of the foregoing sort have led to serious controversy over the appropriate role of philosophy of science. On the one hand, it is sometimes claimed that philosophy of science must necessarily be a historically oriented empirical study of the methods scientists of the past and present have actually used and of the canons they have accepted. On the other hand, it is sometimes maintained that such factual studies of the methods of science belong to the domain of the historian and that the philosopher of science is concerned exclusively with logical and epistemological questions. Proponents of the latter view—which is essentially the reconstructionist approach—may appear quite open to the accusation that they are engaged in some sort of scholastic symbol mongering that has no connection whatever with actual science. To avoid this undesirable state of affairs, it may be suggested, we ought to break down the distinctions between history and philosophy, between psychology and logic, and ultimately between discovery and justification.

There is, I believe, a better alternative. Although basically concerned with abstract logical relations, the philosopher of science can hardly afford to ignore the actual methods that scientists have found acceptable. If a philosopher expounds a theory of the logical structure of science according to which almost all of modern physical science is methodologically unsound, it would be far more reasonable to conclude that the philosophical reasoning had gone astray than to suppose that modern science is logically misconceived. Just as certain empirical facts, such as geometrical diagrams or soap film experiments, may have great heuristic value for mathematics, so too may the historical facts of scientific development provide indispensable guidance for the formal studies of the logician.[4] In spite of this, the philosopher of science is properly concerned with issues of logical correctness that cannot finally be answered by appeal to the history of science. One of the problems with which the philosopher of science might grapple is the question of what grounds we have for supposing scientific knowledge to be superior to other alleged types of knowledge, for example, alchemy, astrology, or divination. The historian may be quick to offer an answer to that question in terms of the relative success of physics, chemistry, and astronomy. It required the philosophical subtlety of David Hume ([1739–1740] 1888) to realize that such an answer involves a circular argument.[5] The philosopher of science, consequently, is attempting to cope with problems for which the historical data may provide enormously useful guidance, but the solutions, if they are possible at all, must be logical, not historical, in character. The reason, ultimately, is that justification is a normative concept, whereas history provides only the facts.

I have been attempting to explain and defend the distinction between discovery and justification largely by answering objections to it, rather than by offering positive arguments. My attitude, roughly, is that it is such a plausible distinction to begin with, and its application yields such rich rewards in understanding, that it can well stand without any further justification. Like any useful tool, however, it must be wielded with some finesse; otherwise the damage it does may far outweigh its utility.

2. Bayes's Theorem and the Context of Justification

It would be a travesty to maintain, in any simpleminded way, that the historian of science is concerned only with matters of discovery and not with matters of justification. In dealing with any significant case, say the replacement of an old theory by a new hypothesis, the historian will be deeply interested in such questions as whether, to what extent, and in what manner the old theory has been disconfirmed, and similarly, what evidence is offered in support of the new hypothesis and how adequate it is. How strongly are factors such as national rivalry among scientists, esthetic disgust with certain types of theories, personal idiosyncrasies of influential figures, and other nonevidential factors operative? Since science aspires to provide objective knowledge of the world, it cannot be understood historically without taking very seriously the role of evidence in scientific development and change. Such historical judgments—whether a particular historical development was or was not rationally justified on the basis of the evidence available at the time—depend crucially upon the historian's understanding of the logic of confirmation and disconfirmation. The historian who seriously misunderstands the logic of confirmation runs the risk of serious historical misevaluation. And to the possible rejoinder that any competent historian has a sufficiently clear intuitive sense of what constitutes relevant scientific evidence and what does not, I must simply reply that I am not convinced. Some basis for my doubt is contained in chapter 6; as we shall see, the theorem of Bayes plays a crucial role.

Having compared the H-D account of confirmation with the Bayesian analysis at length in the preceding chapter (pars. 1–3), we can see an obvious way in which the historian's problem could arise. A historian who accepts the H-D analysis of confirmation sees no place for plausibility judgments in the logic of science—at least not in the context of justification. Finding that plausibility considerations play an important role historically in the judgments scientists render upon hypotheses, such a historian will be forced to exclude them from the context of justification and may conclude that the course of scientific development is massively influenced by nonrational or nonevidential considerations. Such an "H-D historian" might well decide, along with the editors of *Harper's Magazine*, that it was scientific prejudice, not objective evaluation, that made the scientific community largely ignore Velikovsky's (1950) views.[6] Such a historian might similarly conclude that Einstein's commitment to the "principle of relativity" on the basis of plausibility arguments shows his views to have been based more upon preconceptions than upon objective evidence (see Einstein 1949, 2–95). A "Bayesian historian," in contrast, will see these plausibility considerations as essential parts of the logic of confirmation and will place them squarely within the context of justification. The consequence is, I would say, that the historian of science who regards the H-D schema as a fully adequate characterization of the logical structure of scientific inference is in serious danger of erroneously excluding from the context of justification items that genuinely belong within it.

The moral for the historian should be clear. There are considerations relating to the acceptance or rejection of scientific hypotheses that, on the H-D account, must be judged evidentially irrelevant to the truth or falsity of the hypothesis, but which are, nevertheless, used by scientists in making decisions about such acceptance or rejection. These same items, on the Bayesian account, become evidentially relevant. Hence, the judgment of whether scientists are making decisions on the basis of evidence, or on the basis of various psychological or social factors that are *evidentially* irrelevant, hinges crucially upon the question of whether the H-D or the Bayesian account of scientific inference is more nearly correct. It is entirely conceivable that one historian might attribute acceptance of a given hypothesis to nonrational considerations, whereas another might judge the same decision to have an entirely adequate rational basis. Which account is historically more satisfactory will depend mainly upon which account of scientific inference is more adequate. The historian can hardly be taken to be unconcerned with the context of justification and with its differences from the context of discovery; indeed, to do the job properly the historian must understand them very well.

3. The Status of Prior Probabilities

It would be rather easy, I imagine, for the historian, and others who are not intimately familiar with the technicalities of inductive logic and confirmation theory, to suppose that the H-D account of scientific inference is the correct one. This view is frequently expressed in the opening pages of introductory science texts and in elementary logic books.[7] At the same time, it is important to raise the question of whether scientists in general—including the authors of introductory texts—actually comply with the H-D method in practice or whether, in fact, they use something similar to the Bayesian approach sketched in the preceding section. I am strongly inclined to believe that the Bayesian schema comes closer than the H-D schema to capturing actual scientific practice, for it seems to me that scientists do make substantial use of plausibility considerations, even though they may feel somewhat embarrassed to admit it. I believe also that practicing scientists have excellent intuitions regarding what constitutes sound scientific methodology but that they may not always be especially adept at fully articulating them. If we want the soundest guidance on the nature of scientific inference, we should look carefully at scientific practice rather than at the methodological pronouncements of scientists. An excellent example can be found in the preface of Jean Perrin's *Atoms* ([1913] 1916)—which we discussed in detail in chapter 3—in which he argues for the plausibility of a discontinuous (atomistic) conception of physical reality instead of a continuous (energeticist) view.

It is, moreover, the almost universal judgment of contemporary inductive logicians—the experts who concern themselves explicitly with the problems of confirmation of scientific hypotheses—that the simple H-D schema presented above is incomplete and inadequate. Acknowledging the well-known fact that there is very little agreement on which particular formulation among many available ones is most nearly a correct inductive logic, we can still see that among a wide variety of influential current approaches to the problems of confirmation there is at least agreement in rejecting the H-D method. This is not the place to go into detailed discussions of the alternative theories, but I should like to mention five leading candidates, indicating how each demands something beyond what is contained in the H-D schema. In each case, I think, what needs to be added is closely akin to the plausibility considerations mentioned in the preceding section.

1. The most fully developed explicit confirmation theory available is that of Rudolf Carnap (1950b). In the systems of inductive logic elaborated in his book, one begins with a formalized language and assigns a priori weights to all statements in that language, including, of course, all hypotheses. It is easy to show that Carnap's theory of confirmation is thoroughly Bayesian, with the a priori weights functioning precisely as the prior probabilities in Bayes's theorem. Although these systems have the awkward feature that all general hypotheses have prior probabilities equal to zero—and, consequently, posterior probabilities, on any finite amount of evidence, equal to zero—Jaakko Hintikka (1966) has shown how this difficulty can be circumvented without fundamentally altering Carnap's conception of confirmation as a logical probability.[8]

2. Although not many exponents of the frequency theory of probability will agree that it even makes sense to talk about the probability of scientific hypotheses, those who do so explicitly invoke Bayes's theorem for that purpose. Reichenbach ([1935] 1949, sec. 85) is the leading figure in this school, although his treatment of the probability of hypotheses is unfortunately quite obscure in many important respects. I have tried (Salmon 1967b, 115–132) to clarify some of the basic points of misunderstanding.

3. The important "Bayesian" approach to the foundations of statistics has become increasingly influential since the publication of L. J. Savage's 1954 work;[9] it has gained many adherents among philosophers, as well as statisticians. This view is based upon a subjective interpretation of probability ("personal probability," as Savage prefers to say, in order to avoid confusion with earlier subjective interpretations), and it makes extensive use of Bayes's theorem. The prior probabilities are simply degrees of prior belief in the hypothesis, before the concrete evidence is available. The fact that the prior probabilities are so easily interpreted on this view means that Bayes's theorem is always available for use. Savage, himself, is not especially concerned with probabilities of general hypotheses, but those who are interested in such matters have a ready-made Bayesian theory of confirmation (see Jeffreys 1939, 1957). On this view, the prior probabilities are subjective plausibility judgments.

4. Nelson Goodman (1955), whose influential work poses and attempts to resolve "the new riddle of induction," clearly recognizes that there is more to confirmation than mere confirming instances. He attempts to circumvent the difficulties, which are essentially those connected with the H-D schema, by introducing the notion of "entrenchment" of terms that occur in hypotheses. Recognizing that a good deal of the experience of the human race becomes embedded in the languages we use, he brings this information to bear upon hypotheses that are candidates for confirmation. Although he never mentions Bayes's theorem or prior probabilities, the chapter in which he presents his solution can be read as a tract on the Bayesian approach to confirmation.

5. Sir Karl Popper ([1935] 1959) rejects entirely the notions of confirmation and inductive logic.[10] His concept of corroboration, however, plays a central role in his

theory of scientific methodology. Although corroboration is explicitly regarded as nonprobabilistic, it does offer a measure of how well a scientific hypothesis has stood up to tests. The measure of corroboration involves such factors as simplicity, content, and testability of hypotheses, as well as the seriousness of the attempts made to falsify them by experiment. Although Popper denies that a highly corroborated hypothesis is highly probable, the highly corroborated hypothesis does enjoy a certain status: it may be chosen over its less corroborated fellows for further testing, and if practical needs arise, it may be used for purposes of prediction.[11] The important point, for the present discussion, is that Popper rejects the H-D schema, and he introduces additional factors into his methodology that play a role somewhat analogous to our plausibility considerations.

The foregoing survey of major contemporary schools of thought on the logic of scientific confirmation strongly suggests not only that the naive H-D schema is not *universally accepted* nowadays by inductive logicians as an adequate characterization of the logic of scientific inference but also that it is not even a serious candidate for that role.[12] Given the wide popular acceptance of the H-D method, it seems entirely possible that significant numbers of historians of science may be accepting a view of confirmation that is known to be inadequate, one that differs from the current serious contending views in ways that can have a profound influence upon historical judgments. It seems, therefore, that the branch of contemporary philosophy of science that deals with inductive logic and confirmation theory may have some substantive material that is highly relevant to the professional activities of the historian of science.

It is fair to say, I believe, that one of the most basic points on which the leading contemporary theories of confirmation differ from one another is with regard to the nature of prior probabilities. As already indicated, the logical theorist takes the prior probability as an a priori assessment of the hypothesis, the personalist takes the prior probability as a measure of subjective plausibility, and the frequentist must look at the prior probability as some sort of success frequency for a certain type of hypothesis, Goodman (1955) would regard the prior probability as somehow based upon linguistic usage, and Popper ([1935] 1959), though he violently objects to regarding it as a prior probability, needs something like the potential explanatory value of the hypothesis. In addition, I should remark, N. R. Hanson (1961) held plausibility arguments to belong to the logic of discovery, but I have argued (1967b, 111–118) that, *on his own analysis*, they have an indispensable role in the logic of justification.

This is not the place to go into a lengthy analysis of the virtues and shortcomings of the various views on the nature of prior probabilities. Rather, I should like merely to point out a consequence of my view that is germane to the topic of relations between history of science and philosophy of science. If one adopts a frequency view of probability, and attempts to deal with the logic of confirmation by way of Bayes's theorem (as I do), then one is committed to regarding the prior probability as some sort of frequency—for example, the frequency with which hypotheses relevantly similar to the one under consideration have enjoyed significant scientific success. Surely no one would claim that we have reliable statistics on such matters or that we can come anywhere near assigning precise numerical values in a meaningful way. Fortunately, that turns out to be unnecessary. As we noted in the preceding chapter, it is enough to have very, very rough estimates.[13] But this approach does suggest that the question of the plausibility of a scientific hypothesis has something to do with our experience in dealing with scientific hypotheses of similar types. Thus, I should say, the reason I would place a rather low plausibility value on teleological hypotheses is closely related to our experience in the transitions from teleological to mechanical explanations in the physical and biological sciences and, to some extent, in the social sciences. To turn back toward teleological hypotheses would be to go against a great deal of scientific experience about what kinds of hypotheses work well scientifically. Similarly, when Watson (1969) and Crick were enraptured with the beauty of the double helix hypothesis for the structure of the DNA molecule, I believe their reaction was more than purely esthetic.[14] Experience indicated that hypotheses of that degree of simplicity tend to be successful, and they were inferring that it had not only beauty but also a good chance of being correct. Additional examples could easily be exhibited; several are discussed in the preceding chapter.

If I am right in claiming, not only that prior probabilities constitute an indispensable ingredient in the confirmation of hypotheses and the context of justification, but also that our estimates of them are based upon empirical experience with scientific hypothesizing, then it is evident that the history of science plays a crucial, but largely unheralded, role in the current scientific enterprise. The history of science is, after all, a chronicle of our past experience with scientific hypothesizing and theorizing—with learning what sorts of hypotheses work and what sorts do not. Without the Bayesian analysis, one could say that the study of the history of science might have some (at least marginal) heuristic value for the scientist and philosopher of science; but with the Bayesian analysis, the data provided by the history of science constitute, *in addition*, an essential segment of the evidence relevant to the confirmation or disconfirmation of hypotheses. Philosophers of science and creative scientists ignore this fact at their peril.

6

Rationality and Objectivity
in Science

Thomas S. Kuhn's (1962) *The Structure of Scientific Revolutions* has been an extraordinarily influential book.[1] Coming at the height of the hegemony of *logical empiricism*—as espoused by such figures as R. B. Braithwaite, Rudolf Carnap, Herbert Feigl, Carl G. Hempel, and Hans Reichenbach—it posed a severe challenge to the logistic approach that they practiced.[2] It also served as an unparalleled source of inspiration to philosophers with a historical bent. For a quarter of a century there has been a deep division between the logical empiricists and those who adopt the historical approach, and Kuhn's book was undoubtedly a key document in the production and preservation of this gulf.[3]

At a 1983 meeting of the American Philosophical Association (Eastern Division), Kuhn and Hempel—the most distinguished living spokesmen for their respective viewpoints—shared the platform in a symposium devoted to Hempel's philosophy.[4] I had the honor to participate in this symposium. On that occasion Kuhn chose to address certain issues pertaining to the rationality of science that he and Hempel had been discussing for several years. It struck me that a bridge could be built between the differing views of Kuhn and Hempel if Bayes's theorem were invoked to explicate the concept of scientific confirmation. At the time it seemed to me that this maneuver could remove a large part of the dissensus between standard logical empiricism and the historical approach to philosophy of science on this fundamental issue.

I still believe that we have the basis for a new consensus regarding the choice among scientific theories. Although such a consensus, if achieved, would not amount to total agreement on every problem, it would represent a major rapprochement on an extremely fundamental issue. The purpose of the present chapter is to develop this approach more fully. As it turns out, the project is much more complex than I thought in 1983.

1. Kuhn on Scientific Rationality

A central part of Kuhn's challenge to the logical empiricist philosophy of science concerns the nature of theory choice in science. The choice between two fundamental theories (or paradigms), he maintains, raises issues that "cannot be resolved by proof." To see how they are resolved we must talk about "techniques of persuasion" or about "argument and counterargument in a situation in which there can be no proof." Such choices involve the exercise of the kind of judgment that cannot be rendered logically explicit and precise. These statements, along with many others that are similar in spirit, led a number of critics to attribute to Kuhn the view that science is fundamentally irrational and lacking in objectivity.

Kuhn was astonished by this response, which he regarded as a serious misinterpretation. In his "Postscript—1969" in the second edition of *The Structure of Scientific Revolutions* (1970) and in "Objectivity, Value Judgment, and Theory Choice" (1977)[5] he replies to these charges. What he had intended to convey was the claim that the decision by the community of trained scientists *constitutes* the best criterion of objectivity and rationality we can have. To better understand the nature of such objective and rational methods we need to look in more detail at the considerations that are actually brought to bear by scientists when they endeavor to make comparative evaluations of competing theories.

For purposes of illustration, Kuhn (1977) offers a (nonexhaustive) list of characteristics of good scientific theories that are, he claims,

> individually important and collectively sufficiently varied to indicate what is at stake.... These five characteristics—accuracy, consistency, scope, simplicity, and fruitfulness—are all standard criteria for evaluating the adequacy of a theory.... Together with others of much the same sort, they provide the shared basis for theory choice. (331–332)

Two sorts of problems arise when one attempts to use them:

> Individually the criteria are imprecise: individuals may legitimately differ about their applicability to concrete cases. In addition, when deployed together, they repeatedly prove to conflict with one another; accuracy may, for example, dictate the choice of one theory, scope the choice of its competitor. (322)

For reasons of these sorts—and others as well—individual scientists may, at a given moment, differ about a particular choice of theories. In the course of time, however, the interactions among individual members of the community of scientists produce a consensus for the group. Individual choices inevitably depend upon idiosyncratic and subjective factors; only the outcome of the group activity can be considered objective and fully rational.

One of Kuhn's major claims seems to be that observation and experiment, in conjunction with hypothetico-deductive reasoning, do not adequately account for the choice of scientific theories. This has led some philosophers to believe that theory choice is not rational. Kuhn, in contrast, has tried to locate the additional factors that are involved. These additional factors constitute a crucial aspect of scientific rationality.

2. Bayes's Theorem

The first step in coming to grips with the problem of evaluating and choosing scientific hypotheses or theories[6] is, as we saw in the two preceding chapters, recognition of the inadequacy of the traditional hypothetico-deductive (H-D) schema as a characterization of the logic of science. According to this schema, we confirm a scientific hypothesis by deducing from it, in conjunction with suitable initial conditions and auxiliary hypotheses, an observational prediction that turns out to be true. The H-D method has a number of well-known shortcomings: (1) It does not take account of alternative hypotheses that might be invoked to explain the same prediction. (2) It makes no reference to the initial plausibility of the hypothesis being evaluated. (3) It cannot accommodate cases, such as the testing of statistical hypotheses, in which the observed outcome is not deducible from the hypothesis (in conjunction with the pertinent initial conditions and auxiliary hypotheses) but only rendered more or less probable.

In view of these and other considerations, many logical empiricists agreed with Kuhn about the inadequacy of hypothetico-deductive confirmation. A number— including Carnap and Reichenbach—appealed to Bayes's theorem, which may be written in the following form:

$$P(T|E \cdot B) = \frac{P(T|B)\ P(E|B \cdot T)}{P(T|B)\ P(E|B \cdot T) + P(\sim T|B)\ P(E|B \cdot \sim T)} \tag{1}$$

Let T stand for the theory or hypothesis being tested, B for our background information, and E for some new evidence we have just acquired.[7] Then the expression on the left-hand side, the *posterior probability*, represents the probability of our hypothesis on the basis of the background information and the new evidence. The right-hand side contains the *prior probabilities*, $P(T|B)$ and $P(\sim T|B)$; they represent the probability, on the basis of background information alone, without taking account of the new evidence E, that our hypothesis is true or false, respectively. Obviously they must add up to 1; if the value of one of them is known, the value of the other can be inferred immediately. The right-hand side also contains the *likelihoods*, $P(E|T \cdot B)$ and $P(E|\sim T \cdot B)$; they are, respectively, the probability that the new evidence would occur if our hypothesis were true and the probability that it would occur if our hypothesis were false. The two likelihoods, in contrast to the two prior probabilities, must be established independently; the value of one does not automatically determine the value of the other.

In chapter 4 we saw how Bayes's theorem could be applied to simple artificial examples. When we come to more realistic scientific cases, it is not so easy to see how to apply it; the prior probabilities may seem particularly difficult. In that chapter I claim that, in fact, they reflect the plausibility arguments scientists often bring to bear in their deliberations about scientific hypotheses. I shall discuss this issue in section 4; indeed, in subsequent sections we shall have to take a close look at all of the probabilities that enter into Bayes's theorem.

Before moving on to that discussion, however, I want to present two other useful forms in which Bayes's theorem can be given. First, as we have seen, equation (1) can obviously be rewritten as

$$P(T|E \cdot B) = \frac{P(T|B) \; P(E|B \cdot T)}{P(E|B)} \tag{2}$$

Second, equation (1) can be generalized to handle several alternative hypotheses, instead of just one hypothesis and its negation, as follows:

$$P(T_i|E \cdot B) = \frac{P(T_i|B) \; P(E|B \cdot T_i)}{\displaystyle\sum_{j=1}^{k} [P(T_j|B) \; P(E|B \cdot T_j)]} \tag{3}$$

where T_1–T_k are mutually exclusive and exhaustive alternative hypotheses and $1 \leq i \leq k$.

Strictly speaking, (3) is the form that is needed for realistic historical examples — such as the corpuscular (T_1) and wave (T_2) theories of light in the nineteenth century. In that case, although we could construe T_1 and T_2 as mutually exclusive, we could not legitimately consider them exhaustive, for we cannot be sure that one or the other is true. Therefore, we would have to introduce T_3 — what Abner Shimony (1970) has called the *catchall hypothesis* — which says that T_1 and T_2 are both false. Thus T_1–T_3 constitute a mutually exclusive and exhaustive set of hypotheses. This is the sort of situation that obtains when scientists are attempting to choose a correct hypothesis from among two or more serious candidates.

3. Kuhn and Bayes

For purposes of discussion, Kuhn (1977, 328) is willing to admit that "each scientist chooses between competing theories by deploying some Bayesian algorithm which permits him to compute a value for P(T|E), i.e., for the probability of the theory T on the evidence E available both to him and the other members of his professional group at a particular period of time." He then formulates the crucial issue in terms of the question of whether there is one unique algorithm used by all rational scientists, yielding a unique value for P, or whether different scientists, though fully rational, may use different algorithms yielding different values of P. I want to suggest a third possibility to account for the phenomena of theory choice — namely, that many different scientists might use the same algorithm but nevertheless arrive at different values of P.

When one speaks of a Bayesian algorithm, the first thought that comes to mind is Bayes's theorem itself, as embodied in any of the equations (1), (2), and (3). We have, for instance,

$$P(T|E \cdot B) = \frac{P(T|B) \; P(E|B \cdot T)}{P(E|B)} \tag{2}$$

which constitutes an algorithm in the most straightforward sense of the term. P(E|B) is the *expectedness* of the evidence. Given values for the prior probability, likelihood, and expectedness, the value of the posterior probability can be computed by trivial arithmetical operations.[8]

If we propose to use equation (2) as an algorithm, the obvious question is how to get values for the expressions on the right-hand side. Several answers are possible in principle, depending on what interpretation of the probability concept is espoused. If one adopts a Carnapian approach to inductive logic and confirmation theory, all of the probabilities that appear in Bayes's theorem can be derived a priori from the structure of the descriptive language and the definition of degree of confirmation. Since it is extremely difficult to see how any genuine scientific case could be handled by means of the highly restricted apparatus available within that approach, not many philosophers are tempted to follow this line. Moreover, even if a rich descriptive language were available, it is not philosophically tempting to suppose that the probabilities associated with serious scientific theories are a priori semantic truths.

Two major alternatives remain. First, one might maintain that the probabilities on the right-hand side of (2)—especially the prior probability P(T|B)—are objective and empirical. I have attempted to defend the view that they refer, at bottom, to the frequencies with which various kinds of hypotheses or theories have been found successful (Salmon 1967b, chap. 7). Clearly, enormous difficulties are involved in working out that alternative; I shall return to the issue (see also chapter 8, "Dynamic Rationality"). In the meantime, let us consider the other—far more popular—alternative.

The remaining alternative approach involves the use of personal probabilities. Personal probabilities are subjective in character; they represent subjective degrees of conviction on the part of the person who has them, provided that they fulfill the condition of coherence.[9] Consider a somewhat idealized situation. Suppose that, in the presence of background knowledge B (which may include initial conditions, boundary conditions, and auxiliary hypotheses), theory T deductively entails evidence E. This is the situation to which the hypothetico-deductive method appears to be applicable. In this case, P(E|T·B) must equal 1, and equation (2) reduces to

$$P(T|E \cdot B) = P(T|B)/P(E|B) \tag{4}$$

One might then ask a particular scientist for his or her plausibility rating of theory T on background knowledge B alone, quite irrespective of whether evidence E obtains or not. Likewise, the same individual may be queried regarding the degree to which evidence E is to be expected irrespective of the truth or falsity of T. According to the personalist, it should be possible—by direct questioning or by some less direct method—to elicit such *psychological* facts regarding a scientist involved in investigations concerning the theory in question. This information is sufficient to determine the degree of belief this individual should have in the theory T given the background knowledge B and the evidence E, namely, the posterior probability P(T|E·B).

In the more general case, when T and B do not deductively entail E, the procedure is the same, except that the value of $P(E|T \cdot B)$ must also be ascertained. In many contexts, where statistical significance tests can be applied, a value of the likelihood $P(E|T \cdot B)$ can be calculated, and the personal probability will coincide with the value thus derived. In any case, whether statistical tests apply or not, there is no *new* problem in principle involved in procuring the needed degree of confidence. This reflects the standard Bayesian approach in which all of the probabilities are taken to be personal probabilities.

Whether one adopts an objective or a personalistic interpretation of probability, equation (2)—or some other version of Bayes's theorem—can be taken as an algorithm for evaluating scientific hypotheses or theories. Individual scientists, using the same algorithm, may arrive at different evaluations of the same hypothesis because they plug in different values for the probabilities. If the probabilities are construed as objective, different individuals may well have different estimates of these objective values. If the probabilities are construed as personal, different individuals may well have different subjective assessments of them. Bayes's theorem provides a mechanical algorithm, but the judgments of individual scientists are involved in procuring the values that are to be fed into it. This is a general feature of algorithms; they are not responsible for the data they are given.

4. Prior Probabilities

In chapter 4, I argued that the prior probabilities in Bayes's theorem can best be seen as embodying the kinds of plausibility judgments that scientists regularly make regarding the hypotheses with which they are concerned. Einstein (1949), who was clearly aware of this consideration, contrasted two points of view from which a theory can be criticized or evaluated:

> The first point of view is obvious: the theory must not contradict empirical facts.... It is concerned with the confirmation of the theoretical foundation by the available empirical facts.
>
> The second point of view is not concerned with the relation of the material of observation but with the premises of the theory itself, with what may briefly but vaguely be characterized as the "naturalness" or "logical simplicity" of the premises.... The second point of view may briefly be characterized as concerning itself with the "inner perfection" of a theory, whereas the first point of view refers to the "external confirmation." (21–22)

Einstein's second point of view is the sort of thing I have in mind in referring to plausibility arguments or judgments concerning prior probabilities.

Plausibility considerations are pervasive in the sciences; they play a significant— indeed, *indispensable*—role. This fact provides the initial reason for appealing to Bayes's theorem as an aid to understanding the logic of evaluating scientific hypotheses. Plausibility arguments serve to enhance or diminish the probability of a given hypothesis prior to—that is, without reference to—the outcome of a particular observation or experiment. They are designed to answer the question "Is this the kind of hypothesis that is likely to succeed in the scientific situation in which the scientist finds himself or herself?" On the basis of their training and experience, scientists are qualified to make such judgments. This point can best be explained, I believe, in terms of concrete examples. Some have been given in the preceding two chapters, and I shall offer some others below.

In order to come to a clearer understanding of the nature of prior probabilities, it will be necessary to look at them from the point of view of the personalist and that of the objectivist (frequency or propensity theorist).[10] The frightening thing about pure unadulterated personalism is that nothing prevents prior probabilities (and other probabilities as well) from being determined by all sorts of idiosyncratic and objectively irrelevant considerations. A given hypothesis might get an extremely low prior probability because the scientist considering it has a hangover, has had a recent fight with his or her lover, is in passionate disagreement with the politics of the scientist who first advanced the hypothesis, harbors deep prejudices against the ethnic group to which the originator of the hypothesis belongs, and so forth. What we want to demand is that the investigator make every effort to bring all of his or her *relevant* experience in evaluating hypotheses to bear on the question of whether the hypothesis under consideration is of a type likely to succeed, and to leave aside emotional irrelevancies.

It is rather easy to construct really perverse systems of belief that do not violate the coherence requirement. But we need to keep in mind the objectives of science. When we have a long series of events, such as tosses of fair or biased coins or radioactive decays of unstable nuclei, we want our subjective degrees of conviction to match what either a frequency theorist or a propensity theorist would regard as the objective probability. Carnap was profoundly correct in his notion that inductive or logical or epistemic probabilities should be reasonable estimates of relative frequencies.

A sensible personalist, I would suggest, is someone who wants his or her personal probabilities to reflect objective facts. Betting on a sequence of tosses of a coin, a personalist wants not only to avoid Dutch books[11] but also to stand a reasonable chance of winning (or of not losing too much too fast). As I read it, the whole point of F. P. Ramsey's (1950) famous article on degrees of belief is to consider what you get if your subjective degrees of belief match the relevant frequencies. One of the facts recognized by the sensible personalist is that whether the coin lands heads or tails is not affected by which side of the bed he or she got out on that morning. If we grant that the personalist's aim is to do as well as possible in betting on heads and tails, it would be obviously counterproductive to allow the betting odds to be affected by such irrelevancies.

The same general sort of consideration should be brought to bear on the assignment of probabilities to hypotheses. Whether a particular scientist is dyspeptic on a given morning is irrelevant to the question of whether a physical hypothesis that is under consideration is correct or not. Much more troubling, of course, is the fact that any given scientist may be inadvertently influenced by ideological or metaphysical prejudices. It is obvious that an unconscious commitment to capitalism or racism might seriously affect theorizing in the behavioral sciences.

Similar situations may arise in the physical sciences as well; a historical example will illustrate the point. In 1800 Alessandro Volta invented the battery, thereby providing scientists with a way of producing steady electrical currents. It was not until 1820 that Hans Christian Oersted discovered the effect of an electrical current on a magnetic needle. Why was there such a delay? One reason was the previously established fact that a static electric charge has no effect on a magnetic needle. Another reason that has been mentioned is the fact that, contrary to the expectation, if there were such an effect it would align the needle perpendicular to the current carrying wire. As Holton and Brush (1973, 416; emphasis in original) remark, "But even if one has currents and compass needles available, one does not observe the effect unless the compass is placed in the right position so that the needle can respond to a force that seems to act in a direction *around* the current rather than *toward* it". I found it amusing when, on one occasion, a colleague set up the demonstration with the magnetic needle oriented at right angles to the wire to show why the experiment fails if one begins with the needle in that position. When the current was turned on, the needle rotated through 180°; he had neglected to take account of polarity. How many times, between 1800 and 1820 had the experiment been performed without reversing the polarity? Not many. The experiment had apparently not been tried by others because of Cartesian metaphysical commitments. It was undertaken by Oersted as a result of his proclivities toward *naturphilosophie*.

How should scientists go about evaluating the prior probabilities of hypotheses? In elaborating a view he calls *tempered personalism*—a view that goes beyond standard Bayesian personalism by placing further constraints on personal probabilities—Shimony (1970) points out that experience shows that the hypotheses seriously advanced by serious scientists stand some chance of being successful. Science has, in fact, made considerable progress over the past four or five centuries, which constitutes strong empirical evidence that the probability of success among members of this class is nonvanishing. Likewise, hard experience has also taught us to reject claims of scientific infallibility. Thus, we have good reasons for avoiding the assignment of extreme values to the priors of the hypotheses with which we are seriously concerned. Moreover, Shimony reminds us, experience has taught that science is difficult and frustrating; consequently, we ought to assign fairly low prior probabilities to the hypotheses that have been explicitly advanced, allowing a fairly high prior for the catchall hypothesis—the hypothesis that we have not yet thought of the correct hypothesis. The history of science abounds with situations of choice among theories in which the successful candidate has not even been conceived at the time.

In chapter 4 of this book (and in Salmon 1967b, chap. 7) I proposed that the problem of prior probabilities be approached in terms of an objective interpretation of probability, and I suggested that three sorts of criteria can be brought to bear in assessing the prior probabilities of hypotheses: formal, material, and pragmatic.

Pragmatic criteria have to do with the circumstances in which a new hypothesis originates. We have just seen an example of a pragmatic criterion in Shimony's (1970) observation that hypotheses advocated by serious scientists have nonvanishing chances of success. We have also mentioned the opposite side of the same coin, as provided by Martin Gardner (1957, 7–15) in his enlightening characterization of scientific cranks. Since it is doubtful that a single useful scientific suggestion has ever been originated by anyone in that category, hypotheses advanced by people of that ilk have negligible chances of being correct.

The *formal criteria* have to do not only with matters of internal consistency of a new hypothesis but also with relations of entailment or incompatibility of the new hypothesis with accepted laws and theories. As we have seen, Velikovsky's discussion (1950) stands as a prime example in this category. It should be recalled that among his five considerations for the evaluation of scientific theories—mentioned above—Kuhn includes consistency of the sort we are discussing. I take this as a powerful hint that one of the main issues Kuhn has raised about scientific theory choice involves the use of prior probabilities and plausibility judgments.

The *material criteria* concern the actual structure and content of the hypothesis or theory under consideration. The most obvious example is simplicity—another of Kuhn's five items. Simplicity strikes me as singularly important, for it has often been treated by scientists and philosophers as an a priori criterion. It has been suggested, for example, that the hypothesis that quarks are fundamental constitutents of matter loses plausibility as the number of different types of quarks increases since it becomes less simple as a result (see Harari 1983, 56–68). It has also been advocated as a universal methodological maxim: *Search for the simplest possible hypothesis*. Only if the simpler hypotheses do not stand up under testing should one resort to more complex hypotheses.

Although simplicity has obviously been an important consideration in the physical sciences, its applicability in the social/behavioral sciences is problematic. In a recent article, "Slips of the Tongue," Michael T. Motley (1985) criticizes Freud's theory for being too simple—an oversimplification:

> Further still, the categorical nature of Freud's claim that all slips have hidden meanings makes it rather unattractive. It is difficult to imagine, for example, that my six-year-old daughter's mealtime request to "help cut up my meef" was the result of repressed anxieties or anything of that kind. It seems more likely that she simply merged "meat" and "beef" into "meef." Similarly, about the only meaning one can easily read into someone's saying "roon mock" instead of "moon rock" is that the *m* and *r* got switched. Even so, how does it happen that words can merge or sounds can be switched in the course of speech production? And in the case of my "pleased to beat you" error [to a competitor for a job], might Freud have been right? (116)[12]

First, the most reasonable way to look at simplicity, I think, is to regard it as a highly relevant characteristic, but one whose applicability varies from one scientific context to another. Specialists in any given branch of science make judgments about the degree of simplicity or complexity that is appropriate to the context at hand, and they do so on the basis of extensive experience in that particular area of scientific investigation. Since there is no precise measure of simplicity as applied to scientific hypotheses and theories, scientists must use their judgment concerning the degree of simplicity a given hypothesis or theory possesses and concerning the degree of simplicity that is desirable in the given context. The kind of judgment to which I refer is not spooky; it is the kind of judgment that arises on the basis of training and experience. This experience is far too rich to be the sort of thing that can be spelled out explicitly. As Patrick Suppes (1966, 202–203) has pointed out, the assignment of prior probability by the Bayesian can be regarded as the best estimate of the chances of success of the hypothesis or theory on the basis of all relevant experience in that particular scientific domain. The personal probability represents, not an effort to contaminate science with subjective irrelevancies, but rather an attempt to facilitate the inclusion of all relevant evidence.

Simplicity is only one among many material criteria. A second, closely related criterion—frequently employed in contemporary physics—is symmetry. Perhaps the most striking historical example is de Broglie's hypothesis regarding matter waves. Since light exhibits both particle and wave behavior that are linked in terms of linear momentum, he suggested, why should not material particles, which obviously possess linear momentum, also have such wave characteristics as wavelength and frequency? Unknown by de Broglie, experimental work by Davisson was, at that very time, providing positive evidence of wavelike behavior of electrons. And Davisson was deeply puzzled by the results of these experiments (see Sobel 1987, 91–95).

A third, widely used material criterion is analogy; a famous Canadian study of the effects of the consumption of large doses of saccharin provides an excellent example (see Giere 1991, 222–223). A statistically significant association between heavy saccharin consumption and bladder cancer in a controlled experiment with rats lends considerable plausibility to the hypothesis that use of saccharin as an artificial sweetener in diet soft drinks increases the risk of bladder cancer in humans. This example, unlike the preceding one, is inherently statistical and does not have even the prima facie appearance of a hypothetico-deductive inference.

I suspect that the use of arguments by analogy in science is almost always aimed at establishing prior probabilities. The formal criteria enable us to take account of the ways in which a given hypothesis fits *deductively* with what else we know. Analogy helps us to assess the degree to which a given hypothesis fits *inductively* with what else we know.

The moral I would draw concerning prior probabilities is that they can be understood as our best estimates of the frequencies with which certain kinds of hypotheses succeed. These estimates are rough and inexact; some philosophers might prefer to think of them in terms of intervals. If, however, one wants to construe them as personal probabilities, there is no harm in it, as long as we attribute to the subject who has them the aim of bringing to bear all his or her experience that is relevant to the success or failure of hypotheses similar to that being considered. The personalist and the frequentist need not be in any serious disagreement over the construal of prior probabilities (see chapter 8).

One point is apt to be immediately troublesome. If we are to use Bayes's theorem to compute values of posterior probabilities, it would appear that we must be prepared to furnish numerical values for the prior probabilities. Unfortunately, it seems preposterous to suppose that plausibility arguments of the kind we have considered could yield exact numerical values. The usual answer is that, because of a phenomenon known as "washing out of the priors" or "swamping of the priors," even very crude estimates of the prior probabilities will suffice for the kinds of scientific judgments we are concerned to make. Obviously, however, this sort of convergence depends upon agreement regarding the likelihoods.

5. The Expectedness

The term "$P(E|B)$," occurring in the denominator of equation (2), is called the *expectedness* because it is the opposite of surprisingness. The smaller the value of $P(E|B)$, the more surprising E is; the larger the value of $P(E|B)$, the less surprising, and hence, the more expected E is. Since the expectedness occurs in the denominator, a smaller value tends to increase the value of the fraction. This conforms to a widely held intuition that the more surprising the predictions a theory can make, the greater is their evidential value when they come true.

A classic example of a surprising prediction that came true is the Poisson bright spot. If we ask someone who is completely naive about theories of light how probable it is that a bright spot appears in the center of the shadow of a brightly illuminated circular object (ball or disk), we would certainly anticipate the response that it is very improbable indeed. There is a good inductive basis for this answer. In our everyday lives we have all observed many shadows of opaque objects, and they do not contain bright spots at their centers. Once, when I demonstrated the Poisson bright spot to an introductory class, one student carefully scrutinized the ball bearing that cast the shadow because he strongly suspected that I had drilled a hole through it.

Another striking example, to my mind, is the Cavendish torsion balance experiment. If we ask someone who is totally ignorant of Newton's theory of universal gravitation how strongly he or she expects to find a force of attraction between a lead ball and a pith ball in a laboratory, I should think the answer, again, would be that it is very unlikely. There is, in this example as well, a sound inductive basis for the response. We are all familiar with the gravitational attraction of ordinary-sized objects to the earth, but we do not have everyday experience of an attraction between two such relatively small (electrically neutral and unmagnetized) objects as those Cavendish used to perform his experiment. Newton's theory predicts, of course, that there will be a gravitational attraction between any two material objects. The trick was to figure out how to measure it.

As the foregoing two examples show, there is a possible basis for assigning a low value to the expectedness; it was made plausible by assuming that the subject was completely naive concerning the relevant physical theory. The trouble with this approach is that a person who wants to use Bayes's theorem—in the form of equation (2), say—cannot be totally innocent of the theory T that is to be evaluated since the other terms in the equation refer explicitly to T. Consequently, we have to recognize the relationship between $P(E|B)$ and the prior probabilities and likelihoods that appear on the right-hand side in the theorem on total probability:

$$P(E|B) = P(T|B)\ P(E|T \cdot B) + P(\sim T|B)\ P(E|\sim T \cdot B) \qquad (5)$$

Suppose that the prior probability of T is not negligible and that T, in conjunction with suitable initial conditions, entails E. Under these circumstances E cannot be totally surprising; the expectedness cannot be vanishingly small. Moreover, to evaluate the expectedness of E we must also consider its probability if T is false. By focusing on the expectedness, we cannot really avoid dealing with likelihoods.

There is a further difficulty. Suppose, for example, that the wave theory of light is true. It is surely *true enough* in the context of the Poisson bright spot experiment. If we want to evaluate P(E|B) we must include in B the initial con-ditions of the experiment—the circular object illuminated by a bright light in such a way that the shadow falls upon a screen. Given the truth of the wave theory, the *objective probability* of the bright spot is 1, for whenever those initial conditions are realized, the bright spot appears. It makes no difference whether we know that the wave theory is true, believe it, reject it, or have ever thought of it. Under the conditions specified in B the bright spot invariably occurs. Interpreted either as a frequency or a propensity, P(E|B) = 1. If we are to avoid trivialization in many important cases, the expectedness must be treated as a personal probability. To anyone who, like me, wants to base scientific theory preference or choice on objective considerations, this result poses a serious problem.

The net result is a twofold problem. First, by focusing on the expectedness, we do *not* escape the need to deal explicitly with the likelihoods. In section 6 I shall discuss the difficulties that arise when we focus on the likelihoods, especially the problem of the likelihood on the catchall hypothesis. Second, the expectedness defies interpretation as an objective probability. In section 7 I shall propose a strategy for avoiding involvement with either the expectedness or the likelihood on the catchall. That maneuver will, I hope, keep open the possibility of an objective basis for the evaluation of scientific hypotheses.

6. Likelihoods

Equations (1), (2), and (3) are different forms of Bayes's theorem, and each of them contains a likelihood, P(E|T·B), in the numerator. Two trivial cases can be noted at the outset. First, if the conjunction of theory T and background knowledge B are logically incompatible with evidence E, the likelihood equals zero, and the pos-terior probability, P(T|E·B), automatically becomes zero.[13] Second, as we have already noticed, if T·B entails E, that likelihood equals 1, and consequently drops out, as in equation (4).

Another easy case occurs when the hypothesis T involves various kinds of randomness assumptions, for example, the independence of a series of trials on a chance setup.[14] Consider the case of a coin that has been tossed 100 times with the result that heads showed in 63 cases and tails in 37. We assume that the tosses are independent, but we are concerned about whether the system consisting of the coin and tossing mechanism is biased. Calculation shows that the probability, given an unbiased coin and tossing mechanism, of the actual frequency of heads differing from 1/2 by 20% or more on 100 tosses (i.e., falling outside of the range 40–60) is about 0.05. Thus, the likelihood of the outcome on the hypothesis that the coin and mechanism are fair is less than 0.05. On the hypothesis that the coin has a 60–40 bias for heads, by contrast, the probability that the number of heads in 100 trials differs from 6/10 by less than 20% (that is, lies within the 48–72) is well above 0.95. These are the kinds of likelihoods that would be used to compare the *null hypothesis* that the coin is fair with the hypothesis that it has a certain bias.[15] This example typifies a wide variety of cases, including the above-mentioned controlled experiment on rats and saccharin, in which statistical significance tests are applied. These yield a comparison between the probability of the observed result if the hypothesis is correct and the probability of the same result on a null hypothesis.

In still another kind of situation the likelihood $P(E|T \cdot B)$ is straightforward. Consider, for example, the case in which a physician takes an X ray for diagnostic purposes. Let T be the hypothesis that the patient has a particular disease and let E be a certain appearance on the film. From long medical experience it may be known that E occurs in 90% of all cases in which that disease is present. In many cases, as this example suggests, there may be accumulated frequency data from which the value of $P(E|T \cdot B)$ can be derived.

Unfortunately, life with likelihoods is not always as simple as the foregoing cases suggest. Consider an important case, which I will present in a highly unhistorical way. In comparing the Copernican and Ptolemaic cosmologies, it is easy to see that the phases of Venus are critical. According to the Copernican system, Venus should exhibit a broad set of phases, from a narrow crescent to an almost full disk. According to the Ptolemaic system, Venus should always present nearly the same crescent-shaped appearance. One of Galileo's celebrated telescopic observations was of the phases of Venus. The likelihood of such evidence on the Copernican system is unity; on the Ptolemaic it is zero. This is the decisive sort of case that we cherish.

The Copernican system did, however, face one serious obstacle. On the Ptolemaic system, because the earth does not move, the fixed stars should not appear to change their positions. On the Copernican system, because the earth makes an annual trip around the sun, the fixed stars should appear to change their positions in the course of the year. The very best astronomical observations, including those of Tycho Brahe, failed to reveal any observable stellar parallax.[16] However, it was realized that, if the fixed stars are at a very great distance from the earth, stellar parallax, though real, would be too small to be observed. Consequently, the likelihood $P(E|T \cdot B)$, where T is the Copernican system and E the absence of observable stellar parallax, is not zero. At the time of the scientific revolution, prior to the advent of Newtonian mechanics, there seemed no reasonable way to evaluate this likelihood. The assumption that the fixed stars are almost unimaginably distant from the earth was a highly ad hoc, and consequently implausible, auxiliary hypothesis to adopt just to save the Copernican system. Among other things, Christians did not like the idea that heaven was so very far away.

The most reasonable resolution of this anomaly was offered by Tycho Brahe, whose cosmology placed the earth at rest, with the sun and moon moving in orbits around the earth, but with all of the other planets moving in orbits around the sun. In this way both the observed phases of Venus and the absence of observable stellar parallax could be accomodated. Until Newton's dynamics came upon the scene, it seems to me, Tycho's system was clearly the best available theory.

In section 2 I suggested that the following form of Bayes's theorem is the most appropriate for use in actual scientific cases in which more than one hypothesis is available for serious consideration:

$$P(T_i|E \cdot B) = \frac{P(T_i|B) \ P(E|B \cdot T_i)}{\sum\limits_{j=1}^{k} [P(T_j|B) \ P(E|B \cdot T_j)]} \tag{3}$$

It certainly fits the foregoing example, in which we compared the Ptolemaic, Copernican, and Tychonic systems. This equation involves a mutually exclusive and exhaustive set of hypotheses $T_1, \ldots, T_{k-1}, T_k$, where T_1–T_{k-1} are seriously entertained and T_k is the catchall. Thus, the scientist who wants to calculate the posterior probability of one particular hypothesis T_i on the basis of evidence E must ascertain likelihoods of three types: (1) the probability of evidence E given T_i, (2) the probability of that evidence on each of the other seriously considered alternatives T_j ($j \neq i$, $j \neq k$), and (3) the probability of that evidence on the catchall T_k.

In considering the foregoing example, I suggested that, although likelihoods in the first two categories are sometimes straightforward, there are cases in which they turn out to be problematic. We shall look at more examples in which they present difficulties as our discussion proceeds, but the point to be emphasized right now is the utter intractability of the likelihood on the catchall. The reason for this difficulty is easy to see. Whereas the seriously considered candidates are bona fide hypotheses, the catchall is a hypothesis only in a Pickwickian sense. It refers to all of the hypotheses we are *not* taking seriously, including all those that have not been thought of as yet; indeed, the catchall is logically equivalent to their disjunction. These will often include brilliant discoveries in the future history of science that will eventually solve our most perplexing problems.

Among the hypotheses hidden in the catchall are some that, in conjunction with present available background information, entail the present evidence E. On such as-yet-undiscovered hypotheses the likelihood is 1. Obviously, however, the fact that its probability on one particular hypothesis is unity does not entail anything about its probability on some disjunction containing that hypothesis as one of its disjuncts. These considerations suggest to me that the likelihood on the catchall is totally intractable. To try to evaluate the likelihood on the catchall involves, it seems to me, an attempt to guess the future history of science. That is something we cannot do with any reliability.

In any situation in which a small number of theories are competing for ascendency it is tempting, though quite illegitimate, simply to ignore the likelihood on the catchall. In the nineteenth century, for instance, scientists asked what the probability of a given phenomenon is on the wave theory of light and what it is on the corpuscular theory. They did not seriously consider its probability if neither of these theories is correct. Yet we see, from the various forms in which Bayes's theorem is written, that either the expectedness or the likelihood on the catchall is an indispensable ingredient. In the next section I shall offer a *legitimate* way of eliminating those probabilities from our consideration.

7. Choosing between Theories

Kuhn (1962, 1970, 1977) has often maintained that in actual science the problem is never to evaluate one particular hypothesis or theory in isolation; it is always a matter of choosing from among two or more viable alternatives. He has emphasized that an old theory is never completely abandoned unless there is currently available a rival to take its place. Given that circumstance, it is a matter of choosing between the old and the new. On this point I think that Kuhn is quite right, especially as regards reasonably mature sciences. And this insight provides a useful clue on how to use Bayes's theorem to explicate the logic of scientific confirmation.

Suppose that we are trying to choose between T_1 and T_2, where there may or may not be other serious alternatives in addition to the catchall. By letting $i = 1$ and $j = 2$, we can proceed to write equation (4) for each of these candidates. Noting that the denominators of the two are identical, we can form their ratio as follows:

$$\frac{P(T_1|E\cdot B)}{P(T_2|E\cdot B)} = \frac{P(T_1|B)}{P(T_2|B)} \frac{P(E|T_1\cdot B)}{P(E|T_2\cdot B)} \tag{6}$$

No reference to the catchall hypothesis appears in this equation. Since the catchall is not a bona fide hypothesis, it is not a contender, and we need not try to calculate its posterior probability. The use of equation (6) frees us from the need to deal either with the expectedness of E or with its probability on the catchall.

Equation (6) yields a relation that can be regarded as a *Bayesian algorithm for theory preference*. Suppose that, prior to the emergence of evidence E, you prefer T_1 to T_2; that is, $P(T_1|B) > P(T_2|B)$. Then E becomes available. You should change your preference in the light of E if and only if $P(T_2|E\cdot B) > P(T_1|E\cdot B)$. From (6) it follows that

$$P(T_2|E\cdot B) > P(T_1|E\cdot B) \quad \text{iff} \quad P(E|T_2\cdot B)/P(E|T_1\cdot B) > P(T_1|B)/P(T_2|B) \tag{7}$$

In other words, you should change your preference to T_2 if the ratio of the likelihoods is greater than the reciprocal of the ratio of the respective prior probabilities. A corollary is that, if both $T_1\cdot B$ and $T_2\cdot B$ entail E, so that

$$P(E|T_1\cdot B) = P(E|T_2\cdot B) = 1,$$

the occurrence of E can never change the preference rating between the two competing theories.

At the end of section 4, I made reference to the well-known phenomenon of washing out of priors in connection with the use of Bayes's theorem. One might well ask what happens to this swamping when we switch from Bayes's theorem to the ratio embodied in equation (6).[17] The best answer, I believe, is this. If we are dealing with two hypotheses that are serious contenders in the sense that they do not differ too greatly in plausibility, the ratio of the priors will be of the order of unity. If, as the observational evidence accumulates, the likelihoods come to differ greatly, the ratio of the likelihoods will swamp the ratio of the priors. Recall the example of the tossed coin. Suppose we consider the prior probability of a fair device to be ten times as large as that of a biased device. If about the same proportion of heads occurs in 500 tosses as occurred in the aforementioned 100, the likelihood on the null hypothesis would be virtually zero and the likelihood on the hypothesis that the device has a bias approximating the observed frequency would be essentially indistinguishable from unity. The ratio of prior probabilities would obviously be completely dominated by the likelihood ratio.

8. Plausible Scenarios

Although, by appealing to equation (6), we have eliminated the need to deal with the expectedness or the likelihood on the catchall, we cannot claim to have dealt adequately with the likelihoods on the hypotheses we are seriously considering, for their values are not always straightforwardly ascertainable. We have already mentioned one example, namely, the probability of no observable stellar parallax on the Copernican hypothesis. We noted that, by adding an auxiliary hypothesis to the effect that the fixed stars are located an enormous distance from the earth, we could augment the Copernican hypothesis in such a way that the likelihood, on this augmented hypothesis, is one. But, for many reasons, this auxiliary assumption could hardly be considered plausible in that historical context. By now, of course, we have measured the parallax of relatively nearby stars, and from those values have calculated these distances. They are extremely far from us in comparison to the familiar objects in our solar system.

Consider another well-known example. During the seventeenth and eighteenth centuries the wave and corpuscular theories of light received considerable scientific attention. Each was able to explain certain important optical phenomena, and each faced fundamental difficulties. The corpuscular hypothesis easily explained how light could travel vast distances through empty space, and it readily explained sharp shadows. The theory of light as a longitudinal wave explained various kinds of diffraction phenomena, but failed to deal adequately with polarization. When, early in the nineteenth century, light was conceived as a transverse wave, the wave theory explained polarization as well as diffraction quite straightforwardly. And Huyghens had long since shown how the wave theory could handle rectilinear propagation and sharp shadows. For most of the nineteenth century the wave theory dominated optics.

The proponent of the particle theory could still raise a serious objection. What is the likelihood of a wave propagating in empty space? Lacking a medium, the answer is zero. So wave theorists augmented their theory with the auxiliary assumption that all of space is filled with a peculiar substance known as the *luminiferous aether*. This substance was postulated to have precisely the properties required to transmit light waves.

The process I have been describing can appropriately be regarded as the discovery and introduction of *plausible scenarios*. A theory is confronted with an *anomaly*—a phenomenon that appears to have a small, possibly zero, likelihood given that theory. Proponents of the theory search for some auxiliary hypothesis that, if conjoined to the theory, renders the likelihood high, possibly unity. This move shifts the burden of the argument to the plausibility of the new auxiliary hypothesis. I mentioned two instances involved in the wave theory of light. The first was the auxiliary assumption that the wave is transverse. This modification of the theory was sufficiently plausible to be incorporated as an integral part of the theory. The second was the luminiferous aether. The plausibility of this auxiliary hypothesis was debated throughout the nineteenth, and into the twentieth, century. The aether had to be dense enough to transmit transverse waves (which require a denser medium than do longitudinal waves) and thin enough to allow astronomical bodies to move through it without noticeable diminution of speed. Attempts to detect the motion of the earth relative to the aether were unsuccessful. The Lorentz-Fitzgerald contraction hypothesis was an attempt to save the aether theory—that is, another attempt at a plausible scenario—but it was, of course, abandoned in favor of special relativity.

I am calling these auxiliaries *scenarios* because they are stories about how something could have happened, and *plausible* because they must have some degree of acceptability if they are to be of any help in handling problematic phenomena. The wave theory could handle the Poisson bright spot by deducing it from the theory. There seemed to be no plausible scenario available to the particle theory that could deal with this phenomenon. The same has been said with respect to Foucault's demonstration that the velocity of light is greater in air than it is in water (see Holton and Brush 1973, 392–93).

One nineteenth-century optician of considerable importance who did not adopt the wave theory, but remained committed to the Newtonian emission theory, was David Brewster (see Worrall 1990). In a "Report on the Present State of Physical Optics," presented to the British Association for the Advancement of Science in 1831, he maintained that the undulatory theory is "still burthened with difficulties and cannot claim our implicit assent" (quoted by Worrall 1990, 321). Brewster freely admitted the unparalleled explanatory and predictive success of the wave theory; nevertheless, he considered it false.

Among the difficulties Brewster found with the wave theory, two might be mentioned. First, he considered the wave theory implausible, for the reason that it required "an *ether* invisible, intangible, imponderable, inseparable from all bodies, and extending from our own eye to the remotest verge of the starry heavens" (ibid., 322). History has certainly vindicated him on that issue. Second, he found the wave theory incapable of explaining a phenomenon that he had discovered himself, namely, *selective absorption*—dark lines in the spectrum of sunlight that has passed through certain gases. Brewster points out that a gas may be opaque to light of one particular index of refraction in flint glass, while transmitting freely light whose refractive indices in the same glass are only the tiniest bit higher or lower. Brewster maintained that there was no plausible scenario the wave theorists could devise that would explain why the aether permeating the gas transmits two waves of very nearly the same wave length but does not transmit light of a very precise wave length lying in between:

There is no fact analogous to this in the phenomena of sound, and I can form no conception of a simple elastic medium so modified by the particles of the body which contains it, as to make such an extraordinary selection of the undulations which it stops or transmits. (ibid., 323)

Brewster never found a plausible scenario by means of which the Newtonian theory he favored could cope with absorption lines, nor could proponents of the wave theory find one to bolster their viewpoint. Dark absorption lines remained anomalous for both the wave and particle theories; neither could see a way to furnish them with high likelihood.

With hindsight we can say that the catchall hypothesis was looking very strong at this point. We recognize that the dark absorption lines in the spectrum of sunlight are closely related to the discrete lines in the emission spectra of gases, and that they, in turn, are intimately bound up with the problem of the stability of atoms. These phenomena played a major role in the overthrow of classical physics at the turn of the twentieth century.

I have introduced the notion of a plausible scenario to deal with problematic likelihoods. Likelihoods can cause trouble for a scientific theory for either of two reasons. First, if you have a pet theory that confers an extremely small—for all practical purposes, zero—likelihood on some observed phenomenon, that is a problem for that favored theory. You try to come up with a plausible scenario according to which the likelihood will be larger—ideally, unity. Second, if there seems to be no way to evaluate the likelihood of a piece of evidence with respect to some hypothesis of interest, that is another sort of problem. In this case, we search for a plausible scenario that will make the likelihood manageable, whether this involves assigning it a high, medium, or low value.

What does this mean in terms of the Bayesian approach I am advocating? Let us return to

$$\frac{P(T_1|E \cdot B)}{P(T_2|E \cdot B)} = \frac{P(T_1|B) \; P(E|T_1 \cdot B)}{P(T_2|B) \; P(E|T_2 \cdot B)} \tag{6}$$

which contains two likelihoods. Suppose, as in nineteenth-century optics, that both likelihoods are problematic. As we have seen, we search for plausible scenarios A_1 and A_2 to augment T_1 and T_2, respectively. If the search has been successful, we can assess the likelihoods of E with respect to the augmented theories $A_1 \cdot T_1$ and $A_2 \cdot T_2$. Consequently, we can modify (6) to yield

$$\frac{P(A_1 \cdot T_1|E \cdot B)}{P(A_2 \cdot T_2|E \cdot B)} = \frac{P(A_1 \cdot T_1|B) \; P(E|A_1 \cdot T_1 \cdot B)}{P(A_2 \cdot T_2|B) \; P(E|A_2 \cdot T_2 \cdot B)} \tag{8}$$

In order to use this equation to compare the posterior probabilities of the two augmented theories, we must assess the plausibilities of the scenarios, for the prior probabilities of both augmented theories—$A_1 \cdot T_1$ and $A_2 \cdot T_2$—appear in it. In section 4 I tried to explain how prior probabilities can be handled, that is, how we can obtain at least rough estimates of their values. If, as suggested, the plausible scenarios have made the likelihoods ascertainable, then we can use them in conjunction with our determinations of the prior probabilities to assess the ratio of the posterior probabilities. We have, thereby, handled the central issue raised by Kuhn, namely, what is the basis for preference between two theories. . . .[18] Equation (8) is a Bayesian algorithm.

If either augmented theory, in conjunction with background knowledge B, entails E, then the corresponding likelihood is 1 and it drops out of equation (8). If both likelihoods drop out we have the special case in which

$$\frac{P(A_1 \cdot T_1 | E \cdot B)}{P(A_2 \cdot T_2 | E \cdot B)} = \frac{P(A_1 \cdot T_1 | B)}{P(A_2 \cdot T_2 | B)} \tag{9}$$

thereby placing the *whole* burden on the prior probabilities—the plausibility considerations. Equation (9) represents a simplified Bayesian algorithm that is applicable in this type of special case.

Another type of special case was mentioned above. If, as in our coin-tossing example, the values of the prior probabilities do not differ drastically from one another, but the likelihoods become widely divergent as the observational evidence accumulates, there will be a washing out of the priors. In this case, the ratio of the posterior probabilities equals, for practical purposes, the ratio of the likelihoods.

The use of either equation (8) or (9) as an algorithm for theory choice does not imply that all scientists will agree on the numerical values or prefer the same theory. The evaluation of prior probabilities clearly demands the kind of scientific judgment whose importance Kuhn has rightly insisted upon. It should also be clearly remembered that these formulas provide no evaluations of individual theories; they furnish only comparative evaluations. Thus, instead of yielding a prediction regarding the chances of one particular theory being a component of 'completed science,' they compare existing theories with regard to their present merits.

9. Kuhn's Criteria

Early in this chapter I quoted five criteria that Kuhn (1977) mentioned in connection with his views on the rationality and objectivity of science. The time has come to relate them explicitly to the Bayesian approach I have been attempting to elaborate. In order to appreciate the significance of these criteria it is important to distinguish three aspects of scientific theories that may be called *informational virtues, confirmational virtues,* and *economic virtues*. Up to this point we have concerned ourselves almost exclusively with confirmation, for our use of Bayes's theorem is germane only to the confirmational virtues. But since Kuhn's criteria patently refer to the other virtues as well, we must also say a little about them.

Consider, for example, the matter of *scope*. Newton's three laws of motion and his law of universal gravitation obviously have greater scope than the conjunction of Galileo's law of falling bodies and Kepler's three laws of planetary motion. This means, simply, that Newtonian mechanics contains more information than the laws of Kepler and Galileo taken together. Given a situation of this sort, we prefer the more informative theory because it is a basic goal of science to increase our knowledge as much as possible. We might, of course, hesitate to choose a highly informative theory if the evidence for it were extremely limited or shaky because the desire to be right might overrule the desire to have more information content. But in the case at hand that consideration does not arise.

In spite of its intuitive attraction, however, the appeal to scope is not altogether unproblematic. There are two ways in which we might construe the Galileo-Kepler-Newton example of the preceding paragraph. First, we might ignore the small corrections mandated by Newton's theory in the laws of Galileo and Kepler. In that case we can clearly claim greater scope for Newton's laws than for the conjunction of Galileo's and Kepler's laws since the latter is entailed by the former but not conversely. Where an entailment relation holds we can make good sense of comparative scope.

Kuhn, however, along with most of the historically oriented philosophers, has been at pains to deny that science progresses by finding more general theories that include earlier theories as special cases. Theory choice or preference involves *competing* theories that are *mutually incompatible* or *mutually incommensurable*. To the best of my knowledge Kuhn has not offered any precise characterization of scope; Karl Popper, in contrast, has made serious attempts to do so. In response to Popper's efforts, Adolf Grünbaum (1976a) has effectively argued that none of the Popperian measures can be usefully applied to make comparisons of scope among mutually incompatible competing theories. Consequently, the concept of scope requires fundamental clarification if we are to use it to understand preferences among competing theories. However, since scope refers to information rather than confirmation, it plays no role in the Bayesian program I have been endeavoring to explicate. We can thus put aside the problem of explicating that difficult concept.

Another of Kuhn's (1977) criteria is *accuracy*. It can, I think, be construed in two different ways. The first has to do with informational virtues; the second, with economic. On the one hand, two theories might both make true predictions regarding the same phenomena, but one of them might give us precise predictions whereas the other gives only predictions that are less exact. If, for example, one theory enables us to predict that there will be a solar eclipse on a given day and that its path of totality will cross North America, it may well be furnishing correct information about the eclipse. If another theory gives not only the day but also the time, and not only the continent but also the precise boundaries, the second provides much more information, at least with respect to this particular occurrence. It is not that either is incorrect; rather, the second yields more knowledge than the first. However, it should be clearly noted—as it was in the case of scope—that these theories are not incompatible or incommensurable competitors (at least with respect to this eclipse), and hence do not illustrate the interesting type of theory preference with which Kuhn is primarily concerned.

On the other hand, one theory may yield predictions that are nearly, but not quite, correct, whereas another theory yields predictions that are entirely correct— or, at least, more nearly correct. Newtonian astrophysics does well in ascertaining the orbit of the earth, but general relativity introduces a correction of 3.8 seconds of arc per century in the precession of its perihelion.[19] Although the Newtonian theory is literally false, it is used in contexts of this sort because its inaccuracy is small, and the economic gain involved in using it instead of general relativity (the saving in computational effort) is enormous.

The remaining three criteria are *simplicity, consistency,* and *fruitfulness;* all of them have direct bearing upon the confirmational virtues. In the treatment of prior probabilities in section 4, I briefly mentioned simplicity as a factor having a significant bearing upon the plausibility of theories. More examples could be added, but I think the point is clear.

In the same section I also made passing reference to consistency, but more can profitably be said on that topic. Consistency has two aspects, internal consistency of a theory and its compatibility with other accepted theories. Although scientists may be fully justified in *entertaining* collections of statements that contain contradictions, the goal of science is surely to *accept* only logically consistent theories (see Smith 1987). The discovery of an internal inconsistency has a distinctly adverse effect on the prior probability of that theory; to wit, it must go straight to zero.

When we consider the relationships of a given theory to other accepted theories we again find two aspects. There are *deductive* relations of entailment and incompatibility, and there are *inductive* relations of fittingness and incongruity. The deductive relations are quite straightforward. Incompatibility with an accepted theory makes for implausibility; being a logical consequence of an accepted theory makes for a high prior probability. Although deductive subsumption of narrower theories under broader theories is probably something of an oversimplification of actual cases, nevertheless, the ability of an overarching theory to deductively unify diverse domains furnishes a strong plausibility argument.

When it comes to the inductive relations among theories, analogy is, I think, the chief consideration. I have already mentioned the use of analogy in inductively transferring results of experiments from rats to humans. In archaeology, the method of ethnographic analogy, which exploits similarities between extant preliterate societies and prehistoric societies, is widely used. In physics, the analogy between the inverse square law of electrostatics and the inverse square law of gravitation provides an example of an important plausibility consideration.

Kuhn's criteria of consistency (broadly construed) and simplicity seem clearly to pertain to assessments of the prior probabilities of theories. They cry out for a Bayesian interpretation.

The final criterion in Kuhn's (1977) list is *fruitfulness;* it has many aspects. Some theories prove fruitful by unifying a great many apparently different phenomena in terms of a few simple principles. The Newtonian synthesis is, perhaps, the outstanding example; Maxwellian electrodynamics is also an excellent case. As I suggested above, this ability to accomodate a wide variety of facts tends to enhance the prior probability of a given theory. To attribute diverse success to happenstance, rather than basic correctness, is implausible.

Another sort of fertility involves the predictability of theretofore unknown phenomena. We might mention as familiar illustrations the prediction of the Poisson bright spot by the wave theory of light and the prediction of time dilation by special relativity. These are the kinds of instances in which, in an important sense, the expectedness is low. As we have noted, a small expectedness tends to increase the posterior probability of a hypotheses.

A further type of fertility relates directly to plausible scenarios; a theory is fruitful in this way if it successfully copes with difficulties with the aid of suitable auxiliary assumptions. Newtonian mechanics again provides an excellent example. The perturbations of Uranus were explained by postulating Neptune. The perturbations of Neptune were explained by postulating Pluto.[20] The motions of stars within galaxies and of galaxies within clusters are explained in terms of *dark matter*, concerning which there are many current theories. A theory that readily gives rise to plausible scenarios to deal with problematic likelihoods can boast this sort of fertility.

The discussion of Kuhn's (1977) criteria in this section is intended to show how adequately they can be understood within a Bayesian framework—insofar as they are germane to confirmation. If it is sound, we have constructed a fairly substantial bridge connecting Kuhn's views on theory choice with those of the logical empiricists—at least, those who find in Bayes's theorem a suitable schema for characterizing the confirmation of hypotheses and theories.

10. Rationality vs. Objectivity

In the title of this chapter I have used both the concept of *rationality* and that of *objectivity*. It is time to say something about their relationship. Perhaps the best way to approach the distinction between them is to enumerate various grades of rationality. In a certain sense one can be rational without paying any heed at all to objectivity. It is essentially a matter of good housekeeping as far as one's beliefs and degrees of confidence are concerned. As Bayesians have often emphasized, it is important to avoid logical contradictions in one's beliefs and to avoid probabilistic incoherence in one's degrees of conviction. If contradiction or incoherence are discovered they must somehow be eliminated; the presence of either constitutes a form of irrationality. But the removal of such elements of irrationality can be accomplished without any appeal to facts outside of the subject's corpus of beliefs and degrees of confidence. To achieve this sort of rationality is to achieve a minimal standard that I call *static* rationality.[21]

One way in which additional facts may enter the picture is via Bayes's theorem. We have a theory T in which we have a particular degree of confidence. A new piece of evidence turns up—some objective fact E of which we were previously unaware—and we use Bayes's theorem to calculate a posterior probability of T. To accept this value of the posterior probability as one's degree of confidence in T is known as *Bayesian conditionalization*. Use of Bayes's theorem does not, however, guarantee objectivity. If the resulting posterior probability of T is one we are not willing to accept, we can make adjustments elsewhere to avoid incoherence. After all, the prior probabilities and likelihoods are simply personal probabilities, so they can be adjusted to achieve the desired result. If, however, the requirement of *Bayesian conditionalization* is added to those of static rationality we have a stronger type of rationality that I have called *kinematic*.

The highest grade of rationality—what I have called *dynamic rationality*—requires much fuller reference to objective fact than is demanded by advocates of personalism. The most obvious way to inject a substantial degree of objectivity into our deliberations regarding choices of scientific theories is to provide an objective interpretation of the probabilities in Bayes's theorem. Throughout this discussion I have adopted that approach as thoroughly as possible. For instance, I have argued that prior probabilities can be given an objective interpretation in terms of frequencies of success. I have tried to show how likelihoods could be objective—by virtue of entailment relations, tests of statistical significance, or observed frequencies. When the likelihoods created major difficulties I have appealed to plausible scenarios. The result was that an intractable likelihood could be exchanged for a tractable prior probability—namely, the prior probability of a theory in conjunction with an auxiliary assumption.

We noted that the denominators of the right-hand sides of the various versions of Bayes's theorem—equations (1), (2), and (3)—contain either an expectedness or a likelihood on the catchall. It seems to me futile to try to construe either of these probabilities objectively. Consequently, in section 7 I introduced equation (6), which involves a ratio of two instances of Bayes's theorem and from which the expectedness and the likelihood on the catchall drop out. Confining our attention, as Kuhn (1970) recommends, to comparing the merits of competing theories, rather than offering absolute evaluations of individual theories, we were able to eliminate the probabilities that most seriously defy objective interpretation.

11. Conclusions

For many years I have been convinced that plausibility arguments in science have constituted a major stumbling block to an understanding of the logic of scientific inference. Kuhn was not alone, I believe, in recognizing that considerations of plausibility constitute an essential aspect of scientific reasoning, without seeing where they fit into the logic of science. If one sees confirmation solely in terms of the crude hypothetico-deductive method, there is no place for them. There is, consequently, an obvious incentive for relegating plausibility considerations to heuristics. If one accepts the traditional distinction between the *context of discovery* and the *context of justification* it is tempting to place them in the former context. But Kuhn recognized, I think, that plausibility arguments enter into the justifications of choices of theories, with the result that he became skeptical of the value of that distinction. If, as I argued in chapter 4, plausibility considerations are simply evaluations of prior probabilities of hypotheses or theories, then it becomes apparent via Bayes's theorem that they play an indispensable role in the context of justification. As I maintained in chapter 5, we do not need to give up that important distinction.

At several places in this chapter I have spoken of Bayesian algorithms, mainly because Kuhn introduced that notion into the discussion. I have claimed that such algorithms exist—and have attempted to exhibit them—but I accord *very little* significance to that claim. The algorithms are trivial; what is important is the scientific judgment involved in assessing the probabilities that are fed into the equations. The algorithms give frameworks in which to understand the role of the sort of judgment upon which Kuhn rightly placed great emphasis.

The history of science chronicles the successes and failures of attempts at scientific theorizing. If the Bayesian analysis I have been offering is at all sound, history of science—in addition to contemporary scientific experience, of course—provides a rich source of information relevant to the prior probabilities of the theories among which we are at present concerned to make objective and rational choices. This viewpoint captures, I believe, the point Kuhn (1957) made at the beginning of his first book:

> But an age as dominated by science as our own does need a perspective from which to examine the scientific beliefs which it takes so much for granted, and history provides one important source of such perspective. If we can discover the origins of some modern scientific concepts and the way in which they supplanted the concepts of an earlier age, we are more likely to evaluate intelligently their chances for survival. (3-4)

I suggested at the outset that an appeal to Bayesian principles could provide some aid in bridging the gap between Hempel's logical empiricist approach and Kuhn's historical approach. I hope I have offered a convincing case. However that may be, there remain many unresolved issues. For instance, I have not even broached the problem of incommensurability of paradigms or theories. This is a major issue. For another example, I have assumed uncritically throughout the discussion that the various parties to disputes about theories share a common body B of background knowledge. It is by no means obvious that this is a tenable assumption—though an individual scientist would presumably use the same background knowledge B in comparing two different theories. No doubt other points for controversy remain. I do not for a moment maintain that complete consensus would be in the offing even if both camps were to buy the Bayesian line I have been peddling. But I do hope that some areas of misunderstanding have been clarified.[22]

Revisions of Scientific Convictions

As the title of *The Structure of Scientific Revolutions* suggests, a chief focus—perhaps *the* chief focus—of Kuhn's (1962) book is the nature of scientific change. It is also the chief focus of most philosophers who adopt a historical orientation. Sir Karl Popper and his followers have seen the problem of growth of scientific knowledge as a major issue in philosophy of science. Logical positivists and logical empiricists have often been accused of holding a totally unrealistic view of the nature of scientific development. Bayesians have devoted considerable attention to the ways in which degrees of conviction should be modified, but they have encountered fundamental troubles along the way.[1] This chapter attempts to resolve some of most important and difficult ones.

1. Dutch Book Arguments

The majority of Bayesians adopt the subjective (or personalistic) interpretation of probability. They attribute great importance to the requirement of probabilistic coherence because any violation of the restrictions imposed by the calculus of probability could lead to a Dutch book. A Dutch book is a system of bets on an event in which, no matter what happens, the victim loses. For example, if I had a degree of belief of 2/3 that a toss of a coin will result in heads and a degree of belief of 2/3 that the same toss of the same coin will result in tails, my beliefs would be incoherent, regardless of whether the coin is biased or fair. Since these two results are incompatible, the probability of the result heads-or-tails is the sum of the individual probabilities. In this case the sum of the two values is greater than 1, but according to the calculus of probability it is impossible for any probability to exceed 1; hence, the constraints of the calculus have been violated. If I used those values to make bets—offering odds of 2:1 on each outcome—it is obvious that I would inevitably lose.

From this point of view, the calculus of probability can be considered as a system of logic. When one discovers an incoherence, this signals the necessity for a revision of the degrees of belief, but the calculus of probability cannot determine what changes would be correct. In deductive logic, analogously, if I find a contradiction in my beliefs—for example, that it is raining and not raining at the same place at the same time—I would know that I had made an error. Nevertheless, logic cannot tell me which belief is false. These considerations suggest a concept of *static rationality*—a state of mind at any given moment that is free from logical contradictions and probabilistic incoherence. However, neither logic nor the calculus of probability can mandate particular changes of belief concerning facts if one or the other of these conditions is not satisfied.[2]

Bayesians have devoted a good deal of attention to the problem of how degrees of belief are to be changed in the presence of a new piece of evidence. In general, when one receives new evidence it is an occasion for review of one's degrees of belief, and the Bayesians advise us to use *Bayesian conditionalization* in such circumstances. Consider any of the versions of Bayes's theorem we have used, for example,

$$P(T|E \cdot B) = \frac{P(T|B) \, P(E|B \cdot T)}{P(E|B)} \tag{1}$$

Suppose that an observation is about to be made that will eventuate in the evidence E, but we do not know as yet what the outcome will be. Nevertheless, we have values for the prior probability of T, for the probability that the evidence E obtains if T is true (the likelihood), and for the expectedness of evidence E. These are the probabilities that appear on the right-hand side of the equation. Suppose we now learn that E is true. Bayesian conditionalization requires us to accept the value of the posterior probability, as calculated by (1), as our new degree of belief in T.[3] This advice requires a justification of its own; it is not a consequence of the principles of static rationality. Suppose that I have announced my personal values for the terms on the right, but when E emerges I adopt a probability for T that differs from the resulting posterior probability as given by (1). You point out that I am being incoherent, and I acknowledge the fact. I realize that I must make some modification in my degrees of belief (for I want to be probabilistically coherent). However, instead of accepting the posterior probability derived from (1) I decide to change my prior probability P(T|B) to conform to equation (1). In this way I have again become coherent.[4]

In fact, another argument, based on considerations of coherence, has been given for the justification of Bayesian conditionalization. It demonstrates that, if one uses any rule other than Bayesian conditionalization, one can become the victim of a Dutch book, provided that one's opponent knows what rule is to be used. Although the kinematic Dutch book argument is a matter of some controversy,[5] I will accept Bayesian conditionalization as an appropriate response when one is presented with new evidence. This is a principle of *kinematic rationality*. Both static and kinematic rationality are discussed at greater length in the chapter 8, "Dynamic Rationality."

Bas van Fraassen (1989) considers the nature of Bayesian conditionalization and draws the conclusion—by an argument that seems valid to me—that this type of reasoning is nonampliative; that is, it does not increase our knowledge. If we have some new evidence, the prior probability of the hypothesis, the probability of the new evidence given the truth of the hypothesis, and the probability of the new evidence given the falsity of the hypothesis, the rest is just arithmetic; it adds nothing to our store of knowledge. It follows, then, that any *formal* rule of *ampliative* reasoning is threatened by an argument based on a kinematic Dutch book.[6]

2. Revision of Opinion

Let us agree, if only for the sake of the argument, that Bayesian conditionalization is the correct method for the revision of degrees of belief in the light of new evidence. It does not follow that this is the only situation in which it is necessary to revise degrees of conviction. On the contrary, the invention of a new concept or the discovery of a new theory or hypothesis is a paradigmatic occasion for the revision of degrees of belief.[7]

In science we typically have several hypotheses or theories, $T_1, T_2, T_3, \ldots$, that have been advanced to explain a phenomenon. A given scientist attributes prior probabilities to them. Later, someone proposes a new hypothesis, T_4, that is not too implausible. It is entirely possible for T_4 to acquire some of the plausibility formerly attributed to T_1–T_3, with the result that the prior probabilities of the original hypotheses have to be revised. Consider an example. Several years ago archaeologists discovered in Alaska an instrument made of bone that had clearly been created by a human artisan. The age of this object, determined by radiocarbon dating, was approximately 30,000 years. This artifact presented a problem because, at the time of its discovery, there was no site of human habitation in the entire New World with a well-confirmed age greater than 12,000 years.

Let us consider three hypotheses: (T_1) The New World has been continuously inhabited by humans for 30,000 years. This hypothesis could be true, but it would then seem strange that archaeologists had not discovered any evidence of it for the entire period between 30,000 and 12,000 years ago. (T_2) Humans had made a brief incursion into the New World around 30,000 years ago, but they did not remain for very long. Either they returned to the Old World, or they soon became extinct. This hypothesis is obviously ad hoc. (T_3) The determination of the age of the artifact by radiocarbon dating was in error. This is always a possibility. In this situation we assign prior probabilities to the three hypotheses in whatever way seems reasonable. Later we discover that in Siberia the bodies of mammoths that have remained frozen for many millenia are heaved up to the surface by processes of freezing and thawing. When humans find these bodies they use the bones to make implements. This discovery suggests another hypothesis regarding the instrument that had been found in Alaska: (T_4) That instrument was fashioned relatively recently—not more than 12,000 years ago—from the bone of a mammoth that died 30,000 years ago.[8] This new hypothesis would bring about a striking revision of degrees of belief, especially those representing the prior probabilities.[9]

One might be tempted to say that this example is a case of Bayesian conditionalization, in which the new evidence consists of the use of the bones of long-frozen bodies for the fashioning of instruments. But this is not how it seems to me. Unless it has a bearing on the assignment of prior probabilities, it is not clear how this so-called evidence is even relevant to any of the three hypotheses with which we began. The important fact is that a new hypothesis has been introduced into the situation. It seems quite unreasonable to suppose that the invention or discovery of a new idea or hypothesis can be governed by formal rules—at least in the interesting cases. The formulation of a new hypothesis does not involve the use of formal rules of reasoning; therefore it does not carry any danger of a kinematic Dutch book.

The assignment of a prior probability to the new hypothesis and the re-assignment of prior probabilities to the original hypotheses present a problem, namely, the problem of characterizing the type of reasoning on the basis of which these assignments are made. According to Carnap (1950b), the assignment of prior probabilities is always done a priori; therefore, in all cases there is a formal rule for that purpose. According to Reichenbach ([1935] 1949)—at least in ideal cases—one can use the *rule of induction* to this end. Both Carnap and Reichenbach believed that it is possible in principle to formulate in explicit statements all of our knowledge. For many years I held the same belief, but no longer. This idea now seems quite mad—on a par with the notion that it is possible to construct ordinary material objects out of sensations.

Patrick Suppes (1966) offered a different approach. According to him, the judgment of the plausibility of a hypothesis or the assignment of a prior probability is a method that enables us to bring all of the scientific experience of the researcher to bear on the likelihood of success or failure of the hypothesis. This enormous body of experience cannot be written out explicitly and completely. I agree, and therefore conclude that the assignment of prior probabilities to a hypothesis is a type of informal reasoning that cannot be governed by formal rules.

3. The Problem of Old Evidence

For quite some time Bayesians have been concerned about the so-called *problem of old evidence*. For example, as my colleague John Earman (1992, chap. 5) has pointed out, when Einstein formulated his theory of general relativity, the anomalous precession of the perihelion of Mercury had been known for many years. Since this fact constituted part of the general background knowledge available at the time at which the theory was discovered, Bayesian conditionalization on the basis of this evidence would not produce any change in the probability of the theory; the posterior probability would be just the same as the prior probability. Therefore, what is considered one of the three classic tests of general relativity would not produce any confirmation at all. Nevertheless, Einstein and many other scientists considered the behavior of Mercury as very strong evidence—perhaps the strongest of the three. As Stephen Brush (1989) has indicated, prior to the advent of general relativity astronomers had tried to explain the anomalous precession in many ways, but without success. With regard to this phenomenon, classical physics had exhausted its resources. In contrast, prior to the famous total solar eclipse of 1919, no one knew that light from stars curves when it passes close to the sun. Therefore, when this phenomenon was discovered, astronomers tried to see whether it was possible to explain it by means of classical physics. Only after classical methods had failed to explain the curvature of the path of light did this fact become strong evidence for the general theory of relativity. Here, then, is the problem for the Bayesians: Notwithstanding that according to the theorem of Bayes old evidence seems unable to confirm anything, according to the history of science it seems that the old evidence concerning the anomalous precession of Mercury strongly confirmed the general theory of relativity.

Richard Jeffrey (1983) and Daniel Garber (1983) tried to resolve the problem posed by this example in terms of the fact that Einstein, during the time he was formulating his theory, did not know that it would explain the anomalous precession. Therefore, when he demonstrated that the general theory of relativity provided a satisfactory explanation of the anomalous precession, the probability of the theory was increased. According to Jeffrey and Garber, strict Bayesianism requires logical omniscience to preserve rationality, but this is an absurd requirement. No one, not even Einstein, is logically omniscient. In order to give Bayesianism a more "human face," they suggest abandoning this requirement, even though taking this measure exposes us to probabilistic incoherence and to the danger of a Dutch book. However, Earman has noted that the approach of Jeffrey and Garber does not resolve the problem. Even if, when Einstein formulated his theory, he had known that it would resolve the problem of the anomalous precession, that fact still would have constituted evidence for the theory.

Earman (1992) has pointed out a simple and useful fact regarding the problem of old evidence, that is, that it coincides exactly with the problem of new hypotheses. This was the problem we were discussing just before we undertook discussion of the problem of old evidence. Thus, what may have appeared to be a digression was no such thing. The anomalous precession of the perihelion of Mercury was a phenomenon that had been present for a long time *as an anomaly* of classical physics. A new hypothesis was needed. When a new hypothesis that explains that anomaly is finally discovered, this fact must be considered as giving it a strong claim to plausibility. Thus, the phenomenon of anomalous precession makes its contribution to the prior probability of the hypothesis. Given that this evidence—together with other considerations—had already been used to establish the prior probability of general relativity, it would not be justifiable to use that same evidence again to confirm that theory. Let us keep clearly in mind that "to confirm" means "to raise the posterior probability with respect to the prior probability." To use the evidence in this way would amount to counting it twice.

Another example may help to illustrate the point. When Louis de Broglie formulated his theory of matter waves, he exploited an analogy between the well-established wave-particle duality for light, proposing a similar wave-particle duality for material entities such as electrons. This analogy obviously provides a plausibility argument for de Broglie's hypothesis. At the time de Broglie formulated his new hypothesis, Niels Bohr's model of the hydrogen atom, with its discrete stationary electron orbits, had been around for quite a while and was considered extremely puzzling. When de Broglie showed that his matter-wave hypothesis, in conjunction with the assumption that the wave of the electron in a stationary orbit had to be a standing wave, entailed precisely the orbits in the Bohr model, that constituted a powerful plausibility consideration. The experimental confirmation came initially in the results of the Davisson-Germer experiment. In this case, unlike the example of the anomalous precession of the perihelion of Mercury, the formulator of the hypothesis was fully aware of the fact that his hypothesis would explain the "old evidence" before promulgating his new hypothesis, but that result was not introduced again as experimental evidence for the matter-wave theory (see Sobel 1987, 91–95).

The question of old evidence would constitute a problem for the Bayesian approach only if one were to adopt the position that Bayesian conditionalization is the only legitimate method for modifying one's degrees of belief. Since, as I have already said, I reject this limitation, I do not find the problem of old evidence a grave one. However, if one insists on the formalization of scientific confirmation, the argument of van Fraassen (1989) can appear to be serious; that is, one could then be worried about revisions of probabilities of hypotheses that are not governed by rules like Bayesian conditionalization.

With some trivial exceptions that have no significance for the present discussion, the calculus of probability is a mathematical instrument for the manipulation of given values to compute other values. Thus, van Fraassen's (1989) characterization of Bayesian conditionalization as a type of nonampliative reasoning is justified. Given that fact, the obvious question is where do the probabilities that were introduced for the purpose of the calculation come from? In some cases, the result of one calculation can furnish a value that can be used in another calculation. Suppose, for example, that we have a hypothesis whose posterior probability relative to a given amount of evidence has been calculated. That probability could be taken as the prior probability with reference to another piece of evidence. However, to avoid an infinite regress we cannot suppose that all of the values of probabilities are introduced in this way. If we want to use Bayes's theorem systematically, we need another way of assigning prior probabilities. Whether these are the a priori probabilities of Carnap (1950b), the evaluations of frequencies of Reichenbach ([1935] 1949), or the subjective degrees of belief of the (majority of) Bayesians, we have to obtain them in some fashion. My proposal, following Suppes (1966), is to consider the assignment of prior probabilities as the result of our most refined scientific judgments, taking account of all of our objective scientific knowledge and avoiding, insofar as possible, our prejudices. Obviously, it is not possible to force this approach into a set of formal rules.

4. Kuhn, Bayes, and Numbers

In chapter 6, "Rationality and Objectivity in Science," I tried to show how recourse to the theorem of Bayes can shed light on some of Kuhn's (1962) famous doctrines concerning choices among scientific theories. As we noted, in *The Structure of Scientific Revolutions* he asserted that the choice among scientific theories involves considerations that go beyond observation and logic, even when logic is taken to include scientific confirmation in addition to deductive logic. I argued that this doctrine is based on an inadequate theory of confirmation—something like the traditional hypothetico-deductive (H-D) method—which is a serious oversimplification. Bayes's theorem, I contend, provides a far more satisfactory schema.

In the same book, Kuhn (1962) offered a reason for his thesis that the choice among scientific theories involves considerations that go beyond observation and logic—that is, because they involve judgment and persuasion. In another work Kuhn (1977) claimed that practicing scientists have recourse to such factors as simplicity and consistency, where "consistency" represents not only internal logical consistency of the theory but also its coherence with other accepted theories. As I said, it seems reasonable to interpret these considerations as assignments of plausibility, that is, as assignments of prior probabilities. Kuhn (1977) has not rejected this suggestion; in fact, in commenting on the question he has spoken of "Bayesian algorithms," and he has asserted that different scientists use different Bayesian algorithms to arrive at choices among diverse theories.

I must state immediately and emphatically that *I do not believe for one moment* that scientists use algorithms for purposes of choosing among scientific theories, but if there were any such algorithms the theorem of Bayes would be the prime candidate. However, no algorithm, even if we admitted that one existed, could be used without having available the probabilities that must be introduced into the formula in order to calculate other probabilities. Judgment of the sort mentioned by Kuhn is necessary to arrive at the evaluations of plausibility required for the evaluations of the prior probabilities. The differences of opinion among scientists would be based, not upon the use of different algorithms, but upon different values for the probabilities introduced into them.

Even if these suggestions regarding prior probabilities are correct, we observed, the problem of the "catchall" still remains. We noted, however, there is a simple way to evade the problem of the catchall—an approach actually suggested by Kuhn (1977). Instead of trying to furnish an absolute evaluation of a particular hypothesis T_1 we can make a comparative evaluation between two hypotheses, T_1 and T_2. In order to do this we can write Bayes's theorem twice, once for T_1 and again for T_2. We can divide the first by the second, with the following result:

$$\frac{P(T_1|E \cdot B)}{P(T_2|E \cdot B)} = \frac{P(T_1|B)}{P(T_2|B)} \frac{P(E|T_1 \cdot B)}{P(E|T_2 \cdot B)} \tag{2}$$

This equation enables us to compare the posterior probabilities of T_1 and T_2 with respect to the general background knowledge B and a new piece of evidence E. The catchall hypothesis has vanished. What remains, however, is a mathematical equation that seems to call for precise numerical values.

For many years I have regarded the theorem of Bayes as the correct schema for scientific confirmation. A question that frequently arises is where can we get the numerical values of probabilities that are present in the equation? This reaction is easily understandable. The subjective Bayesians can reply that these probabilities are simply personal degrees of credence regimented by the calculus of probability. This response gives rise to several problems. For example, many philosophers and scientists think that it injects so much subjectivity into science that *objective* scientific knowledge would be impossible. Another doubt arises from the fact that such precise numerical degrees of belief seem not to be present in the mind of any real scientist. One possible answer frequently advanced—one that I have often used, for example, in chapter 4, "Plausibility Arguments in Science"—makes reference to the phenomenon known as the "washing out" or "swamping" of the prior probabilities. It shows that the resulting posterior probabilities are not very sensitive to the actual values assigned to the prior probabilities. Although I consider this answer extremely important, in this chapter I would prefer to take a different tack.

The approach to the problem that I would like to propose consists in holding that we have no need for exact values of probabilities. It seems to me possible to learn a great deal about *informal* reasoning from examination of the mathematical calculus of probability—including, of course, Bayes's theorem. My purpose is *not* to develop a formal calculus of qualitative probability; instead I want to show how the *quantitative* theory of probability can shed light on informal reasoning. I did not lay my cards on the table from the outset because I wanted to illustrate the idea. The preceding discussion seems to me to suggest that the principal points (with the exception of the Dutch book argument) do not depend on the availability of exact numerical values. The mathematical calculus of probability can be regarded as an abstract idealization that furnishes important indications, even in the cases in which precise values are not available. There is certainly nothing new in the idea that abstract mathematics can furnish useful instruments for dealing with reality.

The idea of the relation between quantitative reasoning and qualitative reasoning that I am suggesting can be illustrated by examples treated by exact science. During a flight from Pittsburgh to Bologna, I was thinking that the pilot of an airplane has to understand that, other things being equal, the higher the temperature and the relative humidity, the greater the distance required for takeoff. It is not necessary to set forth the precise equations or the precise values of the temperature and the relative humidity to apply this important and useful information. Similar remarks are applicable to the fact that, for a fixed volume of gas, an increase in temperature produces an increase in pressure. Before the invention of the safety valve many people died horrible deaths because of a lack of understanding of this physical relationship.[10] In addition, a player of tennis (and other sports) needs to know that, because of the Bernoulli effect, the spin on the ball can make it follow a curving trajectory. Again, it is not necessary to know the precise theory or the exact values of the variables. It is easy to think of many other examples.

Applying this idea to the use of the mathematical theory of probability as a guide for informal reasoning, we can derive many results similar to those we have seen in relation to the theorem of Bayes. And in this context we should keep in mind that, apart from actual mathematical calculation, scientific reasoning is almost entirely informal. In fact, in my elementary logic text I presented a schema for scientific confirmation based on Bayes's theorem (Salmon 1963a, 85–86; 1973b, 114; 1984a, 137):

The hypothesis has a nonnegligible prior probability.
If the hypothesis is true, then the observational prediction will be true.

The observational prediction is true.
No other hypothesis is so highly confirmed as this one by the truth of the observational prediction; that is, the other hypotheses that are confirmed by the same observational prediction have lower prior probabilities.

Therefore, the hypothesis is true.[11]

Although this schema is not perfect, I believe that it is a notable improvement over the hypothetico-deductive schema. Like the H-D schema, it is limited to cases in which the observational prediction is a deductive consequence of the hypothesis (together with initial conditions and auxiliary hypotheses). A further refinement is necessary for cases in which the probability of the observational prediction, given the hypothesis, is neither zero nor 1. Moreover, the conclusion is too strong. We should say, instead, that the hypothesis has a presumption of truth. Nevertheless, Bayes's theorem tells us what things need to be considered for purposes of evaluating scientific hypotheses. This consideration provides an answer to critics of Bayesianism who claim that Bayes's theorem 'does no work' in science because scientists seldom resort to the calculation of the probability or degree of confirmation of the actual scientific hypotheses with which they are concerned. Indeed, I think that chapter 9, "Hume's Arguments on Cosmology and Design," shows how Bayes's theorem can do important philosophical work as well.

8

Dynamic Rationality

S ince the time of Bishop Butler we have known that *probability is the very guide of life*—at least it should be if we are to behave rationally. The trouble is that we have had an extraordinarily difficult time trying to figure out what kind of thing—or things—probability is. I shall argue that causality is a key piece in the puzzle, and consequently, an indispensable aspect of rationality.

1. The Problem

Example 1. In the autumn of 1944 I went as a student to the University of Chicago. The dormitory in which I lived was not far from Stagg Field, and I often walked by the entrance to the West Stands where a sign was posted: "Metallurgical Laboratories." Nothing ever seemed to be going on there. Later, when I got a job as a technician at the Met Labs, I learned that this was the place where, about two years previously, the first humanly created, self-sustaining nuclear chain reaction had taken place. An atomic pile had been assembled and the decision was taken to make a trial run on 2 December 1942. The control rods were gradually withdrawn, and the chain reaction occurred as predicted. It was an undertaking fraught with high risk; as one commentator has put it, "They were facing either the successful birth of the atomic age—or a catastrophic nuclear disaster within a packed metropolis" (Wyden 1984, 50). I was glad I had not been there at the time. No one could be absolutely certain beforehand that the chain reaction would not spread, involving other substances in the vicinity and engulfing Chicago—or, perhaps, the entire earth—in a nuclear holocaust. On second thought, it did not much matter where one was in case of such an eventuality.

I once read that before this pile was put into operation, the probability of the spreading of the chain reaction had been calculated. The result was that this probability was no more than three in a million. I do not recall the source of this bit of information, but I do remember wondering how the probability had been calculated and how it had been decided that the risk was acceptable. I was even more curious about the meaning of the probability concept in this context.[1]

Example 2. The foregoing perplexity was brought vividly back to mind by an article in *Science* in June 1986 about the Rogers Commission Report on the *Challenger* space shuttle disaster.[2] In particular, the article reports on a press conference held by Richard Feynman, the famous and colorful physicist who was a member of that commission. Although Feynman did not refuse to sign the commission's report, he did make a supplemental statement to the press in which he sharply criticized NASA's methods of assessing risks:

> Feynman objects most strongly to NASA's way of calculating risks. Data collected since the early days of the program...show that about one in every 25 solid rocket boosters has failed. About 2900 have been launched with 121 losses. Feynman says it is reasonable to adjust the anticipated crash rate a bit lower (to 1 in 50) to take account of today's better technology. He would even permit a little more tinkering with the numbers (to 1 in 100) to take credit for exceptionally high standards of part selection and inspection. In this way, the Challenger accident, the first solid rocket failure in 25 shuttle launches (with two boosters each), fits perfectly into Feynman's adjusted rate of one crash per 50 to 100 rocket firings.[3]
>
> But Feynman was stunned to learn that NASA rejects the historical data and claims the actual risk of crash is only 1 in 100,000. This is the official figure as published in "Space Shuttle Data for Planetary Mission RTG Safety Analysis" on 15 February 1985.[4]...Feynman searched for the origin of this optimism and found that it was "engineering judgment," pure and simple. Feynman concluded that NASA "for whatever purpose...exaggerates the reliability of its product to the point of fantasy."

The article goes on to report a conversation with NASA's chief engineer, Milton Silveira, who stated that NASA does not "use that number as a management tool":

> The 1 in 100,000 figure was hatched for the Department of Energy (DOE), he says, for use in a risk analysis DOE puts together on radioactive hazards on some devices carried aboard the shuttle....
>
> DOE and General Electric, supplier of the power units, write up a detailed risk analysis before launch. They are accustomed to expressing risk in statistical terms. NASA is not, but it must help them prepare the analysis. To speak in DOE's language, NASA translates its "engineering judgment" into numbers. How does it do this? One NASA official said, "They get all the top engineers together down at Marshall Space Flight Center and ask them to give their best judgment of the reliability of all the components involved." The engineers' adjectival descriptions are then converted to numbers. For example..."frequent" equals 1 in 100; "reasonably probable" equals 1 in 1000; "occasional" equals 1 in 10,000; and "remote" equals 1 in 100,000.
>
> When all the judgments were summed up and averaged, the risk of a shuttle booster explosion was found to be 1 in 100,000. That number was then handed over to DOE for further processing....

Among the things DOE did with the number was to conclude that "the overall risk of a plutonium disaster was found to be terribly, almost inexpressibly low. That is, 1 in 10,000,000, give or take a syllable."

> "The process," says one consultant who clashed with NASA, "is positively medieval." He thinks Feynman hit the nail exactly on the head.... Unless the risk estimates are based on some actual performance data, he says, "it's all tomfoolery."

Feynman is quoted as saying, "When playing Russian roulette, the fact that the first shot got off safely is little comfort for the next."

Example 3. Then, there is the old joke about precipitation probabilities. We have seen them in weather forecasts in newspapers and on television; sometimes they serve as some sort of guide of life. They are always given in multiples of ten—for example, a 30 percent chance of showers. How are they ascertained and what do they mean? Well, there are ten meteorologists, and they take a vote. If three of the ten vote for showers, that makes a 30 percent chance.

2. Grades of Rationality

When one speaks of rationality, it may refer to rational action or to rational belief. Rational action must somehow take account of probabilities of various outcomes. I am inclined to think that the policy of maximizing expected utilities—though perhaps not precisely correct—is satisfactory in most practical situations (see Machina 1987). Expected utilities are defined in terms of probabilities. I want to discuss the nature of these probabilities. There is a long tradition that identifies them with degrees of belief. I am not going to deny that answer, but I want to look at it with some care. First, however, a matter of terminology.

Patrick Maher (1986) has argued—quite cogently, in my opinion—that belief has nothing to do with rational *behavior.* He takes belief to be an all or nothing affair; one either believes a given proposition or does not believe it. Beliefs have the purely cognitive function of constituting our picture of the world. When it comes to rational behavior, he argues, we should use what he calls *degrees of confidence* to evaluate expectations of utility. Within Bayesian decision theory we can identify coherent degrees of confidence with personal probabilities. He prefers to avoid using such terminology as "degree of belief" or "degree of partial belief" in order to minimize confusion of the practical function of degrees of confidence (which can be evaluated numerically) with the cognitive function of beliefs (which are only qualitative). I have a slight preference for the term *degree of conviction* over *degree of confidence* (because "confidence" has a technical meaning in statistics that I do not want to attach to degrees of conviction), but both terms are acceptable. In addition, I shall sometimes use the more neutral term *degree of credence*, which should serve well enough for those who accept Maher's view and those who do not.[5]

There are two immediate objections to taking mere subjective degrees of conviction as probabilities. The first is that, in general, we cannot expect them to constitute an admissible interpretation of the probability calculus. The second is that we are not prepared to take raw subjective feelings as a basis for *rational* action. So, one standard suggestion is to require that the probabilities constitute *rational* degrees of conviction.

There are, I think, various grades of rationality. At the lowest level, philosophers have, for the most part, taken logical consistency as a basic rationality requirement. If one believes that no mammals hatch their young from eggs, that the spiny anteater (echidna) is a mammal, and that it hatches its young from eggs, standard deductive logic tells us that something is wrong—that some modification should be made, that at least one of these beliefs should be rejected. But logic, by itself, does not tell us what change should be made. Logical consistency can be regained—with regard to this set of beliefs—by rejecting any one of the three beliefs. Logical consistency provides one minimal sort of rationality.[6] In traditional logic, the price of adopting an inconsistent set of beliefs is that everything follows from them. Let us say that anyone whose total set of beliefs is logically consistent qualifies for *basic deductive rationality*.

A minimal kind of probabilistic rationality is represented by the subjective Bayesians, who require only *coherence*. Coherence is a sort of probabilistic consistency requirement. A set of degrees of conviction that violate the relationships embodied in the standard axioms of probability is incoherent. If, for example, I hold a probability of 2/3 that this coin will land heads up on the next toss, and a probability of 2/3 that it will land tails up on the next toss, and that these two outcomes are mutually exclusive, then I am being incoherent. This shows that there is something fundamentally wrong with the foregoing set of degrees of conviction. As is well known, the price of incoherence is that a so-called Dutch book can be made against anyone who holds incoherent degrees of conviction and is prepared to use them as fair betting quotients. A set of bets constitutes a Dutch book if, no matter what the outcome, the bettor loses. It is obvious how this happens in the case in which 2/3 is assigned as the probability of heads and also of tails. Adopting the terminology of L. J. Savage (1954), we usually refer to subjective probabilities that are coherent as *personal probabilities*. It follows immediately from their definition that personal probabilities constitute an admissible interpretation of the probability calculus, and thus overcome the first objection to subjective probabilities.

Violation of the coherence requirement shows that something is wrong with a set of degrees of conviction. Probability theory, however, does not tell us what modification to make in order to repair the difficulty, any more than pure deductive logic tells us which statement or statements of the foregoing inconsistent triad should be rejected. In the probabilistic example, we could achieve coherence by deciding that there is a 1/3 probability of heads, a 1/3 probability of tails, and a 1/3 probability that the coin will come to rest standing on edge, where these three alternatives are mutually exclusive and exhaustive. L. J. Savage (1954) held an official view that probabilistic coherence is the only requirement of rationality, but unofficially, so to speak, he adopted others.

Abner Shimony (1970) defined a notion of *strict coherence*, and he proved a kind of Dutch book result. Strict coherence demands, in addition to coherence, that no logically contingent molecular statement (statement not involving quantifiers or binary connectives) should have a probability of zero or 1. He has shown that anyone who violates strict coherence can be put in a position of making a set of bets such that, no matter what happens, the bettor cannot win, though he or she will break even on some outcomes. Carnap has always considered strict coherence a basic rationality requirement.

A fundamental distinction between the requirements of coherence and strict coherence should be noted. Almost without exception, coherence requirements impose restrictions on relationships among probability values, as found, for instance, in the addition rule, the multiplication rule, and Bayes's theorem. They do not generally yield individual probability values. The two exceptions are that logical truths have probability 1 and logical contradictions have probability zero. Moreover, the requirement of coherence does not constrain individual probability values beyond the universal constraint that they must lie within the closed unit interval from zero to 1. Strict coherence imposes a hefty restriction on a large class of probabilities, where the statements to which they apply (contingent molecular statements) are *not* logical truths or falsehoods. We shall look at the rationale for the strict coherence requirement below.

Strict coherence is a sort of *openmindedness* requirement; various authors have demanded other sorts of *openmindedness* requirements. In characterizing a view he called *tempered personalism*, Shimony (1970) argues that, in evaluating and comparing scientific hypotheses, we should allow hypotheses seriously proposed by serious scientists to have nonnegligible prior probabilities. We should also accord a substantial prior probability to the *catchall hypothesis*—the "none of the above" supposition—the notion that we have not yet thought of the correct hypothesis. It is important to note that Shimony offers empirical justifications for these openmindedness conditions. He observes historically that science has progressed remarkably by taking seriously the hypotheses proposed by qualified scientists. He also points out that in many historical situations the set of hypotheses under consideration did not include some hypothesis, proposed only later, that turned out to be successful.

The principles I have mentioned thus far pertain to what might be called the *statics* of degrees of conviction. You look at your total body of knowledge at a given moment, so to speak, and try to discover whether it contains any logical inconsistencies or probabilistic incoherencies (or violations of strict coherence). If such defects are discovered some changes in degrees of conviction are required, but we have not provided any basis for determining what alterations should be made. For the inconsistent triad a modest amount of research reveals that there are several species of egg-laying mammals and that the echidna is one of them. Hence, we should reject the first statement while retaining the other two. Additional empirical evidence has provided the solution. In the case of the coin, a modification was offered, but it was not a very satisfactory one. The reasoning might have been this. There are three possibilities: the coin could land heads up, tails up, or on edge. Applying some crude principle of indifference, we assigned equal probabilities to all of them. But silly as this may be, we did eliminate the incoherence.

Clearly we have not said nearly enough about the *kinematics* of degrees of conviction—the ways our personal probabilities should change over time. There are, I believe, three sorts of occasion for modification of personal probabilities.

First, there is the kind of situation in which an incoherency is discovered, and some change is made simply to restore coherence. As we have seen, this can be accomplished without getting further information. When an alteration of this type occurs, where modifications are made willy-nilly to restore coherence (or strict coherence), there is change of degrees of conviction but no kinematics thereof. We can speak properly of a kinematics of degrees of conviction only when we have kinematic principles that somehow direct the changes.

Second, personal probabilities are sometimes modified as a result of a new idea. Suppose we have been considering two rival hypotheses, H_1 and H_2, and the catchall H_c. We have a probability distribution over this set. Someone thinks of a new hypothesis H_3 that has never before been considered. We now have the set H_1, H_2, H_3, and a new catchall H_c', where H_c is equivalent to $H_3 \vee H_c'$. It is possible, of course, to assign new probabilities in a way that does not involve changing any of the old probabilities, but that is not to be expected in all cases. Indeed, a new idea or a new hypothesis may change the whole picture as far as our plausibility considerations are concerned.[7] From what has been said so far, however, we do not have any principles to guide the way such modifications take place. At this point it looks like no more nor less than a psychological reaction to a new factor. We still have not located any kinematical principle.

Third, personal probabilities must often be modified in the light of new evidence. In many such cases the natural approach is to use Bayes's theorem. This method is known as *Bayesian conditionalization*. It is important to realize that coherence requirements alone do not mandate Bayesian conditionalization. Consider a trivial example. Suppose someone is tossing a coin in the next room. There are only two possibilities, namely, that the coin is two-headed or it is standard. I cannot examine the coin, but I do get the results of the tosses. My prior probability for the hypothesis that a two-headed coin is being tossed is 1/10; that is, I have a fairly strong conviction that it is not two-headed. Now I learn that the coin has been tossed four times and that each toss resulted in a head. If the coin is two-headed that result is inevitable; if the coin is standard the probability of four heads in a row is 1/16. Using Bayes's theorem you calculate the posterior probability of the two-headed hypothesis and find that is 0.64. You tell me that I must change my mind and consider it more likely than not that the coin is two-headed. I refuse. You show me the calculation and tell me that I am being incoherent. I see that you are right, so I must somehow modify my degrees of conviction. I tell you that I was wrong about the prior probability, and I change it from 1/10 to 1/100. Now I calculate the posterior probability of the two-headed hypothesis, using *that* value of the prior, and come out with the result that it is just about 0.14. My system of personal probabilities is now coherent, and I still consider it more likely than not that the coin is a standard one. I satisfied the coherence requirement but failed to conform to Bayesian conditionalization.

To many Bayesians, Bayesian conditionalization is a cardinal kinematical principle of rationality. It says something like this: Suppose you are considering hypothesis H. Before you collect the next bit of evidence, announce your prior probabilities and your likelihoods. When the next bit of evidence comes in, calculate the posterior probability of H by using those priors and likelihoods.

Change your degree of conviction in H from the prior probability to the posterior probability.

Bayesian conditionalization works effectively only if the prior probabilities do not assume the extreme values of zero or 1. A cursory inspection of Bayes's theorem reveals the fact that no evidence can change the probability of a hypothesis whose prior probability has one of those extreme values. As L. J. Savage (1954) once said about hypotheses being evaluated, one should have toward them a mind that is, if not open, at least slightly ajar. In addition to the requirement of coherence (or strict coherence), many Bayesians accept both Bayesian conditionalization and some sort of openmindedness condition. We shall construe the openmindedness condition as a principle of statics since it applies to the degrees of conviction that exist at any one time. Nevertheless, we should keep in mind that the main motivation is kinematical—to enable Bayesian conditionalization to work effectively. Bayesian conditionalization is a method for modifying degrees of conviction as a result of considering new evidence.

We are now in a position to distinguish three different grades of rationality in terms of considerations already adduced. Each succeeding grade is to be understood as incorporating the principles involved in the preceding grades.

> Basic Deductive Rationality
> Logical Consistency
> Static Probabilistic Rationality
> Coherence
> Strict Coherence
> Openmindedness
> Kinematic Probabilistic Rationality
> Bayesian Conditionalization

Later in this chapter I shall add a fourth grade, to be known as *dynamic rationality*.

The division between the kinematic and static grades of rationality marks a difference between rationality principles that do, and those that do not, involve any relation to the real world. The fact that the static principles do not relate to the real world is not a fault, however, for one sort of rationality consideration is quite independent of objective fact. The basic idea is that one aspect, at least, of rationality involves simply the management of one's body of opinion in terms of its inner structure. It has no concern with the objective correctness of opinions; it deals with the avoidance of blunders of various sorts within the body of opinion— for example, the kind of mistake that occurs when logical inconsistency or probabilistic incoherence is present.

The requirement of strict coherence nicely illustrates the point. Why should we adopt a prohibition against assigning probability zero or 1 to every molecular statement? Do we really believe that there is a probability greater than zero, for example, of finding a piece of pure copper that is not an electrical conductor? I am inclined to believe that we could search the universe thoroughly from the big bang to the present without finding a nonconducting piece of copper, and that a search from now until the end of time (if such there is) would yield the same result.

However, the requirement of strict coherence is not concerned with considerations of this sort. It is designed to prevent a person from having a set of degrees of conviction that would lead to such disadvantageous bets as wagering at odds of a million dollars to zero against someone finding a piece of nonconducting copper. It does not matter what the facts of nature are; such a bet would be at best pointless and at worst disastrous. Good housekeeping of one's stock of degrees of conviction recommends against allowing probabilities that, if they are to be used as fair betting quotients, could place one in that kind of situation.[8]

I have no desire to denigrate this type of consideration; it constitutes an important aspect of rationality. But it also seems to fall short of giving us a full-blooded concept of rationality. Surely *rational behavior* demands attention to the objective facts that are relevant to the action one contemplates or undertakes. Examples 1–3 in the first section of this chapter were intended to illustrate this point dramatically.

Bayesian conditionalization does, to some extent, provide a connection between opinion and objective fact. It specifies what degrees of conviction should be held constant and which allowed to change under specified circumstances. It tells us how to respond to new items of evidence.

Nevertheless, Bayesian conditionalization cannot provide a sufficient connection between our degrees of conviction and the objective facts. One obvious restriction is that it says nothing about changing our degrees of conviction in the absence of new evidence. It could be construed as prohibiting any change of opinion without new evidence, but that would be unduly restrictive. As we have already seen, there are at least two types of occasions for revising degrees of conviction even if there is no new evidence. The first is required to restore coherence to an incoherent set of degrees of conviction. It might be said, in response, that Bayesian conditionalization presupposes a coherent set of degrees of conviction. If we have an incoherent set, any change whatever that will restore coherence (without violating any other rationality condition listed above) will do. Then Bayesian conditionalization should be used.

The second type of situation that calls for revision of degrees of conviction in the absence of new evidence is the occurrence of a new idea. Surely the Bayesian does not want to bar the sort of revision that is based on thought and reflection. However, if that sort of revision is permitted, it can be used to sever connections between degrees of conviction and objective facts. Whenever accumulating evidence—in the form of observed relative frequencies, for example—seems to mandate a certain degree of conviction, a redistribution of personal probabilities can preclude that result. It is because of this weakness of connection between degree of conviction and objective fact, given only Bayesian conditionalization, that I wish to pursue higher grades of rationality.

3. Two Aspects of Probability

Hacking (1975) maintained that the concept of probability could not appear upon the scene until two notions could be brought together, namely, objective chance and degree of credence.[9] The first of these concepts is to be understood in terms of relative frequencies or propensities; the second is an epistemic notion that has to do with the degree to which one is justified in having confidence in some proposition. In the twentieth century Rudolf Carnap (1950b, 1962b) codified this general idea in his systems of inductive logic in which there two probability concepts—probability$_1$, inductive probability or degree of confirmation; probability$_2$, relative frequency. The major part of Carnap's intellectual effort for many years was devoted to his various attempts to characterize degree of confirmation. Because his systems of inductive logic require a priori probability measures, which strike me as unavoidably arbitrary, I have never been able to accept his approach.

It is interesting to note, however, that Carnap (1962a) approaches the problem by beginning with raw subjective degrees of credence and imposing rationality requirements upon them. He adopts coherence, strict coherence, and Bayesian conditionalization. He goes beyond these by imposing additional symmetry conditions. In Carnap (1963) he offers a set of fifteen axioms, all of which are to be seen as rationality conditions that narrow down the concept of degree of confirmation. These axioms, in other words, are intended to beef up the rather thin notion of rationality characterized wholly by coherence, strict coherence, and Bayesian conditionalization. It is a motley assortment of axioms that require a strange collection of considerations for their justification. The problem of arbitrary a priorism remains in all of his subsequent work.[10] Carnap has stated that the ultimate justification of the axioms is inductive intuition. I do not consider this answer an adequate basis for a concept of rationality. Indeed, I think that *every* attempt, including those by Jaakko Hintikka (1966) and his students, to ground the concept of rational degree of credence in logical probability suffers from the same unacceptable a priorism.

If the conclusion of the previous paragraph is correct, we are left with subjective probabilities and physical probabilities. By imposing coherence requirements on subjective probabilities we transform them into personal probabilities, and (trivially) these satisfy the standard probability calculus. Moreover, I am convinced—by arguments advanced by F. P. Ramsey (1950), L. J. Savage (1954), and others—that there *are* psychological entities that constitute degrees of conviction. We can find out about them in various ways, including the observation of betting behavior. What is required if they are to qualify as *rational* degrees of conviction is the question we are pursuing.

Turning to the objective side, we find propensities and relative frequencies. Although the so-called *propensity interpretation of probability* has enjoyed considerable popularity among philosophers for the last couple of decades, it suffers from a basic defect. As Paul Humphreys (1985) pointed out, the probability calculus contains *inverse probabilities*, as in Bayes's theorem, but there are no corresponding inverse propensities. Consider a simple quality-control situation. A certain factory manufactures floppy disks. There are several different machines in this factory, and these machines produce disks at different rates. Moreover, each of these machines produces a certain proportion of defective disks; suppose, if you like, that the proportions are different for different machines. If these various frequencies are given it is easy to figure the probability that a randomly selected disk will be defective. It is perfectly sensible to speak of the propensity of a given machine to produce a defective disk and of the propensity of the factory to produce defective disks. Moreover, if we pick out a disk at random and find that it is defective, it is easy, by applying Bayes's theorem, to calculate the probability that it was produced by a particular machine. It is not sensible, however, to speak of the propensity of this disk to have been produced by a given machine. Consequently, propensities do not even provide an admissible interpretation of the probability calculus. The problem is that the term "propensity" has a causal aspect that is not part of the meaning of "probability" (see Salmon 1979 for further discussion of this topic).

I have no objection to the concept of propensity as such. I believe there are probabilistic causes in the world, and they are appropriately called "propensities." There is a propensity of an atom to decay, a propensity of a tossed die to come to rest with side 6 uppermost, a propensity of a child to misbehave, a propensity of a plant sprayed with an herbicide to die, and so on. Such propensities produce relative frequencies. We can find out about many propensities by observing frequencies.

Indeed, it seems to me that the propensity concept may play quite a useful role in quantum theory. In that context wave equations are often employed, and references to amplitudes are customary. To calculate a probability one squares the absolute value of the amplitude. We can, of course, speak formally about wave equations as mathematical entities, but when they are applied to the description of the physical world it is reasonable to ask what the waves are undulations of. There seems to be no answer that is generally recognized as adequate. In other realms of physics we study sound waves, vibrating strings, light waves, water waves, and so forth. People sometimes refer to quantum mechanical waves as waves of probability, but that is not satisfactory, for what the wave gives has to be squared to get a probability. In dealing with other kinds of waves we can say what the wave amplitude is an amplitude of: displacement of water molecules, changes in electromagnetic field strength, fluctuations of air density, and so forth. My suggestion is that the quantum mechanical wave is a wave of propensity—propensity to interact in certain ways, given appropriate conditions. The results of such interactions are frequencies, and observed frequencies give evidence as to the correctness of the propensity we have attributed in any given case. If we adopt this terminology we can say that *propensities* exhibit interference behavior; it is no more peculiar to talk about the interference of waves of propensity than it is to speak of the interference of electromagnetic waves. In this way we can avoid the awkward necessity of saying that *probabilities* interfere with one another.

It is my view that—in quantum mechanics and everywhere else—physical probabilities are somehow to be identified with frequencies. One reason for this is that relative frequencies constitute an admissible interpretation of the probability calculus if the axioms require only finite additivity. Hans Reichenbach ([1935] 1949) demonstrated that the axioms of his probability calculus are logical consequences of the definition of probabilities as limiting frequencies. Van Fraassen (1980, 184) has pointed out that the limiting frequency interpretation is not an admissible interpretation of a calculus that embodies countable additivity, but I do not consider that an insuperable objection. Complications of this sort arise when we use such mathematical idealizations as infinite sequences and limiting frequencies to describe the finite classes of events and objects with which we deal in the real world. Similar problems arise when we use geometrical descriptions of material objects or when we use the infinitesimal calculus to deal with finite collections of such discrete entities as atoms or electric charges.

The conclusion I would draw is that there are two kinds of probabilities, personal probabilities and relative frequencies. Perhaps we can legitimately continue thinking of physical probabilities as limits of relative frequencies in infinite sequences; perhaps it will turn out that the concept has to be finitized. This is a deep and difficult issue that I shall not pursue further in this chapter. For purposes of this discussion I shall speak vaguely about long-run relative frequencies, and I shall assume that observed relative frequencies in samples of populations provide evidence regarding the relative frequencies in the entire population. The question to which I want now to turn is the relationship between personal probabilities and frequencies.

4. On Respecting the Frequencies

Section 3 of F. P. Ramsey's (1950) article "Truth and Probability" is entitled "Degrees of Belief,"[11] and in it he develops a logic of partial belief. This article is rightly regarded as a landmark in the history of the theory of subjective probability. Ramsey mentions two possible approaches to the topic: (1) as a measure of intensity of belief, which could be established introspectively, and (2) from the standpoint of causes of action. He dismisses the first as irrelevant to the topic with which he is concerned and proceeds to pursue the second. The way he does so is noteworthy:

> I suggest that we introduce as a law of psychology that [the subject's] behaviour is governed by what is called the mathematical expectation; that is to say that, if p is a proposition about which he is doubtful, any goods or bads for whose realization p is in his view a necessary and sufficient condition enter into his calculations multiplied by the same fraction, which is called the 'degree of his belief in p'. We thus define degree of belief in a way which presupposes the use of the mathematical expectation.[12]
>
> We can put this in a different way. Suppose his degree of belief in p is m/n; then his action is such as he would choose it to be if he had to repeat it exactly n times, in m of which p was true, and in the others false. . . .
>
> This can also be taken as a definition of degree of belief, and can easily be seen to be equivalent to the previous definition. (174).

In the time since Ramsey (1926) composed this article, a good deal of empirical work in psychology has been devoted to the behavior of subjects in gambling situations, and it seems to yield an unequivocal verdict of *false* on Ramsey's proffered law. There is, however, good reason to regard it as a normative principle of rational behavior, and that is entirely appropriate if we regard logic — including the logic of partial belief — as a normative subject.

The situation is relatively straightforward, as Ramsey (1950) shows by example. If a given type of act is to be performed n times, and if on m of these occasions it results in a good *g*, whereas in the other *n* − *m* it results in a bad *b* (where the sign of *b* is taken to be negative), then the frequency *m/n* provides a weighting factor that enables us to calculate the average result of undertaking that action. The total outcome for *n* occasions is obviously

$$m \times g + (n - m) \times b$$

and the average outcome is

$$[m \times g + (n - m) \times b]/n$$

If an individual knows the rewards (positive and negative) of the possible outcomes of an act, as well as the frequencies of these outcomes, he or she can tell exactly what the total result will be. In that case planning would be altogether unproblematic, for the net result of performing any such act a given number of times could be predicted with certainty. The problem is that we do not know beforehand what the frequency will be.

In view of our advance ignorance of the frequency, we must, as Carnap (1962b) repeatedly reminded us, use an estimate of the relative frequency, and we should try to obtain the best estimate that is available on the basis of our present knowledge. For Carnap, probability₁ fulfils just this function; it is the best estimate of the relative frequency. However, inasmuch as I have rejected Carnap's inductive logic on grounds of intolerable a priorism, his approach is unavailable to me.

Ramsey's approach to this problem is quite different from Carnap's. Ramsey (1950) wants to arrive at an understanding of the nature of degree of belief, and he shows that we can make sense of it if (and only if?) we identify it with the betting quotient of the subject. Using the betting quotient in this way makes sense because of its relation to frequencies. The degree of partial belief, which is to be identified with the subject's betting quotient, can thus be regarded as the subject's best guess or estimate of the relative frequency. If one repeats the same type of act in the same sort of circumstances *n* times, then, in addition to the actual utilities that accrue to the subject in each type of outcome, the actual frequencies with which the various outcomes occur determine the net gain or loss in utility for the subject. Because of the crucial role played by actual frequencies in this theory, I would say that Ramsey's account of degree of belief is *frequency-driven*. The whole idea is to get a handle on actual frequencies because, given the utilities, the frequency determines what you get.

Ramsey's (1950) treatment of the nature of subjective probabilities and their relations to objective probabilities stands in sharp contrast to D. H. Mellor's (1971) theory as set forth in his book.[13] In that work he explicitly adopts the strategy of basing his "account of objective probability on a concept of partial belief" instead of going at it the other way around. We can say that Mellor offers a *credence-driven* account of objective probability.

On Mellor's (1971) view, propensities are not probabilities, and they are not to be identified with chances. A propensity is a disposition of a chance setup to display a *chance distribution* under specifiable conditions—for example, upon being flipped in the standard way, a fair coin displays the distribution (chance of heads = 1/2; chance of tails = 1/2). The chance setup displays this *same distribution* on *every* trial. Obviously the chance distribution, which is always the same for the same chance setup, is not to be identified with the outcomes, which may vary from one trial to another. In addition, the chance distribution is not to be identified with the relative frequencies of outcomes generated by the chance setup.

It might seem odd to say that chance distributions are "displayed" when clearly it is the outcome, not the distribution, that is observable. But obviously, as Mellor (1971) is clearly aware, there is nothing in the notion of a disposition that prevents it, when activated, from manifesting something that is not directly observable. A hydrogen atom, for example, has a disposition to emit photons of particular frequencies under certain specifiable circumstances. *To find out what chance distribution a given chance setup displays*, Mellor maintains, *we must appeal to our warranted partial beliefs*. Chances are probabilities, but they are not to be identified with relative frequencies. Relative frequencies are generated by the operation of chance setups having chance distributions. At the foundation we have warranted partial beliefs, which determine chance distributions, which in turn yield relative frequencies.

A fuller elaboration of the credence-driven point of view, and one that differs from Mellor's in fundamental respects, is provided by David Lewis (1980):[14]

> We subjectivists conceive of probability as the measure of reasonable partial belief. But we need not make war against other conceptions of probability, declaring that where subjective credence leaves off, there nonsense begins. Along with subjective credence we should believe also in objective chance. The practice and the analysis of science require both concepts. Neither can replace the other. Among the propositions that deserve our credence we find, for instance, the proposition that (as a matter of contingent fact in our world) any tritium atom that now exists has a certain chance of decaying within a year. Why should subjectivists be less able than other folk to make sense of that? (263).

Lewis points out that there can be "hybrid probabilities of probabilities," among them credences regarding chances:

> . . . we have some very firm and definite opinions concerning reasonable credence about chance. These opinions seem to me to afford the best grip we have on the concept of chance. Indeed, I am led to wonder whether anyone *but* a subjectivist is in a position to understand objective chance. (264; emphasis in original)

There is an important sense in which David Lewis's (1980) guide to objective chance is extraordinarily simple. It consists of one principle that seems to him "to capture all we know about chance." Accordingly, he calls it

> The Principal Principle. Let C be any reasonable initial credence function. Let t be any time. Let x be any real number in the unit interval. Let X be the proposition that the chance, at time t, of A's holding equals x. Let E be any proposition compatible with X that is admissible at time t. Then

$$C(A|X\cdot E) = x$$

"That," as Lewis says, "will need a good deal of explaining."[5] But what it says roughly is that the degree to which you believe in A should equal the chance of A. I certainly cannot disagree with that principle. The question is, who is in the driver's seat, subjective credence or objective chance? As we have seen, Lewis has stated unequivocally his view that it is subjective credence; I take myself as agreeing with Ramsey (1950) that we should leave the driving to some objective feature of the situation. Ramsey opts for relative frequencies, and in a sense I think that is correct. In a more fundamental sense, however, I shall opt for propensities.

David Lewis (1980) identifies chance with propensity; I take the notion of propensity as the more basic of the two. As I said above, I do not reject that concept, when it is identified with some sort of probabilistic causality (provided it is not taken to be an interpretation of probability). Moreover, I do not have any serious misgivings about attributing causal relations in single cases. On the basis of my theory of causal processes and causal interactions—spelled out most fully in Salmon (1984b, chaps. 5–7)—I believe that individual causal processes transmit probability distributions and individual causal interactions produce probabilistic outcomes (indeterministically, in some kinds of cases, at least) (Salmon 1984b). Scientific theories, which have been highly confirmed by massive amounts of frequency data, tell us what the values of these propensities are. Propensities are, as James H. Fetzer (1971, 1981) has often maintained, entities that are not directly observable, but about which we can and do have strong indirect evidence. Their status is fully objective.

As David Lewis (1980, 266) formulates his principle, he begins by saying, "Let C be *any* reasonable initial credence function" (my emphasis). This means, in part, that "C is a nonnegative, normalized, finitely additive measure defined on all propositions" (267). In addition, it is *regular*. Regularity is closely akin to strict coherence, but regularity demands that no proposition receives the value zero unless it is logically false. In addition, Lewis requires Bayesian conditionalization as *one* (but not necessarily the *only*) way of learning from experience:

> In general, C is to be reasonable in the sense that if you started out with it as your initial credence function, and if you always learned from experience by conditionalizing on your total evidence, then no matter what course of experience you might undergo your beliefs would be reasonable for one who had undergone that course of experience. I do not say what distinguishes a reasonable from an unreasonable credence function to arrive at after a given course of experience. We do make the distinction, even if we cannot analyze it; and therefore I may appeal to it in saying what it means to require that C be a reasonable initial credence function. (268)

The fact that David Lewis (1980) does not fully characterize reasonable functions poses problems for the interpretation of his view, but the point that is generally acknowledged by subjectivists or personalists is that reasonable credence functions are not unique. One person may have one personal probability distribution, and another in the same situation may have a radically different one. So it appears that, on Lewis's account, there is no such thing as *unique* objective chance.

As I understand the principal principle, it goes something like this. Suppose that I have a credence function C that, in the presence of my total evidence E, assigns the *degree of credence* x to the occurrence (truth) of A. Suppose further that E says nothing to contradict X, the assertion that the *objective chance* of A is x. Then, I can try adding X to my total evidence E to see whether this would change my degree of conviction regarding A. The supposition is that in most circumstances, if my degree of credence in A is x, asserting that it is also the objective chance of A is not going to change my degree of credence to something other than x (provided, of course, that there is nothing in the evidence E that contradicts that statement about the objective chance of A). Under those circumstances I have established the objective chance of A on the basis of my degree of credence in A.

One of the main things that bothers me about this account is that it seems possible that another subject, with a different credence function C' and a different degree of conviction x' in A, will assign a different objective chance to the truth of A. There is, of course, no doubt that different subjects will have different estimates of the chance of A, or different opinions concerning the chance of A. That does not help David Lewis's (1980) account, inasmuch as he is attempting to characterize *objective chance*, not *estimate of objective chance* or *guess at objective chance*. The personalist may appeal to the well-known swamping of the priors—the fact that two investigators with very different prior probabilities will experience convergence of the posterior probabilities as additional evidence accumulates if they share the same evidence and conditionalize upon it. There are two objections to this response. First, and most obviously, the convergence in question secures intersubjective agreement, without any assurance that it corresponds to objective fact. Second, the convergence cannot be guaranteed if the parties do not agree on the likelihoods, and the likelihoods are just as subjective as the prior probabilities are.

Accordingly, I would want to read the principal principle in the opposite direction, so to speak. To begin, we should use whatever empirical evidence is available—either through direct observation of relative frequencies or through derivation from a well-established theory—to arrive at a statement X that the chance (in D. Lewis's 1980 sense) of A is x. Taking that value of the chance, we combine X with the total evidence E, and calculate the subjective degree of conviction on the basis of the credence function C. If $C(A|X \cdot E)$ is not equal to x, then C is not a *rational* credence function. In other words, we should use our knowledge of objective chance to determine what constitutes a reasonable degree of conviction. Our knowledge of objective chance, or propensity, is based ultimately upon observations of relative frequencies. Epistemically speaking, this amounts to a *frequency-driven* account of rational credence. In the following section I shall suggest that, ontically speaking, we should prefer a *propensity-driven* account—in my sense of the term "propensity."

5. Dynamic Rationality

In section 2 of this chapter we looked at various grades of rationality, static and kinematic, that can appropriately be applied to our systems of degrees of conviction. We noted emphatically that static rationality comprises only internal 'housekeeping' criteria—ones that do not appeal to external objective facts in any way. In that connection, I mentioned my view that, from the standpoint of *rational action*, it is necessary to refer to objective fact.

Kinematic rationality, which invokes Bayesian conditionalization, makes a step in that direction—it does make *some* contact with the real world. I tried to show, however, that the connection it effects is too tenuous to provide a full-blooded concept of rationality. In an attempt to see how this connection could be strengthened we considered several theories concerning the relationships between subjective and objective probabilities. We found in Ramsey (1950) a robust account of the connection between partial beliefs and relative frequencies, which led me to characterize his theory as a *frequency-driven* account of subjective probability. In both Mellor (1971) and David Lewis (1980) we also saw strong connections between subjective probabilities and objective chance. Inasmuch as both of these authors base their accounts of objective chance on subjective probabilities, I characterized them as offering a *credence-driven* account of objective probability. Lewis provides a particularly strong connection in terms of his *principal principle*. I am prepared to endorse something like this principle, except that I think it should run in the direction opposite to that claimed by Lewis. He says, in effect, that we can plug in subjective probabilities and get out objective chance (which he identifies with propensity). I think it should be just the other way around.

I should like to confer the title *dynamic rationality* upon a form of rationality that incorporates some sort of requirement to the effect that the objective chances— whether they be interpreted as frequencies or as propensities—must be respected, as well as other such rationality principles as logical consistency, probabilistic coherence, strict coherence, openmindedness, and Bayesian conditionalization. I believe Ramsey (1950) was offering a theory of dynamic rationality because frequencies are its driving force. My version of dynamic rationality will establish one connection between propensities and frequencies, as well as another connection between propensities and personal probabilities. Since I regard propensities as probabilistic causes, the term "dynamic rationality" is especially appropriate.

In order to see how this works, let us look at an example. In the Pennsylvania State Lottery, the "daily number" consists of three digits, each of which is drawn from the chamber of a separate machine containing ten Ping-Pong balls numbered from 0 through 9. It is assumed that all of the balls in any of the machines have equal chances of being drawn and that the draws from the several machines are mutually independent. The winning number must have the digits in the order in which they are drawn. Since there are 1000 numbers between 000 and 999, each lottery ticket has a probability of 1/1000 of winning. The tickets cost $1 each, and the payoff for getting the right number is $500. Thus, the expectation value of each ticket is $0.50.[16] As Ramsey (1950) emphasized, if a person played for 1000 days at the rate of one ticket per day, and if the actual frequency of wins matched exactly the probability of winning, the individual would win once for a total gain of $500 and an average gain of $0.50 per play. The actual result was 666.

Unfortunately, the lottery has not always been honest. In 1980 it was discovered that the April 24 drawing (and perhaps others) had been "fixed." On that occasion, white latex paint had been injected into all of the balls except those numbered 4 and 6, thereby making the injected balls heavier than the others. Since the balls are mixed by a jet of air through the bottom of the chamber and are drawn by releasing one ball through the top, the heavier ones were virtually certain not to be drawn. The probabilities of drawing the untampered balls were thereby greatly increased. Those who knew about the crooked arrangement could obviously take advantage of it. If the only balls that could be drawn were 4 and 6, then each of the eight possible combinations of these two digits had a probability of 1/8. In that case, the expectation value would work out to about $62.50, which is not bad for a ticket that can be purchased for $1.[17]

One reason for going into this example in some detail is the bring out the familiar fact that, when one is going to use knowledge of frequencies in taking some sort of action on a single outcome, it is important to use the appropriate reference class. The class must be epistemically homogeneous—that is, the agent must not know of any way to partition it in a manner that makes a difference to the probabilities of the outcomes. Someone who had no knowledge of the way the lottery had been fixed would have to assign the value 1/1000 to the probability of any given number being the winning number. Someone who knew of the fix could assign a probability of virtually zero to some of the numbers and a probability of 0.125 to the remaining ones. As we have seen, the difference between the two probabilities is sufficient to make a huge difference in the expectations.

Three steps in the decision procedure (the decision whether to purchase a lottery ticket or not) have been taken. First, the event on whose outcome the gamble depends (a particular night's drawing) is referred to a reference class (all of the nightly drawings). Second, the probability that one particular number will win is assessed relative to that reference class. Third, the epistemic homogeneity of that reference class is ascertained. If it turns out to be inhomogeneous, a relevant partition is made, and the process is repeated until an epistemically homogenous reference class is found. This is the probability with respect to the broadest epistemically homogeneous reference class (what Reichenbach ([1935] 1949) and I have called "weight") that is taken as the probability value in calculating the expectation.

In any given situation, the epistemically homogeneous reference class may or may not be objectively homogeneous (see Salmon 1984b, chap. 3). In cases such as the present—where we have a causal or stochastic process generating the outcomes—if the class is objectively homogeneous, I would consider the weight assigned to the outcome the *propensity* or *objective chance* of that mechanism (chance setup) to produce that outcome. A mechanism with this propensity generates relative frequencies, some of which are identified as probabilities if we adopt a frequency interpretation.

If the reference class is merely epistemically homogeneous, I would regard the weight as our *best available estimate* of the propensity or objective chance. That weight can easily be identified as our subjective degree of credence—or, if it is not actual, as the degree of credence we should have.

I have no wish to create the impression that any sort of crude counting of frequencies is the only way to determine the propensities of various chance setups. In the case of the lottery, we have enough general knowledge about the type of machine employed to conclude that the propensities for all of the different numbers from 000 to 999 are equal if all of the balls are of the same size and weight. We also have enough general background knowledge to realize that injecting some but not all of the balls with latex paint will produce a chance setup that has different propensities for different numbers. In theoretical calculations in quantum mechanics one routinely computes propensities in order to get values of probabilities. I do think that, at the most fundamental epistemic level, it is the counting of frequencies that yields knowledge of values of propensities.

On my view, as I have said above, propensities are not probabilities; rather, they are probabilistic causes. They generate some basic probabilities directly, as, for example, when the lottery machines produce sequences of digits. From the probabilities generated directly by propensities we can compute other probabilities, such as the (inverse) probability that a given number was produced by a crooked machine.

In cases like the lottery machines, where there are actually many repetitions of the same type of event to provide a large reference class, it may seem that the foregoing account makes good sense. In many similar situations, where we do not have a large number of trials but are familiar with the sort of mechanism involved, we may justifiably feel confident that we know what kinds of frequencies would result if many trials were made. Again, the foregoing account makes sense. But how, if at all, can we extend the account to deal with nonrepetitive cases?

Let us begin by considering situations like the lottery where long sequences of trials exist but where the subject is going to bet on only one trial in that sequence. Using the mathematical expectation, he or she can ascertain what the average return per trial would be, but what point is there in having that number if we can only apply it in a single instance? The answer is that the *same policy* can be used over and over again in connection with *different sequences*.

Imagine a gambling casino in which there are many (say 100) different games of chance, each of which generates a long sequence of outcomes. In these various games there are many different associated probabilities.[18] Let us assume that our subject knows these probabilities and decides to play each game just once. After deciding the size of the bet in each case, he or she can figure the mathematical expectation for each wager. The situation we have constructed now consists of 100 long sequences of trials, each with an associated expectation value. The subject makes a random selection of one item from each of the sequences; these constitute a new sequence of trials, each with an associated expectation value. The main differences between this new sequence and the original 100 are that each item is produced by a different chance setup, and both the probabilities and the expectation values differ from item to item. In playing through this new sequence of trials the subject will win some and lose some, and in general the tendency will be to win more of the cases with higher probabilities and to lose more of those with low probabilities. The average gain (positive or negative) per play will tend to be close to the average of the 100 expectation values associated with the 100 different basic sequences.

We realize, of course, that our subject might be extremely unlucky one evening and lose every bet placed, but such an overall outcome will be very rare.

If he or she repeats this type of performance night after night, the mathematical expectation of the evening's play will be the average outcome per evening in a sequence of evenings, and the frequency of significant departures from that amount will be small. Reichenbach ([1935] 1949, sec. 56, 72) proved what amounts to the same result to justify his policy for dealing with single cases.

Whether we think of physical probability as propensity or as (finite or infinite) long-run relative frequency, there is always the problem of the short run, as characterized elegantly by Charles Sanders Peirce (1931):

> According to what has been said, the idea of probability belongs to a kind of inference which is repeated indefinitely. An individual inference must be either true of false and can show no effect of probability; and therefore, in reference to a single case considered in itself, probability can have no meaning. Yet if a man had to choose between drawing a card from a pack containing twenty-five red cards and a black one, or from a pack containing twenty-five black cards and a red one, and that of a red one were destined to transport him to eternal felicity, and that of a black one to consign him to everlasting woe, it would be folly to deny that he ought to prefer the pack containing the larger proportion of red cards, although from the nature of the risk, it cannot be repeated. (vol. 2, sec. 2.652).

There is no question about ascertaining the probabilities in Peirce's example; they are assumed to be well known. The problem has to do with the application of this knowledge in a concrete situation. It does not much matter whether the application involves one case or a dozen or a hundred, as long as the number of cases is much smaller than the number of members of the entire population.

In (Salmon 1955), I addressed this problem and suggested a pragmatic vindication of a short-run rule to the effect that one should assume that the short-run frequency will match, as closely as possible, the long-run probability. That vindication did not hold up. Now I would be inclined to give a different argument. Assume that I am right in identifying the probability with the long-run frequency and that the long-run frequency of draws resulting in red cards from the predominately red deck is 25/26. Assume that this value has been established on the basis of extensive observation of frequencies. It is not crucial to Peirce's (1931) problem that only one draw ever be made from the deck; what is essential is that a person's entire fate hinges on the outcome of one draw.

I shall now say that the drawing of a card from the deck is a chance setup whose propensity to yield a red card is 25/26. The statement about the propensity is established on the basis of observed frequencies, as well as such theoretical considerations as the symmetries of the chance setup. As I have said above, I think of propensities as probabilistic causes. If it is extremely important to achieve a particular outcome, such as a red card, one looks for a sufficient cause to bring it about. If no sufficient cause is available, one seeks the most potent probable cause. In Peirce's (1931) example, a draw from the predominately red deck is obviously a more potent cause than is a draw from the predominately black deck. The fact that the outcome cannot be guaranteed is bad luck, but that is the kind of situation we are in all the time. One might call it *the human predicament*. That is why *"to us,* probability is the very guide of life." Taking the strongest measure possible to bring about the desired result is clearly the sensible thing to do. For the moment, at least, I am inclined to regard this as an adequate answer to the problem of the short run.

Having discussed the easy sort of case in which the frequencies, propensities, mathematical expectations, and reasonable degrees of belief are clear, we must now turn to the hard type of case in which the event whose probability we want to assess is genuinely unique in some of its most important aspects. Here the subjective probabilities may be reasonably tractable—especially in New York City, where it seems that most people will bet on anything—but the objective probabilities are much more elusive. I believe, nevertheless, that the notions of propensity and probabilistic cause can be of significant help in dealing with such cases.

Before turning to them, however, we should briefly consider an intermediate kind of case, which appears at first blush to involve a high degree of uniqueness, but where, upon reflection, we see that pertinent frequencies can be brought to bear. The *Challenger* disaster provides a useful instance. The life of Christa McAuliffe, the schoolteacher who joined the crew on that flight, was insured for a million dollars (with her family as beneficiary) by a philanthropic individual. I have no idea what premium was charged.

McAuliffe was the first female who was not a professional astronaut to go on a space shuttle flight. Since she had undergone rigorous screening and training, there was reason to believe that she was not at special risk as compared with the rest of the crew and that she did not contribute a special risk to that particular mission. As Feynman (Marshall 1986) pointed out, however, there was a crucial piece of frequency information available—namely, the frequency of failure of solid rocket boosters. Taking that information into account, and even making the sort of adjustments Feynman suggested, the minimum premium (omitting profit for the insurance company) should have been $10,000.

Feynman's (Marshall 1986) adjustments are worthy of comment. One adjustment was made on the basis of improved technology; whether that is reasonable or not would depend in large measure on whether the relative frequency of failures had actually decreased as time went by. The second had to do with higher standards of quality control. Perhaps there were frequency data to show that such a factor was relevant, but even in their absence one could take it as a probabilistic cause that would affect the frequency of space shuttle disasters.

Another good example comes from the continuing controversy regarding the so-called strategic defense initiative, popularly known as *Star Wars*. In the debate two facts seem incontrovertible. First, enormous technological advances will be required if the project is to be feasible at all.[19] Second, extraordinarily complicated software will be necessary to control the sensors, guidance systems, weapons, and so on. This software can never be fully tested before it is used to implement a (counter?)attack. If the computers function successfully, the system will attack only in case an enemy launches an attack against the United States. The computers will have to make the decision to launch a (counter?)attack, for there will not be time for human decision makers to intervene. What degree of confidence should we have regarding the ability of writers to produce software that is almost unimaginably complex and that will function correctly *the first time*—without any errors that might launch an attack against a nation that had not initiated hostilities against the United States and without any errors that would completely foul up the operation of the system in case of actual war or render it completely nonoperational? We all have enough experience with computers—often in use by government agencies—to make a pretty reasonable guess at *that* probability.

The moral of the consideration of these 'intermediate' cases is simple. It is the point of Feynman's (Marshall 1986) criticism of NASA's risk assessment. Do not ignore the available relevant information about frequencies. *Respect the frequencies.*

Let us now turn to the to the kind of case in which the presumption of the availability of relevant frequency information is much less plausible. Examples from history are often cited to illustrate this situation. Because of the complexity of actual historical cases, I cannot present a detailed analysis, but I hope to be able to offer some suggestive remarks.

One of the most significant occurrences of the present century was the dropping of the atomic bomb on Hiroshima. The background events leading up to the development of the atomic bomb are familiar: the discovery of nuclear fission and the possibility of a self-sustaining chain reaction, the realization that the Nazis were working on a similar project, the fear that they would succeed and—with its aid—conquer the world, and the urging by important scientists of the development of such a weapon. Before the bomb was completed, however, Germany had surrendered, and many scientists felt that the project should halt or, if a bomb was created, it should not be put to military use. Given this situation in the spring of 1945, what were the probabilistic causes that led to the dropping of the bomb on a major Japanese city? What probabilistic causes led to rejection of the proposal by many scientists to demonstrate its power to the Japanese, and perhaps other nations, instead of using it on a civilian population?

This is obviously an extremely complicated matter, but a few factors can be mentioned. One event of great significance was the death of Franklin D. Roosevelt and the elevation of Harry S. Truman to the presidency; this seems clearly to have raised the probability of military use. Another contributing factor was the deep racism in America during World War II, making the lives of Asian civilians less valuable than those of Caucasians. Still another contributing factor was the failure of scientists to anticipate the devastating effects of radiation, and hence, to view the atomic bomb as just another, more powerful conventional weapon. An additional complicating feature of the situation was the great failure of communication between scientists and politicians and between scientists and military people.

Some scientists had suggested inviting Japanese observers to Trinity, the test explosion in Nevada, but others feared the consequences if the first bomb turned out to be a dud. Some had suggested erecting some buildings, or even constructing a model city, to demonstrate the power of the bomb, but various objections were raised. After Trinity, some scientists suggested exploding an atomic bomb over the top of Mount Fujiyama, but some of the military argued that only destruction of a city would provide an adequate demonstration (Wyden 1984). The personalities of key individuals had crucial import.

As one considers a complex historical situation, such as this, it is natural, I believe, to look at the multitude of relevant items as factors that tend to produce or inhibit the outcome in question. These tendencies are propensities—contributing or counteracting probabilistic causes—and we attempt to assess their strengths. In so doing we are relying on a great deal of experience with scientific, political, economic, social, and scientific endeavors. We bring to bear an enormous amount of experience pertaining to human interactions and the effects of various traits of personality. My claim is that, in assigning the personal probabilities that would have been appropriate in a given historical situation, we are estimating or guessing at the strengths of and interactions among probabilistic causes.

In writing about scientific explanation during a period of nearly twenty-five years, Hempel repeatedly addressed the special problems of historical explanation (see especially Hempel 1942, 1962b, and 1965b). He dealt with a number of concrete examples, in an effort to show how such explanations can be understood as scientific explanations conforming to his well-known models of explanation. I would suggest that they can easily be read in terms of attributions of probabilistic causes to complex historical occurrences.

I certainly do not pretend to have numerical values to assign to the propensities involved in the dropping of the bomb on Hiroshima, given one set of conditions or another—or to any other significant historical event—and I do not think anyone else does either. But it does seem reasonable to regard each of the factors we have cited (and others, perhaps) as probabilistic causes that worked together to bring about a certain result. I would suggest that experts who are familiar with the complexities of the situation can make reasonable qualitative assessments of the strengths of the various factors, classifying them as strong, moderate, weak, or negligible. In addition, they can make assessments of the ways in which the various probabilistic causes interact, reinforcing one another or tending to cancel one another out. It is considerations such as these that can be brought to bear in arriving at personal probabilities with respect to the unique events of human history.

6. Causality, Frequency, and Degree of Conviction

If one were to think of causality solely in terms of constant conjunction, then it would be natural to identify probabilisitic causality with relative frequency in some fashion or other. As I tried to show (Salmon 1984b, chaps. 5–6), a far more robust account of causality can be provided in terms of *causal processes* and *causal interactions*. This account is intended to apply to probabilistic causal relations, as well as to causal relations that can be analyzed in terms of necessary causes, sufficient causes, or any combination of them. It is also intended to give meaning to the notions of *production* and *propagation*.

My aim in the present chapter is to bring these causal considerations to bear on the problem of the relationship between objective and subjective probabilities and to relate them to rationality. As we noted above, Ramsey (1950) stressed the crucial role of mathematical expectations of utility in rational decision making. For purposes of this discussion, let us make the grossly false assumption that we know the utilities that attach to the various possible outcomes of the actions we contemplate.[20]

What we would *really* like to have, given that knowledge, is the ability to predict accurately the outcome of every choice we make, but we realize that this is impossible. Given that fact, what we would *really* like to have is knowledge of the actual frequencies with which the outcomes will occur. That would also be nice, but it, too, is impossible. Given that fact, the next best thing, I suggest, would be to know the strengths of the probabilistic causes that produce the various possible outcomes. They are the agencies that produce the actual frequencies. I think of them as actual physical tendencies, and I have called them *propensities*. It is the operations of physical devices having these propensities—chance setups, including our own actions—that produce the actual short-run frequencies on which our fortunes depend, as well as the long-run frequencies that I am calling *probabilities*.[21]

The best estimate of the actual short-run frequency is, I would argue, the possible value closest to the propensity. Recall Peirce's (1931) example. If the agent draws from the deck with the preponderance of red cards, that act has a propensity of degree 25/26 to yield a red card. The actual frequency of red in a class containing only one member must be either 1 or zero; 1 is obviously closer to the propensity than is zero. In the other deck, the propensity for red is 1/26, and this value is closer to zero than it is to 1. Since the agent wants a red card, he or she chooses to draw from the deck whose propensity for red is closer to the desired short-run frequency.

The best way to look at an important subset of subjective or personal probabilities is, I think, to consider them as estimates of the strengths of probabilistic causes. In many cases, several probabilistic causes may be present, and the propensity of the event to occur is compounded out of them. Smoking, exercise, diet, body weight, and stress are, for example, contributing or counteracting causes of heart disease. In such cases we must estimate, not only the strengths of the several causes, but also the interactions among them. Obviously, two causes may operate synergistically to enhance the potency of one another, they may tend to cancel each other out, or they may operate independently. Our assessments of the strengths of the probabilistic causes must take into account their interactions, as well as the strengths of the component causes.

The obvious problem that must now be faced is how we can justifiably estimate, guess, infer, or posit the strengths of the propensities. When we attempt to assign a propensity to a given chance setup, we are dealing with a causal hypothesis, for a propensity is a probabilistic cause. Since Bayes's theorem is, in my view, the appropriate schema for the evaluation of scientific (including causal) hypotheses, I should like to offer a brief sketch of a Bayesian account.

There is a widely held maxim, which I regard as correct, to the effect that the meaningful collection of scientific data can occur only in the presence of one or more hypotheses upon which the data are supposed to have an evidential bearing. It is therefore important, I believe, to take a brief excursion into the context of invention (discovery) in order to consider the generation of hypotheses about propensities—that is, about probabilistic causes.[22] It is a rather complex matter. First, we have to identify the chance setup and decide that it is an entity, or set of entities, worthy of our interest. This is analogous to selecting a reference class as a basis for an objective probability—that is, a long-run relative frequency. We must also identify the outcome (or set of outcomes) with which we are to be concerned. This is analogous to the selection of an attribute class (or sample space) for an objective probability relationship. Without these, we would have no hypothetical probabilistic causal relation to which a value of a propensity could be assigned.

But, as I emphasized above, not all objective probabilities are propensities. Inverse probabilities are not propensities. Many correlations do not stand for propensities. It would be a joke to say that a barometer is a chance setup that exhibits a high propensity for storms whenever its reading drops sharply. It would not necessarily be a joke to say that the atmosphere in a particular locale is a chance setup with a strong propensity for storms when it experiences a sharp drop in pressure. Thus, whenever we hypothesize that a propensity exists, we are involved in a rather strong causal commitment.

When we have identified the chance setup and the outcomes of interest and we hypothesize a probabilistic causal relation between them, we need to make hypotheses about the strength of the propensity. The idea I want to suggest is, roughly, that an observed frequency fulfills one essential function relating to the formulation of a hypothesis about the propensity. Actually, we need more than just a single hypothesized value of the propensity; we need a prior probability distribution over the full range of possible values of the propensity, that is, the closed unit interval. My proposal is that, in the absence of additional theoretical knowledge about the chance setup, the observed frequency determines the value of the propensity for which the prior probability is maximal, namely, the value of the observed frequency itself.[23]

There is an additional assumption that is often made by propensity theorists. A chance setup is defined as a mechanism that can operate repeatedly, such that, on each trial, it has the same propensity to produce the given outcome. This means that the separate instances of its operation are independent of one another; for example, the outcome in any given case does not depend probabilistically upon the outcome of the preceeding trials. This constitutes a strong factual assumption. Another, closely related assumption has to do with the randomness of the sequence of outcomes. There is a strong temptation to make these assumptions. If we do, we can use the Bernoulli theorem to calculate the probabilities of various frequency distributions, given various values for the associated propensity.

Imagine that we have a black box with a button and two lights — one red, the other green — on the outside. We notice that, when the button is pressed, either the red or the green comes on; we have never seen both of them light up simultaneously, nor have we seen a case in which neither comes on when the button has been pushed. However, we have no a priori reason to believe that either of the latter two cases is impossible. This is a generic chance setup. Suppose that we have observed ten cases, in six of which the green light has been illuminated. We now have a prior distribution for green, peaking at 0.6; a prior distribution for red, peaking at 0.4; and prior distributions for both lights and for neither light, each peaking at 0. So far, I would suggest, it is all context of invention (discovery). At the same time, and in the same context, we may make hypotheses — of the sort mentioned above — about such matters as the independence of the outcomes or the randomness of the sequence. All such hypotheses can be tested by additional observational evidence. We can check such questions as whether the red light goes on every time the number of the trial is a prime number (supposing that red was illuminated on the second, third, fifth, and seventh trials), whether green goes on whenever (but not only whenever) there have been two reds in a row, or whether red occurs on every seventh trial. Answers to all of these questions depend upon the frequencies we observe.

Not all philosophers agree that there is a viable distinction between the context of invention (discovery) and the context of appraisal (justification). But those of us who do agree have emphasized the psychological aspects of the former context. Such psychological factors would, I believe, have considerable influence in shaping the distribution of prior probabilities—for example, how flat it is or how sharply peaked at the value of the observed frequency. Prior probabilities represent plausibility judgments; the prior distribution exhibits the plausibilities we assign to the various possible values of the propensity.

The fact that some aspects of the context of invention (discovery) are psychological does not prevent objective features of the situation from entering as well. In particular, I am suggesting, observed frequencies—the paradigm of objective fact—play a decisive role. In this connection, it should perhaps be mentioned that I have long maintained a view regarding the confirmation of scientific hypotheses that might suitably be called *objective Bayesianism* (see, e.g., Salmon 1967b, chap. 7). One key feature of this approach to confirmation is that the prior probabilities of hypotheses are to be construed as frequencies. A plausibility judgment should identify a hypothesis as one of a given type, and it should estimate the frequency with which hypotheses of that type have succeeded. I shall not rehearse the arguments here, but we should recall, as noted above, that Shimony (1970) supports his treatment of prior probabilities in his *tempered personalism* by an explicit appeal to frequency considerations. This is another way in which, I believe, the frequencies should be respected.

7. Conclusion

Ramsey's conception of rationality accords a central role to mathematical expectations of utility. The probabilities occurring in the expression for the expectation are, in Ramsey's terms, degrees of partial belief. His rationale for this approach brings into sharp focus the actual short-run frequencies that determine the net outcomes of our choices. It seems clear that Ramsey regarded his subjective probabilities as estimates of (short- or long-run) frequencies. For this reason I regard his view as a *frequency-driven* conception of rationality.

Dynamic rationality, as I would define it, consists in the attempt to use propensities—that is, probabilistic causes—as the weighting factors that occur in the formula for expected utility. Since we cannot be sure that our choices and decisions will be fully efficacious in bringing about desired results, it is reasonable to rely on the strengths of probabilistic causes. This line of thought treats our voluntary choices, decisions, and actions as probabilistic causes of what happens as a result of our deliberations. Dynamic rationality involves a *propensity-driven* view of objective probabilities and short-run frequencies.

Because the values of propensities are not known with certainty, we have to make do with the best estimates we can obtain. I have been urging that some personal probabilities are, precisely, best estimates of this sort. There are, however, physical probabilities that cannot be identified with propensities; consequently, I believe, there are personal probabilities that cannot straightforwardly be construed as estimates of propensities. These personal probabilities may be taken as estimates of frequencies. Where we do not have access to the pertinent propensities, we do best by using estimates of frequencies. Since, however, I take it that frequencies are generated by propensities, my approach involves a *propensity-driven* account of degree of conviction.

In dealing with universal laws in the sciences we are used to the idea that some laws—for example, the ideal gas law—provide empirical regularities without furnishing any causal underpinning. We look to a deeper theory—the kinetic-molecular theory—for a causal explanation of that regularity. In the domain of statistical regularities a similar distinction can be made. There are statistical laws that express statistical regularities without providing any kind of causal explanation. Such statistical regularities are physical probabilities (long-run frequencies). In some cases a deeper statistical theory exists that embodies the propensities that constitute the causal underpinning. In both the universal and the statistical cases, however, the unexplained empirical regularities can serve as a basis for prediction and decision making.

A major thesis of this chapter is that observed frequencies often provide the best evidence we have concerning long-run frequencies and the strengths of probabilistic causes. Sometimes our evidence is of a less direct and more theoretical variety. At the most primitive level, I believe, observed frequencies constitute the evidential base.

Induction by enumeration has traditionally been regarded as a rule of primitive induction for *justifying* inferences to values of physical probabilities (long-run frequencies). Whether it may be a suitable rule for this aim is an issue I shall not go into here.[24] Since our main concerns in this chapter have been with propensities and probabilistic causes, these concerns have *not* been at the level of primitive induction. Causal attributions are more complicated than mere statistical generalization. For purposes of the present discussion, I have suggested that we consider induction by enumeration instead as a method—in the context of invention (discovery)—that makes a crucial contribution to the generation of hypotheses. Used in this way, it provides, I believe, a concrete method for *respecting the frequencies*. It then serves as a counterpart to David Lewis's *principal principle*—providing the link between objective and subjective probabilities, but going *from* the objective (observed frequency) *to* the subjective (personal probabilities).

Observed frequencies furnish knowledge of physical probabilities and propensities. They also constitute the basis for personal probabilities—both those that correspond to propensities and those that do not. These personal probabilities, in turn, provide the weights that should figure in our calculations of mathematical expectations. The mathematical expectations that result from this process then constitute our *very guide of life*.

Hume's Arguments on Cosmology and Design

M any different kinds of reasons have been given, in various times and places, for belief in the existence of God. Mystics have sometimes claimed direct experiential awareness of the deity. Other believers have maintained that faith, not reason, is the proper foundation for religious conviction. Pascal, a devout adherent of this latter group, also claimed that belief could be fostered by a pragmatic "wager." During David Hume's lifetime—as well as before and after—"natural religion" enjoyed considerable popularity. According to this approach, the existence of a Supreme Being could be established by rational arguments, either a priori or a posteriori. In Hume's *Dialogues* (1970) [1779], the a priori arguments—such as Anselm's ontological argument and Aquinas's first-cause argument—are treated briefly, but they are given rather short shrift. They are defended by Demea, who is clearly the intellectual inferior of Philo and Cleanthes; they are dispatched quickly and without much ceremony. Indeed, Demea is not even able to stick out the discussion to the end.

1. The Design Argument

The main topic of the *Dialogues* is what Philo calls "experimental theism"—the thesis that the existence of God can be approached as a scientific hypothesis and that his existence can be established with a high degree of confirmation by observational evidence. The classic statement of this approach is given by William Paley (1851) in a book entitled *Natural Theology; or, Evidences of the Existence and Attributes of the Deity, Collected from the Appearances of Nature*:

> In crossing a heath, suppose I pitched my foot against a *stone*, and were asked how the stone came to be there, I might possibly answer, that, for any thing I knew to the contrary, it had lain there forever: nor would it perhaps be very easy to show the absurdity of this answer. But suppose I had found a *watch* upon the ground, and it should be inquired how the watch happened to be in that place, I should hardly think of the answer which I had before given, that, for anything I knew, the

watch might have always been there. Yet why should not this answer serve for the watch, as well as for the stone? Why is it not as admissible in the second case, as in the first? For this reason, and no other, viz. that, when we come to inspect the watch, we perceive (what we could not discover in the stone) that its several parts are framed and put together for a purpose, e.g., that they are so formed and adjusted as to produce motion, and that motion is so regulated as to point out the hour of the day; that, if the several parts had been differently shaped from what they are, of a different size from what they are, or placed after any other manner, or in any other order, than that in which they are placed, either no motion at all would have been carried on in the machine, or none which would have answered the use, that is now served by it. (5)

Precisely the same sort of reasoning has, incidentally, been recently applied to Stonehenge. After an elaborate discussion of the workings of the watch, Paley goes on to draw the moral, that "every indication of contrivance, every manifestation of design, which existed in the watch, exists in the works of nature; with the difference, on the side of nature, of being greater and more, and that in a degree which exceeds all computation" (13).[1]

In the second dialogue, Hume ([1779] 1970) puts the same argument in the mouth of Cleanthes:

I shall briefly explain how I conceive this matter. Look round the world: contemplate the whole and every part of it: you will find it to be nothing but one great machine, subdivided into an infinite number of lesser machines, which again admit of subdivisions, to a degree beyond what human senses and faculties can trace and explain. All these various machines, and even their most minute parts, are adjusted to each other with an accuracy, which ravishes into admiration all men, who have ever contemplated them. The curious adapting of means to ends, throughout all nature, resembles exactly, though it much exceeds, the productions of human contrivance; of human design, thought, wisdom, and intelligence. Since therefore the effects resemble each other, we are led to infer, by all the rules of analogy, that the causes also resemble; and that the Author of Nature is somewhat similar to the mind of men; though possessed of much larger faculties proportioned to the grandeur of the work, which he has executed. By this argument *a posteriori*, and by this argument alone, do we prove at once the existence of a Deity, and his similarity to human mind and intelligence. (22)

Throughout the major portions of the *Dialogues*, Cleanthes attempts to defend this type of argument, whereas Philo launches repeated attacks against it. After some polite preliminaries in the first dialogue, Cleanthes advances the argument near the beginning of the second dialogue. Philo persists in making damaging attacks that Cleanthes seems unable to answer until at the opening of the twelfth and final dialogue (after Demea's departure), Philo makes a startling declaration of belief in the argument from design. This sudden apparent reversal of position has led to many disputes about Hume's intention and many controversies as to whether Philo or Cleanthes is Hume's protagonist. I shall postpone discussion of this historical question until the concluding postscript. My purpose for now is simply to attempt a logical analysis of the arguments presented in the *Dialogues*.

As Cleanthes presents the design argument, it is an instance of argument by analogy. Such arguments have traditionally been classified as an important type of inductive argument, and they seem to occur in certain areas of scientific research. It is often said, for example, that the use of medical experiments on animals in order to ascertain the effects of various substances upon humans constitutes an important application of analogical reasoning. Some of the research on carcinogens seems to proceed in this manner.

To characterize these arguments—the medical arguments, as well as the design argument—as mere analogies is, however, to run the risk of seriously misrepresenting them and overlooking much of their force. The strength of a simple argument by analogy depends crucially upon the degree of similarity between the entities with respect to which the analogy is drawn. But mice are strikingly dissimilar to humans, and watches are even more dissimilar to universes. Philo is quick to point this out with devastating force (23–24). By the end of the second dialogue, it seems to me, both Philo and Cleanthes are fully aware that the argument is more complex. The discussion does not, however, end there. Early in the third dialogue, Cleanthes advances a different form of the design argument—one that Philo, remarkably, does not answer. Cleanthes seems to distinguish it from simple analogies by commenting upon its "irregular nature" (37). I shall return to this version of the argument below.

There is one obvious sense in which the arguments are not simple analogies; namely, it is not the degree of similarity that matters most, but rather the *relevance* of the similarities and the *irrelevance* of the dissimilarities. The fact that humans wear clothing and smoke cigarettes, but mice do not, is irrelevant. The fact that the physiological processes in the body of the mouse are similar to those in the human body is highly relevant to the question of whether a substance, such as cigarette tar, which produces cancer in experimental animals, is likely to cause cancer in humans. Such questions of relevance are, obviously, highly theoretical in character—they do not hinge on directly observable similarities or differences. We must conclude that these arguments, initially characterized as analogies, are more subtle and complex. They are arguments whose function is to evaluate *causal hypotheses*; if we wish to understand them, we must try to subject them to the sort of analysis that is appropriate to arguments offered in support of causal hypotheses in science. Hume was aware of this fact, I believe. If we look at the various facets of the discussion between Philo and Cleanthes, we shall find the main constituents of just such arguments.

2. Causal Hypotheses and Bayes's Theorem

The question of what argument form is correct for the establishment of causal hypotheses is, of course, still a matter of great controversy. In preceding articles I have been urging the adoption of Bayes's theorem to play that role. The purpose of this chapter is to carry out a Bayesian analysis of the design argument for the existence of God and to use this framework for a discussion of Hume's *Dialogues*. The reader may decide at the conclusion whether this approach provides an illuminating analysis. I believe it furnishes a useful framework for fitting together the various arguments that occur throughout the *Dialogues*.

Hume makes no explicit reference to Bayes's theorem. While writing an earlier version of this chapter I considered it unlikely that he was aware of the theorem, but in an interesting discussion note David Raynor (1980) wrote,

> [Salmon] made no claim for historical accuracy because he mistakenly believed that Hume probably did not know of Bayes's theorem when he was composing the *Dialogues*. But Hume *was* aware of Bayes's theorem at least as early as 1767. In the light of this hitherto unnoticed historical fact, a Bayesian analysis of the *Dialogues* seems all the more plausible." (105)

Raynor's note contains the historical documentation.

Since Bayes's theorem has already been presented, discussed, and illustrated in chapter 4, I shall simply reiterate it and present a simple example by way of reminder. Here is a basic form of the theorem:

$$P(B|A \cdot C) = \frac{P(B|A)\, P(C|A \cdot B)}{P(B|A)\, P(C|A \cdot B) + P(\sim B|A)\, P(C|A \cdot \sim B)} \tag{1}$$

Example (fictitious): Suppose that a small percentage of pearls have a particular sort of color flaw that renders them worthless. This flaw appears in 1% of all cultured pearls and in 3% of all natural pearls. Let A be the class of pearls, B the class of cultured pearls, and C the class of color-flawed pearls. Assume, moreover, that 90% of all pearls examined are cultured pearls. Now a pearl is found that exhibits this undesirable color flaw. What is the probability that it is a cultured pearl? Using the given values, we have

$$P(B|A \cdot C) = \frac{0.9 \times 0.01}{(0.9 \times 0.01) + (0.1 \times 0.03)} = 3/4 \tag{2}$$

As we have seen, three probabilities—two likelihoods and one prior probability— are required to calculate the posterior probability.[2]

3. Philo's Estimates

My main thesis regarding Hume's *Dialogues Concerning Natural Religion* is that his characters (especially Philo) devote considerable attention to each of these three types of probabilities in their discussion of the hypothesis that a supremely intelligent, powerful, and benevolent deity created the universe. Although the argument in the *Dialogues* is not cast in formal terms, Hume showed a full appreciation of the three types of considerations that must be brought to bear in order to evaluate the theistic causal hypothesis. Whether, in the end, Hume accepts or rejects the design argument, he certainly recognizes it as something deeper and more subtle than a simple appeal to analogy.

The aim of the design argument, as presented by Cleanthes, is to show that the universe, as an object exhibiting a high degree of orderliness, was very probably the result of intelligent design. Since we cannot observe other worlds coming into being—universes are not so plentiful as blackberries, as Peirce (1931) remarked—we cannot perform an induction by simple enumeration on observed births of universes in order to draw a direct conclusion about the creation of our own. "Have worlds ever been formed under your eye," Philo asks at the end of the second dialogue, "and have you had leisure to observe the whole progress of the phenomenon, from the first appearance of order to its final consummation?" (32). The answer, of course, is negative. We must, instead, make an indirect inference from the origins of other types of entities whose beginnings we can observe. Let us, therefore, proceed on that basis.

The first step in our attempt to apply Bayes's theorem is to find meanings for the terms that occur in it. Let A denote the class of instances of entities coming into being. We shall want to consider quite a broad class, so as not to bias our evidence. The class A will include, for example,

(i) Formation of a zygote when a sperm and egg unite.
(ii) Building of a house that was designed by an architect.
(iii) Growth of a tree from a seed.
(iv) Formation of a piece of ice as a lake freezes.
(v) Reduction of buildings to piles of rubble as a tornado tears through a town.
(vi) Carving of a gully by water rushing down a hillside.
(vii) Formation of compost as organic matter decays.

Let B designate the class of instances in which intelligence operates. I shall not include cases of the operation of computing machines since there is no need to raise the question of whether machines have intelligence:

(i) Designing a house.
(ii) Writing a poem.
(iii) Designing and making a telescope.
(iv) Doing a sum.
(v) Proving a mathematical theorem.

In some of these examples a product results, but in others, such as the last, it might be doubtful that anything comes into existence. In the present discussion, however, we will not be concerned with cases in which intelligence does not create an entity—indeed, a material entity.

The letter C will now be taken to represent the class of entities that exhibit order. For example,

(i) A watch.
(ii) A ship.
(iii) A living organism—plant or animal.
(iv) The human eye.
(v) A rainbow.
(vi) The solar system.
(vii) A diamond.

All of the foregoing examples are entities that rather obviously exhibit some sort of order. We shall need to return to the concept of order below, to analyze it with greater precision. With the foregoing meanings assigned to the letters in Bayes's theorem that designate classes, we can now interpret all of the probability expressions that appear in it:

$P(B/A)$ = the probability that a case of coming-into-being is an instance of the operation of intelligence.

$P(\sim B/A)$ = the probability that a case of coming-into-being is an instance of the operation of something other than intelligence.

$P(C/A \cdot B)$ = the probability that something produced by intelligence exhibits order.

$P(C/A \cdot \sim B)$ = the probability that something produced by some agency other than intelligence exhibits order.

$P(B/A \cdot C)$ = the probability that something which comes into being and exhibits order was produced by intelligence.

This last probability, which appears on the left-hand side of Bayes's theorem, is the one we seek. Proponents of the design argument maintain that it is very high.

With the help of Philo, let us now attempt to assess these various probabilities. We may begin with the prior probabilities, $P(B/A)$ and $P(\sim B/A)$. Philo furnishes a brief inventory of "springs and principles" from which a variety of entities arise (see, especially, the seventh and eighth dialogues):

(1) *Two types of biological generation, animal and vegetable.* Whenever any organism comes into being, one or the other is operative.
(2) *Instinct.* When bees make honeycombs, spiders spin webs, birds build nests, or wolves mark out their territories by urinating on trees, this principle is operative.
(3) *Mechanical causation.* The formation of snowflakes, diamonds, and other crystals, as well as the formation of galaxies, solar systems, molecules, and atoms, all illustrate this principle.
(4) *Intelligent design.* All types of human artifacts arise from this principle.

When we consider the number of organisms, animal or vegetable, living on the earth—including the millions of microbes that inhabit each human body—we see immediately that biological generation operates in large numbers of instances. Similar considerations apply to instinct. Mechanical causation obviously operates with great frequency upon the earth; for all we know, it may be the only principle that operates anywhere else in this vast universe. Hume knew that the physical universe is enormous, but he could not have guessed its actual size. We now have evidence to indicate that it contains perhaps 10 billion galaxies and that each of these contains, on the average, from 10 to 100 billion stars. And who can say how many atoms have been formed in the interiors of these stars? Our earth alone contains something like 10^{50} atoms.

Although there are many unanswered questions in cosmology and astrophysics concerning the formation of galaxies, stars, and atoms, we have achieved some scientific understanding of these processes. They appear to be mechanical. There is, consequently, a great deal of evidence pointing to the conclusion that the number of instances in which mechanical causation operates is almost incomprehensibly greater than all of the rest combined. All of the evidence indicates that $P(B/A)$ is low—incredibly small—and that $P(\sim B/A)$ is high. Although he did not have some of the numbers we have today, Hume was clearly aware of the fact. "What peculiar privilege has this little agitation of the brain that we call *thought*," Philo asks in the second dialogue, "that we must thus make it the model of the whole universe?" (28). We now have a rough assessment of the prior probabilities.

The probability $P(C/A\cdot\sim B)$, that the operation of some principle other than intelligence will yield order, cannot be taken as negligible. The examples already cited under biological generation, instinct, and mechanical causation make this point evident. Philo cites such evidence in his discussion of these various "springs and principles."

The probability $P(C/A\cdot B)$, that a result of intelligent planning and design will exhibit order, may be quite high. That fact, in and of itself, does *not* guarantee that the probability we are attempting to evaluate—the posterior probability, $P(B/A\cdot C)$—will be high. If $P(\sim B/A)$ and $P(C/A\cdot\sim B)$ are large enough, the probability that an object that exhibits order was the product of intelligent design may still be quite low. This point was illustrated by the fictitious pearl example. In the course of the argument, Philo brings out these considerations quite explicitly.

Although it may be granted that intelligent planning often gives rise to order, it should be explicitly noted that the probability $P(C/A\cdot B)$ is not equal to unity, and moreover, it may not even be near that value. I do not find this point made explicitly in the *Dialogues*, and it is not essential to the argument, though perhaps it has some bearing. As the Vietnam war and countless other wars remind us, intelligent design may produce disorder: bombs reduce villages to rubble, defoliants destroy plant life, and many other products of human ingenuity maim and kill. Meanwhile, back home, industrial pollution and insecticides destroy many other forms of life as well.

Let us now take stock of the situation. By making plausible assessments of the probabilities that appear on the right-hand side of Bayes's theorem, we are in a position to say, quite confidently, that the probability $P(B/A\cdot C)$—that an unspecified entity, which came into being and exhibited order, was produced by intelligent design—is rather low. This conclusion can be reinforced even more directly by considering all of the entities with which we are acquainted that come into being exhibiting order. If we count the relative frequency with which such entities were the result of intelligent design, we easily see that it is rather low. This conclusion holds even if we exclude such items as galaxies and atoms on the ground that we are not very sure how they are created. There is still a vast numerical preponderance of such occurrences as animal reproduction, growth from seeds, formation of crystals, and spinning of spider webs over the relatively few instances in which watches, houses, and ships are built by people. Why does that not settle the case against the design argument without further ado? Why doesn't everyone agree that it is *improbable* that the universe, as an object exhibiting order, is a result of intelligent design? Is this mere obstinacy and prejudice on the part of upholders of "natural religion"?

4. The Uniqueness of the Universe

There is a sound logical reason why the matter cannot be settled quite so easily. In discussing the design argument, we are attempting to assign a probability to the hypothesis that the universe was created by an intelligent designer. We are dealing with a *single case*, for the creation of the universe is a unique event, at least as far as any possible experience we may have is concerned. The problem of assessing probabilities of single events notoriously involves the problem of deciding to which class the single event is most appropriately to be assigned. If we are trying to find the probability that a particular individual, John Jones, will survive for another decade—perhaps to determine a fair life insurance premium for him to pay—we have to take into account certain relevant factors, such as age, occupation, marital status, state of health, and so forth. The general rule, as I prefer to formulate it, is to refer the individual case to the broadest homogeneous class available—that is, to the broadest class that cannot be relevantly subdivided. Relevance is the key notion. I shall leave it somewhat vague in this context, even though I have tried to explicate it elsewhere (Salmon 1967b, 90–96). But it is, I believe, clear enough intuitively for present purposes. A similar principle, incidentally, is invoked by Hempel (1965b, 397–403) in his inductive-statistical model of scientific explanation, namely, the *requirement of maximal specificity*.

When we apply this consideration to the design argument, we see that we must give a closer specification of the particular case with which we are dealing. We must take into account the type of order that the universe exhibits, and we must also take a closer look at the nature of the intelligent creator hypothesized by the proponent of natural theology.

If we consider the sheer magnitude of the universe, all of our experience suggests that it was not constructed by a unitary being. Where human artifacts are concerned, the larger the project, the more likely it was to have been executed by a group rather than an individual. Moreover, even complicated machines are sometimes made by rather dull artisans who only copy the work of others. Philo, with cutting wit, drives the point home in the fifth dialogue: Taking account of the magnitude of the machine and the imperfections in its construction, for all we know the world may have been created by a juvenile deity who had not yet learned his trade, or a stupid deity who only copies and does not do that very well, or a committee of deities, or a superannuated senile deity who had lost the knack by the time he got around to making our world. If we pay close heed to experience, it is impossible to assign a high probability to the hypothesis that the world, if created as a divine artifact, was the product of a God bearing any resemblance to the theist's conception.

Worse still is the fact that the God of traditional theism has usually been regarded as a pure spirit—a disembodied intelligence. In no instance within our experience, however, has a disembodied intellect produced any kind of artifact, whether or not it might have exhibited any order. Indeed, since disembodied intelligence has never operated in any fashion, to the best of our knowledge, we must conclude from experience that for such an intelligence $P(B/A) = 0$, and in that case, $P(C/A \cdot B)$ is simply undefined.

If we examine more closely the kind of order exhibited by the universe, says Philo in the sixth dialogue, we may find that it more closely resembles that of a vegetable or an animal than it does that of a machine. In that case, *again*, $P(C/A \cdot B)$ is low, whereas $P(C/A \cdot \sim B)$ must be rated high, for animals and vegetables arise from biological generation, not intelligent design. The fact that farmers deliberately plant crops and breed animals does nothing to enhance the value of $P(C/ A \cdot B)$, for the order exhibited by the organisms thus produced arises from genetic principles, not from the intelligent design of the farmer, who merely provides the occasion for their operation.

Philo is not seriously entertaining the hypothesis that the universe arises from biological generation, but he does seem to give purely mechanical principles somewhat more credence. Later in the sixth dialogue he remarks, "And were I obliged to defend any particular system of this nature, which I never willingly should do, I esteem none more plausible than that which ascribes an eternal, inherent principle of order to the world, though attended with great and continual revolutions and alterations" (59). Philo is not, of course, assigning a high prior probability to this hypothesis—quite the contrary. When, in the eighth dialogue, he undertakes a fuller elaboration of this theme, he cautions against such a notion: "Without any great effort of thought, I believe that I could, in an instant propose other systems of cosmogony which would have some faint appearance of truth; though it is a thousand, a million to one if either yours or anyone of mine be the true system" (68).

In this eighth dialogue, Philo brings up "the old Epicurean hypothesis," which he characterizes as "the most absurd system that has yet been proposed" but that "with a few alterations ... might ... be brought to bear a faint appearance of probability" (69). He offers what seems to me a clear anticipation of Darwinian evolution; moreover, other aspects of this hypothesis bear a tenuous connection with more recent cosmological ideas. With the benefit of modern scientific knowledge, we are inclined to endow this sort of hypothesis with a more substantial prior probability, but Philo does not press for this conclusion. He urges only that the prior probability of the mechanical hypothesis is not less than the prior probability of the hypothesis of intelligent design. Moreover, he seems to argue that each of these hypotheses provides the same likelihood of order of the type exhibited by tile universe: "But wherever matter is so poised, arranged, and adjusted, as to continue in perpetual motion, and yet preserve a constancy in the forms, its situation must, of necessity, have all the same appearance of art and contrivance which we observe at present" (70).

At the close of the eighth dialogue, we come to the end of the analysis of one form of the design argument. The conclusion is that the hypothesis of intelligent design, where no "moral attributes" are ascribed to the Creator, is completely on a par with the mechanical hypothesis. Neither is very plausible, and neither is supported by compelling evidence. The prior probabilities and the likelihoods, respectively, are about the same for the two hypotheses. But because of the plethora of alternatives— mentioned, but not elaborated, at the beginning of the eighth dialogue—none can be assigned a high posterior probability. The probabilities of all of the many incompatible alternative hypotheses cannot add up to more than 1, and none of the candidates stands out as commanding a very large share of the total. The hypothesis of intelligent design is quite improbable, but no other hypothesis fares any better.

According to traditional theism, the Creator of the universe possesses the moral attributes of justice, benevolence, mercy, and rectitude. Cleanthes is as anxious to attribute them to the deity as he is to attribute wisdom and power. For religious purposes, the moral attributes are at least as important as the natural; "to what purpose establish the natural attributes of Deity," he asks, "while the moral are still doubtful and uncertain?" (89). Thus, Cleanthes maintains, the universe gives evidence of a Creator who is intelligent, powerful, and benevolent—analogous, but immeasurably superior, to his human counterpart. This is a different theistic hypothesis; it is not seriously discussed until the tenth dialogue.

Philo does not seem to argue that the addition of the moral attributes significantly affects the prior probability of the theistic hypothesis, but he does claim that it has a marked adverse effect upon the likelihood. Other things being equal, a lowering of the likelihood will result in a reduced posterior probability. Following this tack, he notes that the world seems utterly indifferent to values. The rain falls alike upon the just and the unjust. Evil abounds. Untold misery and suffering plague mankind. Hume's eloquence, when he speaks through Philo, creates a vivid picture. And Hume did not even know about nuclear bombs, chemical and biological warfare, and overpopulation problems. It is crucial to realize that Philo is *not* raising the traditional theological problem of evil. That problem is the problem of reconciling the existence of an omniscient, omnipotent, and perfectly benevolent God with the existence of real or apparent evil in the world. The problem of evil is resolved if you can show that there is no contradiction between asserting the existence of such a deity and the occurrence of human suffering and pain.

The problem for the design argument is rather different, as Philo emphatically insists. It is not merely a question of showing that the world as we know it is not incompatible with the existence of God. Rather the proponent of natural religion must maintain that the world as we know it is *positive evidence* for the claim that it was created by a wise, powerful, benevolent being. To test a scientific hypothesis, one derives observational consequences that would be expected to hold if the hypothesis were true and then observes to see whether such consequences obtain. In the eleventh dialogue, Philo puts this question:

> Is the world, considered in general and as it appears to us in this life, different from what a man...would, *beforehand*, expect from a very powerful, wise, and benevolent Deity? It must be strange prejudice to assert the contrary. And from thence I conclude that, however consistent the world may be, allowing certain suppositions and conjectures, with the idea of such a Deity, it can never afford us an inference concerning his existence. The consistency is not absolutely denied, only the inference. (96)

Philo then proceeds to enumerate four rather obvious ways in which an all-powerful Creator could have reduced the amount of evil in the world if he had wanted to (96–102):

(1) Pain need not be inflicted upon man—he need not have been endowed with the capacity for suffering. "Men pursue pleasure as eagerly as they avoid pain; at least, they might have been so constituted. It

seems, therefore, plainly possible to carry on the business of life without any pain" (97).

(2) The deity need not have governed the world by inviolable general laws. A little tinkering with the works now and then could be quite beneficial: an extra large wave upon a warship that is sailing to inflict harm on helpless people; a little calming of the waters for a ship on an errand of mercy.

(3) Mankind and the other species have been so frugally endowed with capacities as to make their existence hazardous and grim. Why could not the Creator have endowed them a bit more lavishly, for example, with industry, energy, and health?

(4) A further source of "misery and ill of the universe is the inaccurate workmanship of all the springs and principles of the great machine of nature... they are, all of them, apt, on every occasion, to run into one extreme or the other. One could imagine that this grand production had not received the last hand of the maker—so little finished is every part, and so coarse are the strokes with which it is executed" (101). It never rains but it pours; it is always feast or famine.

If Philo could so easily conceive of ways in which, for all we can tell, the world might have been vastly improved, why could not a supreme intelligence have done even better? Remember, Philo is not asking whether it is conceivable that God made the best world it was possible to make and this is it. He is asking whether this world is what we would antecedently expect of an omnipotent, omniscient, benevolent being.

> Did I show you a house or palace where there was not one apartment convenient or agreeable; where the windows, doors, fires, passages, stairs, and the whole economy of the building were the source of noise, confusion, fatigue, darkness, and the extremes of heat and cold, you would certainly blame the contrivance, without any further examination. The architect would in vain display his subtilty, and prove to you that, if this door or that window were altered, greater ills would ensue. What he says may be strictly true: The alteration of one particular, while the other parts of the building remain, may only augment the inconveniences. But still you would assert in general that, if the architect had had skill and good intentions, he might have formed such a plan of the whole, and might have adjusted the parts in such a manner as would have remedied all or most of these inconveniences. His ignorance, or even your own ignorance of such a plan, will never convince you of the impossibility of it. If you find any inconveniences and deformities in the building, you will always, without entering any detail, condemn the architect. (95–96)

Whereas the universe might equally be the product of mechanical causes or of an intelligence not assumed to be benevolent, the universe as we observe it can hardly be attributed to a Creator who is benevolent, as well as omniscient and omnipotent. Given the hypothesis that God is supremely wise, good, and powerful, there is a *very low* probability, $P(C/A \cdot B)$, that he would create a world such as this![3] Notice that Philo is not denying that the world exhibits order—one of his complaints, mentioned above, is that the world is too orderly. He is saying that the kind of order it manifests is not what would be expected as the result of creation by a wise, powerful, and benevolent Creator.

Hume evidently felt that the objections to the design argument that are based upon the existence of apparently gratuitous evil were far more potent than any of the other objections; at any rate, he has Philo declare at the close of the tenth dialogue,

> ... when we argued concerning the natural attributes of intelligence and design, I needed all my skeptical and metaphysical subtlity to elude your grasp. But there is no view of human life or of the conditions of mankind from which, without the greatest violence, we can infer the moral attributes or learn that infinite benevolence conjoined with infinite power and infinite wisdom, which we must discover by the eyes of faith alone.... It is your turn now to tug the laboring oar, and to support your philosophical subtilties against the dictates of plain reason and experience. (92)

Although Cleanthes in unwilling, for religious reasons, to abandon the argument concerning moral attributes, and to rest the case entirely upon the natural attributes, some defenders of "natural religion" might consider pursuing this approach.

There is, it seems to me, a logical (as well as a religious) reason why the proponent of the design argument cannot abandon consideration of the moral attributes. If one asks what sort of order the world exhibits that gives such strong evidence of intelligent design, the answer given repeatedly is "the adjustment of means to ends." It is patently impossible to bring such considerations to bear unless we have some conception of what ends the means are designed to serve. If the ends are justice, mercy, and benevolence as we conceive them in our better moments, one sort of order would count as evidence for intelligent design. If the end is to secure from mankind blind unreasoning adulation of and obedience to a deity who says in one breath, "Thou shalt not kill," and in another, "Slay and spare not," another kind of order would be suitable. Still another kind of order would be appropriate if the Creator, as revealed in the book of Job, is mainly concerned to win a bet with his old gaming companion, Satan. Thus, the moral attributes of God are inextricably bound up even in the argument directed toward his natural attributes. The author who wrote in his *A Treatise of Human Nature* (bk. 11, pt. 111, sect. 111), "Reason is, and ought only to be, a slave of the passions, and can pretend to no other office but to serve and obey them," could hardly have overlooked this point.

5. Order and Purpose

We have not yet come sufficiently to grips with the nature of the order that the theist finds in the world and that seems to him to furnish compelling evidence of intelligent design. Philo points out that the order exhibited by living organisms comes from biological generation. This is no news to Cleanthes; he was not under the impression that babies are brought by the stork from the baby-factory, where they are fashioned by artisans. This consideration does not seem to the proponent of the design argument to undermine his position; indeed, such biological wonders seem to him to reinforce his argument, for they are further evidence of ingenious design. Pursuing the example of the watch presented in the earlier quotation, Paley (1851) addresses precisely this issue:

> Suppose, in the next place, that the person, who found the watch, should, after some time, discover, that, in addition to all the properties which he had hitherto observed in it, it possessed the unexpected property of producing, in the course of its movement, another watch like itself; (the thing is conceivable;) that it contained within it a mechanism, a system of parts, a mould for instance, or a complex adjustment of lathes, files and other tools, evidently and separately calculated for this purpose; let us inquire, with effect ought such a discovery to have upon his former conclusion?
>
> The first effect would be to increase his admiration of the contrivance, and his conviction of the consummate skill of the contriver. If that construction *without* this property, or, which is the same thing, before this property had been noticed, proved intention and art to have been employed about it; still more strong would the proof appear, when he came to the knowledge of this further property, the crown and perfection of the rest. (8–9)

Paley proceeds to drive home the argument:

> The conclusion which the *first* examination of the watch, of its works, construction, and movement suggested, was, that it must have had, for the cause and author of that construction, an artificer, who understood its mechanism, and designed its use. The conclusion is invincible. A *second* examination presents us with a new discovery. The watch is found in the course of its movement, to produce another watch, similar to itself; and not only so, but we perceive in it a system of organization, separately calculated for that purpose. What effect would this discovery have, or ought it to have, upon our former inference? What, as hath already been said, but to increase, beyond measure, our admiration of the skill, which had been employed in the formation of such a machine? Or shall it, instead of this, all at once turn us round to an opposite conclusion, viz. that no art or skill whatever has been concerned in the business, although all other evidences of art and skill remain as they were, and this last supreme piece of art be now added to the rest? Can this be maintained without absurdity? Yet this is atheism. (12–13)

In these passages, Paley (1851) articulates what is, perhaps, the most fundamental issue concerning natural religion and the design argument. Hume addresses this issue in the *Dialogues*. For Cleanthes, the order that arises out of biological generation is only further evidence of intelligent design on the part of the Creator of the universe. Philo retorts that, within our experience, all instances of intelligent design issue from biological organisms—biological generation always lies behind intelligence. We do not have experience of intelligent creation as a prior source of biological generation. Why, then, does the defender of the design argument insist, contrary to experience, that intelligent design must lie behind the operation of any other principle that generates order?

The answer lies, of course, in a teleological conception of order. In his fullest and most careful statement of the design argument, Cleanthes describes the universe as a "great machine," composed of a prolific array of "lesser machines," all of which are characterized by "the curious adapting of means to ends." This theme recurs frequently throughout the *Dialogues*, and it is conspicuous in the quotations from Paley (1851), who confines his discussion to instances in which he believes it can be shown that the parts work together to achieve some useful end. Indeed, there is a pervasive tendency in the writings of the defenders of natural religion to equate order and design—to treat them as synonymous. But such a procedure flagrantly begs the question, as Philo is careful to point out in the seventh dialogue:

> To say that all this order in animals and vegetables proceeds ultimately from design is begging the question; nor can that great point be ascertained otherwise than by proving, *a priori*, both that order is, from its nature, inseparably attached to thought, and that it can never of itself or from original unknown principles belong to matter. (65)

Can such a proof be given? The universe, it is admitted on all sides, exhibits order. Is this order evidence of intelligent design? It cannot be shown by experience. Its assertion can only be as the sort of a priori pronouncement the proponent of "experimental theism" eschews. It is, in short, an anthropomorphic conceit.

As I read the eighth dialogue, it seems to me to contain a rather clear anticipation of a nonteleological theory of biological evolution. I do not know whether this reading is correct or whether I am indulging in wishful thinking. Be that as it may, in the nineteenth century Darwin provided just such a theory. To put the matter briefly, Darwin showed how the nonpurposive factors of chance mutations and natural selection could lead to the evolution of the species, and in that evolutionary process, those species that are best adapted to compete for food and reproductive opportunities would evolve. Although evolutionary biology poses many problems, there is little reason to think that a retreat into teleology is the answer. Twentieth-century molecular biology has added to our understanding of the mechanisms of heredity by showing precisely how order is reproduced. It seems to me, indeed, that one of the chief accomplishments of Galileo and Newton was the removal of Aristotelian teleological conceptions from physics. The beautiful order of the solar system, for example, so greatly venerated by Kepler, could be reduced to mechanical principles. The next great scientific revolution was due to Darwin; its main achievement was to rid the biological sciences of their teleological elements. Order in the physical world, and in its biological realms, was shown to be independent of intelligent design. The a priori principle to which the design argument ultimately appeals was thus undermined.

6. The Concept of Order

The net result of the scientific revolution, of which Newton's mechanics was the fundamental ingredient, was to present a picture of the world that exhibits a wonderful mechanical order and simplicity. The world is seen as a collection of material particles that respond to forces in accordance with Newton's three incredibly simple laws of motion. Moreover, one of the forces—gravitation—obeys another very simple law. The further development of classical mechanics, which reached its zenith at the end of the nineteenth century, revealed that other forces — electric and magnetic, in particular — also conform to simple and precisely specifiable laws. At the turn of the century, it appeared that all natural phenomena could be explained in terms of these fundamental principles of classical physics. Although Newton had entertained the idea that God might need to intervene from time to time to keep the cosmic machine in good adjustment, later authors — most notably Laplace — maintained that the universe is entirely governed by mechanical principles. The success of classical mechanics, along with the advent of Darwinian evolutionary theory, banished teleological principles from nature.

The fact that the constituents of the world behave in accordance with simple laws is a kind of order, but it is not the only kind of order nature exhibits. A random and totally disorganized system of material particles would obey the same laws as a finely constructed machine — for example, the heaps of metal in junkyards obey the same physical laws as do the finest newly fabricated motor cars. The universe does not, however, appear to be a cosmic heap of unorganized parts; to the eyes of many observers, it seems more like a finely constructed machine. The manifestation of this latter kind of order seemed to many deists[4] to constitute evidence of intelligent creation, even though all teleological elements had been removed from the operation of nature itself. The story is told of an astronomer who believed in intelligent creation and a friend who did not. One day the friend came to visit the astronomer, and he saw an elaborate model of the entire solar system that the astronomer had just acquired. The model was so constructed that when a crank was turned, the various planets and their satellites moved around a sun at the center, just as the real planets and satellites do. When asked by the friend where he had procured the model and who had constructed it, the astronomer replied simply that no one had made it — the parts had just happened to fall together in that arrangement by chance. Thus, with irony, did he attempt to show the absurdity of his friend's atheism.

One of the severe difficulties faced by eighteenth- and nineteenth-century authors who discussed, pro and con, the design argument was their inability to characterize in any satisfactory way the kind of order they saw in the world. To say that the universe conforms to simple laws is one thing; to say that it also exhibits an orderly configuration is quite another. The best they seemed able to do was to talk in terms of the adjustment of means to ends and the similarity of the universe — or its parts — to humanly constructed machines. Each of these characterizations leads to problems. It is difficult to deal with the means-end relationship if we do not have some independent evidence concerning the nature of the end that is supposed to be served. The similarity of the world to a machine can easily be challenged, as Philo does in the seventh dialogue where he says, "The world plainly resembles more an animal or a vegetable than it does a watch or a knitting loom" (62).

Scientific developments since about the middle of the nineteenth century have significantly clarified the concept of order. In dealing with physical problems closely related to practical concerns about the efficiency of machines, physicists and engineers created the science of thermodynamics. They established a viable concept of energy, discovered laws relating the various forms of energy to each other, and developed the concept of *entropy*. Entropy is essentially a measure of the unavailability of energy to do mechanical work. Given two physical systems, each with the same amount of energy, the one with the lower entropy is the one from which it is possible in principle to extract the greater amount of work. All machines, in the course of their operation, tend to dissipate some of their energy in useless forms—through friction, for example. They are subject to the universal law of degradation of energy, the famous second law of thermodynamics. The discovery of this fundamental law of nature—that entropy tends to increase—has led to considerable speculation about the 'heat death' of the universe. The entire universe seems to be "running down."[5]

Toward the end of the nineteenth century, thermodynamics was given a theoretical foundation in statistical mechanics, and the concept of entropy was given a statistical interpretation.[6] Low entropy was shown to be associated with nonrandom, highly ordered arrangements, which are relatively improbable, whereas high entropy is associated with random, unordered arrangements, which are relatively probable. If, for example, we have a container with 80 molecules in it—40 oxygen molecules and 40 nitrogen molecules—the arrangement would be highly ordered and nonrandom if all of the oxygen molecules happened to be in the left half of the container and all of the nitrogen molecules happened to be in the right half. The probability, incidentally, of the molecules spontaneously sorting themselves out in this way in the course of their random motions is approximately 10^{-24}. Arrangements in which about half of the oxygen molecules are in one-half of the box and about half in the other (and similarly for the nitrogen molecules) are disordered, random, and vastly more probable (see Reif 1965, 4–14). We can now restate the problem of the design argument in terms of these thermodynamic considerations. If (as is admittedly problematic) we allow ourselves to speak of the entropy of the entire universe, it appears that the order exhibited by the universe can be described by saying that the entropy of the universe is now relatively low, and it has been even lower in the past. As we have seen, to say that the entropy is low is tantamount to saying that the universe contains large stores of available energy. The question of creation can be posed again. The universe is a physical system in a relatively low entropy state; is there strong reason to claim, by virtue of this fact, that it must have been produced by intelligent design?

According to the modern statistical interpretation of the second law of thermodynamics, the entropy in a closed physical system is very probably high. It is not, however, impossible for the entropy of such a system to decrease as a result of a mere statistical fluctuation—though such occurrences are exceedingly improbable. It is also possible for a closed physical system to be in a state of low entropy as a result of a recent interaction with its environment. This means, of course, that the system has not been closed for an indefinite period in the past—it has not always been isolated but has, instead, interacted with other parts of the world. For example, a thermos bottle containing tepid water with ice cubes floating in it is in a low-entropy state. In the course of time the ice cubes will melt, the water will cool, and the whole system inside the insulated container will arrive at a uniform temperature. This is a state of higher entropy. Upon finding a thermos containing liquid in the lower of the two states, we would infer without hesitation that it had recently been put into that state by an outside agency—in this case, a person who had removed the ice cubes from a refrigerator and deliberately placed them in the thermos with the water. In this instance, the low-entropy state was a result of an interaction with an intelligent planner who put the ice cubes into to water to fulfill a conscious purpose. Low entropy states that are the results of interaction with the environment do not always involve human intervention. A hailstorm on a summer day may deposit pieces of ice in the lukewarm water of a swimming pool.[7]

It is, I suspect, quite desirable to reformulate the problem of intelligent design in terms of the concept of entropy, rather than the vague concept of order. There will not be as strong a temptation to beg the question by identifying low entropy with conscious design as there was to do so by identifying—almost by definition— order with purpose and "the adjustment of means to ends." We may, once more, look around the world, surveying the physical systems that come into being in low-entropy states, to ascertain what percentage of them are created with conscious design. The results will be similar to those already discussed. An exceedingly small proportion of low-entropy systems—that is, systems that are highly organized and orderly—results from an interaction with the environment that involves any con-scious purpose or design. There is no need to repeat the whole story.

7. Modern Cosmology

In our day, unlike Hume's, we can claim some physical knowledge about the evolution of the universe.[8] There is strong reason to believe that, sometime be-tween 10 and 20 billion years ago, the universe consisted of a compact concen-tration of energy which exploded with incredible violence. This "big bang" conception of the universe is supported, not only by the observed red shifts of light from distant galaxies, but also by the more recently discovered cosmic microwave background radiation. It seems likely that statistical fluctuations in the rapidly expanding fireball gave rise to stable inhomogeneities in the distribution of matter, and that these inhomogeneities, by gravitational attraction of neighboring matter, led to the formation of galaxies. Further concentrations, within the galaxies, led to the formation of stars. Our universe contains roughly 10 billion such galaxies, each of which is an inhomogeneous concentration of energy in the vast expanses of the universe. Each galaxy contains, on the average, perhaps 10 billion stars. These again are inhomogeneous concentrations of energy. Can we not say that we have just noted roughly 100 billion billion systems that came into being in states of low entropy? Where in the annals of human history can we find like numbers of systems created in low-entropy states by conscious human intervention? We be-lieve, moreover, that we have some understanding of the processes by which atoms are formed in the interiors of stars. The earth alone contains on the order of 10^{50} atoms—each a highly organized Low-entropy system—each, to the best of our knowledge, brought into being without conscious intent. Not only do the low-entropy systems created without intelligent purpose outnumber by orders of magnitude those that involve intelligent design, but also the orders of magnitude are greater by orders of magnitude!

But what of the universe itself—that unique system whose mode of creation is at issue? We cannot, at present, say much. It seems possible that our universe will continue to expand for many more billions of years, its rate of expansion continually declining, until eventually it begins to contract in a reverse process, which will inevitably lead to a "big crunch." Perhaps it has a cyclic history of repeated expansion and contraction ad infinitum. This hypothesis seems unlikely as things now stand, for there does not seem to be enough matter in the universe to bring the expansion to a halt and induce a subsequent process of contraction, Perhaps it will simply continue expanding forever. It is difficult to draw any conclusions with much confidence. Perhaps—as Philo might say—the universe was created by a juvenile deity, who, on a supercosmic holiday, set off a supergigantic firecracker. Or, perhaps our universe is a black hole in some incomprehensibly larger universe.

8. Assessment of the Hypothesis

What morals should we draw from our analysis of the design argument? I think we should say that the attempt to deal with the question of God's existence as a scientific hypothesis is quite successful. We have considered the hypothesis that the universe is the product of intelligent design, and we have seen that a very rough estimate, at least, can be made of its probability by applying Bayes's theorem. The result is, of course, disappointing to the theist (or deist), for the conclusion is that the hypothesis is quite *improbable*. This conclusion can, I believe, be drawn unequivocally from the kinds of evidence that Philo adduces in Hume's *Dialogues*, and more recent scientific developments tend only to reinforce this result, pushing the posterior probability of intelligent design even closer to zero. If scientific evidence is relevant, it tends to disprove the existence of God as an intelligent designer of the universe.

Compare this result with a legal situation. Suppose that a suspect in a crime is defended by an attorney who can show the following:

(i) It is antecedently improbable that the client committed the crime since the suspect had no motive.

(ii) If his client had committed the crime, the clues found at the scene of the crime would very probably have been altogether different.

(iii) Someone else, who had motive and opportunity to commit the crime, would very probably have left just the sort of traces which were found at the scene of the crime, if this other person had committed the crime.

Under these circumstances, only an egregious miscarriage of justice could lead to the conviction of the client.

Or suppose that a scientist has an explanatory hypothesis for some phenomenon. Upon investigation, it turns out that

(i) This hypothesis is antecedently implausible because it is in direct conflict with a large body of well-established theory.

(ii) It makes the occurrence of the facts to be explained quite improbable if it is true.

(iii) There is a plausible alternative hypothesis that makes the facts to be explained highly probable.

Under these circumstances, only gross prejudice would make a scientist retain such a hypothesis.

Hume's *Dialogues* constitute, in my opinion, a logical, as well as an artistic, masterpiece. With consummate skill, he put together the diverse elements of a complex argument. It is a striking tribute to his logical acumen that he seemed intuitively aware of the structure of scientific arguments whose explicit characterization is essentially a product of the twentieth century.

9. The Relevance of Scientific Evidence

In this chapter, I have tried to carry out what I consider a legitimate philosophic enterprise. Using materials from Hume's *Dialogues*, I have tried to give a logical analysis of one of the historically and systematically important arguments for the existence of God. Although many such analyses of the ontological and cosmological arguments can be found in the philosophical literature, I am not aware that the same type of analysis of the design argument has been undertaken.[9] The reason may lie in the fact that deductive logic seems to enjoy greater popularity among philosophers than does inductive logic. Whatever the reason for the neglect of the design argument, there is a very significant difference between the outcome of its analysis and the outcomes of the analyses of the others. If an analysis of one of the a priori arguments reveals a fallacy, the most severe consequence is merely that the argument in question may have to be abandoned. Discovery of a fallacy is not apt to transform a putative argument for the existence of God into a valid argument for his nonexistence. Our analysis of the design argument, in contrast, does tend to show that there is no intelligent creator (although it is admittedly irrelevant to other theological hypotheses).

The main reaction of those who have placed some stock in the design argument—if they admit that the foregoing analysis has any validity at all—is likely to be the contention that the design argument should be abandoned and that all such considerations are irrelevant to the existence or nonexistence of God. This position is, I believe, untenable. The result of the analysis is *not* to show that all conceivable empirical evidence is irrelevant to the theological hypothesis. Cleanthes makes this point clear in the third dialogue when he offers a new twist on the design argument:

> Suppose, therefore, that an articulate voice were heard in the clouds, much louder and more melodious than any which human art could ever reach; suppose that this voice were extended in the same instant over all nations and spoke to each nation in its own language and dialect; suppose that the words delivered not only contain a just sense and meaning, but convey some instruction altogether worthy of a benevolent Being superior to mankind could you possibly hesitate a moment concerning the cause of this voice, and must you not instantly ascribe some design or purpose? (34)

Philo does not deny that such an inference would be appropriate in this case. If such evidence did obtain, he seems to concede, it would constitute grounds for believing in an intelligent source. Cleanthes maintains that the argument from design, as originally stated, is just as strong as this. Although Philo leaves the matter hanging there, such a view is unsupportable. The hypothetical evidence just offered involves meaningful linguistic performances of an intricate sort. In our experience, complex linguistic performances take place rather frequently, and they *invariably* arise from intelligent design. When, moreover, such a performance comes from an inanimate object, such as a radio or a phonograph, we do have overwhelmingly strong experiential evidence to support the claim that it can be traced back to an intelligent source. There is conceivable evidence that does not happen to obtain, but which, if it did obtain, would support the hypothesis of a superhuman intelligence. Empirical evidence is not, ipso facto, irrelevant.[10]

Karl Popper (1963) has argued, with considerable justification, that the essence of the scientific approach is to put forth bold conjectures that can be tested and that are capable of being falsified by negative evidence. It is decidedly unscientific to frame hypotheses in a manner that makes them appear confirmable by positive evidence but which, at the same time, renders them immune to falsification by possible negative evidence. In the field of theology, this scientific approach is illustrated by a passage from the Bible:

> Elijah drew near to all the people and said, "How long will you hobble on this faith and that? If the Eternal is God, follow him; if Baal, then follow him." The people made no answer. Then Elijah said to the people, "I, I alone, am left as a prophet of the Eternal, while Baal has four hundred and fifty prophets. Let us have a couple of bullocks; they can choose one bullock for themselves and chop it up, laying the pieces on the wood, but putting no fire underneath it; I will dress the other bullock and lay it on the wood, putting no fire underneath it. You call to your god, I will call to the Eternal, and the God who answers by fire, he is the real God."
>
> "All right," said the people.
>
> So Elijah told the prophets of Baal, "Choose one bullock for yourselves, and dress it first (for you are many), calling to your god, but putting no fire underneath."
>
> They took their bullock, dressed it, and called to Baal from morn to midday, crying, "Baal, answer us!" But not a sound came, no one answered, as they danced about the altar they had reared.
>
> When it came to midday, Elijah taunted them: "Shout," he told them, "for he is a god!" So they shouted, gashing themselves with knives and lances, as was their practice, till the blood poured over their bodies. After noon they raved on till the hour of the evening sacrifice; but not a sound came, there was no one to answer them, no one to heed them.
>
> Then said Elijah to all the people, "Come close to me." All the people came close to him, and he repaired the altar of the Eternal which had been broken down, making a trench around the altar about the space of eighteen hundred square yards. He arranged the wood, chopped up the bullock, and laid the pieces on the wood. "Fill four barrels with water," he said, "and pour them over the sacrifice and over the wood."

"Do it again," he added, and they did it again. "Do it a third time," he said, and they did it a third time, till the water flowed round the altar. He also filled the trench with water.

Then at the hour for the evening sacrifice Elijah the prophet came forward. "Oh Eternal, God of Abraham and Isaac and Israel," he cried, "this day may it be known that thou art God in Israel and that I am *thy* servant, that I have done all this at thy bidding. Hear me, Oh Eternal, hear me, to let this people know that thou the Eternal art God and that thou has made their minds turn to thee again."

Then the Eternal's lightning fell, burning up the sacrifice, the wood, the stones, and the dust, and licking up the water in the trench. At the sight of this, all the people fell on their faces, crying, "The Eternal is God, the Eternal is God!" (*Kings I*, 18:21–40; Moffatt translation).

This, I submit, is an instance in which a religious hypothesis was formulated and put to the test. Cleanthes spoke of "experimental theism," but he proposed no experiments. Elijah practiced real experimental theism. There is, alas, a concluding unscientific postscript: " 'Seize the prophets of Baal,' said Elijah, 'let not a man of them escape.' They seized the prophets, and Elijah, taking them down to the brook Kishon, killed them there."

If a scientific approach to the propositions of theology is to be adopted, the evidence must be considered, whether it be favorable or unfavorable. Negative evidence, as Francis Bacon ([1620] 1889) so forcefully reminds us, cannot be ignored.

> The human understanding when it has once adopted an opinion (either as being the received opinion or as being agreeable to itself) draws all things else to support and agree with it. And though there be a greater number and weight of instances to be found on the other side, yet these it either neglects and despises, or else by some distinction sets aside and rejects; in order that by this great and pernicious predetermination the authority of its former conclusions may remain inviolate. And therefore it was a good answer that was made by one who when they showed him hanging in a temple a picture "of those who had paid their vows as having escaped shipwreck," and would have him say whether he did not now acknowledge the power of the gods. "Aye," asked he again, "but where are they painted that were drowned after their vows?" (aphorism xlvi)

Appendix: Hume's Intentions

Up to this point in the discussion, I have confined my attention to an analysis of the arguments presented in the *Dialogues* and scrupulously avoided the historical question of Hume's own assessment of their merits. As I have said, it seems to me that Philo's objections completely undermine the design argument, as presented by Cleanthes; indeed, Philo's arguments render it highly improbable that the universe is a result of intelligent design. Hume constructed the *Dialogues* with great care, and I am unable to believe that the force of these considerations escaped him. Since other commentators have, however, interpreted Hume differently, I shall now attempt to defend my interpretation.[11]

Three assumptions seem fully justified:

(1) Hume cared deeply about his *Dialogues*. He wanted them to be read and taken seriously. He had ample reason to believe that this aim would be frustrated if he were to present the *Dialogues* as a dispute between a theist and an atheist in which the atheist mops up the floor with the theist. From the outset, therefore, in the prologue, "Pamphilus to Hermippus," he has everyone insisting that atheism is absurd, that there can be no dispute about the existence of God, only about his nature. By presenting the discussion in this guise, he may well have felt that there was some chance that the *Dialogues* would receive some serious attention and not be rejected out of hand as sheer infidel villainy.

(2) Hume was a superb stylist who fully appreciated the importance of literary unity. It would have been poor form to have Philo begin by asserting his unshakable faith in the existence of God, and then end up disavowing that belief. Furthermore, if Hume had allowed that to happen, word would soon have made the rounds that the pious opening meant nothing, and that the atheist won out in the end. This surely would have defeated Hume's purpose.

(3) Hume was a clear-headed philosopher. It is not antecedently likely that he could construct arguments sufficient to demolish the design argument without knowing what he was doing.

The most pressing question we now face is what to make of Philo's astonishing avowal of belief in the design argument in the final dialogue. Let us look at the considerations Philo himself advances. He makes three statements. In each case, it seems to me, we have previously been given adequate grounds for rejecting Philo's reason. First, Philo poses the following question to a hypothetical atheist:

> I would ask him: Supposing there were a God who did not discover himself immediately to our sense, were it possible for him to give stronger proofs of his existence than what appear on the whole face of nature? What indeed could such a Divine Being do but copy to present economy of things, render many of his artifices so plain that no stupidity could mistake them, afford glimpses of still greater artifices which demonstrate his prodigious superiority above our narrow apprehensions, and conceal altogether a great many from such imperfect creatures? (109)

Philo knows quite well what such a being could do; he could do precisely what Cleanthes described in the third dialogue when he offered the example of the voice from heaven. When one reads that third dialogue, it seems strange that that example is introduced and then dropped, virtually without discussion. Cleanthes offered his new version of the design argument with the claim that his earlier version provides as much evidence for the existence of God as would the voice from heaven were it to occur. Is Hume's Philo, the clever skeptic, incapable of concocting an argument to refute this contention of Cleanthes? Considering the ingenuity of some of Philo's other objections, it hardly seems likely. I should like to suggest, rather, that the new twist on the design argument was offered, partly at least, as a prior refutation of Philo's declaration of belief quoted above. It stands as an undisputed answer to the question, what sort of empirical evidence would lend strong support to the theistic hypothesis if it were available?

Shortly thereafter, Philo expresses skepticism about the possibility of a genuine agnosticism:

> So little ... do I esteem this suspense of judgment in the present case to be possible that I am apt to suspect there enters somewhat of a dispute of words into this controversy, more than is usually imagined. That the works of nature bear a great analogy to the productions of art is evident; and, according to all the rules of good reasoning, we ought to infer, if we argue at all concerning them, that their causes have a proportional analogy Here, then, the existence of a Deity is plainly as-
> certained by reason; and if we make it a question whether, on account of these analogies, we can properly call him a *mind* or *intelligence*, notwithstanding the vast difference which may reasonably be supposed between him and human minds; what is this but a mere verbal controversy? (110–111)

Hume has, as I said above, put devastating arguments into Philo's mouth; it is simply incredible to suppose that he underrated or forgot them. On repeated occasions in the earlier dialogues, Philo mercilessly attacked the supposed analogy on which the design argument rests. To be sure, Philo makes light of his objections from time to time, as when he remarks that "the beauty and fitness of final causes strike us with such irresistible force that all objections appear (what I believe they really are) mere cavils and sophisms, nor can we then imagine how it was ever possible for us to repose any weight on them" (92). Hume cannot really believe that they have so little force. Throughout the earlier dialogues, Philo plays with Demea and Cleanthes, and this is part of the game—stringing them along, so to say. Hume must expect the discerning reader to see that the objections cannot be so easily dismissed.

Philo then makes a declaration regarding God's moral qualities:

> And here I must also acknowledge, Cleanthes, that, as the works of nature have a much greater analogy to the effects of *our* art and contrivance than to those of *our* benevolence and justice, we have reason to infer that the natural attributes of the Deity have a greater resemblance to those of men than his moral have to human virtues. But what is the consequence? Nothing but this, that the moral qualities of man are more defective in their kind than his natural abilities. For, as the Supreme Being is allowed to be absolutely and entirely perfect, whatever differs most from him departs the farthest from the supreme standard of rectitude and perfection. (113)

In the preceding dialogue, Philo argued invincibly that, on the basis of the design argument, no conclusion can be drawn attributing moral qualities to God. Now he turns around and says that God is "allowed to be absolutely and entirely perfect." On what basis is this allowed? Certainly not by virtue of the design argument or any other empirical evidence. God's perfection is one of the tenets of traditional theism, which can be held only as a dogma. Such perfection must be accepted, if at all, as an incomprehensible aspect of God's nature. But in the preceding dialogue, both Cleanthes and Philo reject such incomprehensibilities as tools used in "ages of stupidity and ignorance" to "promote superstition" (105). Surely Philo is not sincerely embracing this brand of dogmatism.

These are the considerations Philo offers in support of his "unfeigned sentiments" (114) on the existence and nature of God. Each one, it seems to me, has been totally undercut by strong reasons previously presented in the *Dialogues*. I find it hard to believe that Hume expected his readers to miss this point and to suppose that Philo is *now* expressing his genuine convictions. And if that were not enough, at the very end of the last dialogue Philo *again* (as I shall shortly show) rejects *every one* of these considerations.

If Philo has really believed in the design argument all along, we must suppose that his earlier attacks upon it were made essentially in jest. But be that as it may, Philo has offered *arguments* against natural religion; whether he was being flippant or serious, we must look at the arguments on their own merits. As I have tried to show in this chapter, they cannot be dismissed as mere "cavils and sophisms." I believe Hume expected his readers to pay heed to the arguments and not to be misled by Philo's disclaimers.

From the initial reference to his "careless skepticism" (5), Philo is painted as a shifty character. His protestations of sincere belief are not to be trusted. On several occasions earlier in the *Dialogues*, he had pretended to agree with Demea, but as it turned out later, the agreement was not what it seemed to be (at least to Demea). Hume makes sure the reader does not miss this point: "'And are you so late in perceiving it?' replied Cleanthes. 'Believe me, Demea, your friend Philo, from the beginning, has been amusing himself at both our expense …'" (105). Obviously offended, Demea shortly thereafter departs.

We know by the time we get to the final dialogue that, when Philo adopts a stance, he does so for a reason. What is the point of the one he assumes at the beginning of the final dialogue? Partly, I think, it is simply to carry through to the end the pretense that the *Dialogues* constitute a discussion of the attributes of God — his existence being beyond question — not a dispute between a theist and an atheist. More important, I believe, the purpose is to set the stage for the coup de grâce.

The twelfth dialogue does not end with these pieties; they occupy less than half of it. Philo follows them immediately with a semicandid statement of his aim: "But in proportion to my veneration for true religion is my abhorrence of vulgar superstition …" (114). The remainder of the dialogue is devoted to a discussion of the value of religion. Philo argues eloquently that religion, at least in the forms in which it has always been promulgated, has no social, political, psychological, or moral value. At best, it has no effect; at worst, its effects are positively pernicious. Philo's strategy in the twelfth dialogue seems to be this. Suppose we put aside, he says, all of the arguments of the preceding dialogues as so much worthless logic-chopping. Suppose, flying in the face of reason and experience, we claim that we simply cannot help embracing the theistic standpoint. Will the effects be beneficial? The answer is a resounding no. Religion is not even a useful fiction. In contrast, "the smallest grain of natural honesty and benevolence has more effect on men's conduct than the most pompous views suggested by theological theories and systems" (115).

Although Philo has little trouble assembling an impressive array of evil results of religion, he does admit that "true religion has no such pernicious consequences," but, he hastens to add, we must treat of religion as it has commonly been found in the world" (p. 118). And what is this "true religion"? Nothing that has ever been embodied in any institutionalized religion. It is a rarified thing, given to only a few persons:

> If the whole of natural theology... resolves itself into one simple, though somewhat ambiguous, at least undefined, proposition, *That the cause or causes of order in the universe probably bear some remote analogy to human intelligence* — If this proposition be not capable of extension, variation, or more particular explication; if it affords no inference that affects human life, or can be the source of any action or forbearance; and if the analogy, imperfect as it is, can be carried no further than to the human intelligence, and cannot be transferred, with any appearance of probability, to the other qualities of the mind; if this is really the case, what can the most inquisitive, contemplative, and religious man do more than give a plain philosophical assent to the proposition, as often as it occurs, and believe that the arguments on which it is established exceed the objections which lie against it? (122–123)

Even so, he adds that

> the most natural sentiment which a well-disposed mind will feel on this occasion is a longing desire and expectation that heaven would be pleased to dissipate, at least alleviate, this profound ignorance by affording some more particular revelation to man-kind, and making discoveries of the nature, attributes, and operations of the divine object of our faith. (123)

Notice that, in these passages, Philo retracts all three of the points on which he based his acceptance of the design argument at the beginning of the final dialogue: (1) that one could not imagine stronger evidence for God's existence, (2) that the existence of a deity is plainly shown by reason, and (3) that the moral attributes can be allowed to apply to the Creator. Recalling earlier dialogues, Philo seems to be reiterating his conviction that, if there were an intelligent, powerful, benevolent God, surely he would provide more unambiguous evidence.

Hume's assessment of the design argument is, it seems to me, quite easy to discern. The argument is logically untenable; the evidence adduced to support the theistic conclusion is hopelessly inadequate. In one of the few substantive footnotes, where Hume is presumably speaking for himself (not through the mouth of Philo or Cleanthes), he does, however, indicate that our belief is sometimes compelled in spite of reason:

> No philosophical dogmatist denies that there are difficulties both with regard to the senses and to all science, and that these difficulties are, in a regular, logical method, absolutely insolvable. No skeptic denies that we lie under an absolute necessity, not withstanding these difficulties, of thinking, and believing, and reasoning, with regard to all kinds of subjects, and even of frequently assenting with confidence and security. The only difference... is that the skeptic, from habit, caprice, or inclination insists most on the difficulties, the dogmatist, for like reasons, on the necessity. (113)

Clearly, Hume has little regard for the dogmatist. Demea represents that point of view and comes off badly in the *Dialogues*. If, however, a philosophical skeptic finds himself incapable of dispensing with religious belief—not because of any argument, for the arguments are all inadequate, but because of his natural inclinations—then what should he do? The latter half of the last dialogue has a clear message, namely, to do everything one can to minimize the intellectual content and practical import of that belief. Religious conviction may, for some, be unavoidable; that does not make it desirable.

"A person, seasoned with a just sense of the imperfections of natural reason, will fly to revealed truth with the greatest avidity," says Philo in his summation, "while the haughty dogmatist, persuaded that he can erect a complete system of theology by the mere help of philosophy, disdains any further aid and rejects this adventitious instructor" (123). This statement suggests that the believer faces a dilemma: either systematic theology or revealed religion.

The dogmatic alternative has been found repugnant; but what, precisely, is the other? If "revealed truth" is taken to refer to Christian Scriptures or holy writ of any other established religion, this alternative seems equally undesirable, for it smacks of institutionalized religion, which has been thoroughly condemned. Either way, it seems, the consequences of belief are fervently to be avoided. Perhaps, however, revealed truth is not to be found in religious writings but rather in the world around us. In that case, what is the truth that is revealed? Hume's answer, I believe, is clear: there is no *reason* on earth to believe in an intelligent, powerful, benevolent Creator.

"To be a philosophical skeptic is, in a man of letters, the first and most essential step towards being a sound, believing Christian," Philo concludes. This passage is often cited by those who wish to maintain that Hume was himself, at bottom, a believing theist. I do not believe that such an interpretation is justified. What is Philo supposed to say to Cleanthes, the teacher of Pamphilus, who is himself a believing Christian? Should he say that philosophical skepticism is to be recommended because it is the first and most essential step to becoming a solid atheist?

Hume would have agreed, I believe, that philosophical skepticism is the first and most essential step toward sound religious convictions. It is the only antidote to "sophistry and illusion." Philo could genuinely recommend philosophical skepticism to the young Pamphilus, in the hope that its cultivation would lead him closer to the truth. At the same time, Hume may have been less than fully optimistic about the degree to which philosophic argument is efficacious, for in the end, Pamphilus finds Cleanthes more persuasive than Philo.

Hume thought profoundly about natural religion, and he wanted to convey the fruits of that intellectual labor to a certain audience. He knew that his ideas would be almost universally rejected and despised unless they were suitably "packaged." The wrappings had to be attractive in order to make them appear palatable at first glance to the casual observer and to give no hint of offense. At the same time, the 'product' must not be completely concealed by the wrapper. In his "most artful" philosophical work, Hume presented his devastating critique of natural religion. He wrapped it in cellophane (so that the message could be discerned), but he used red cellophane (so that its character would not be too obvious). Careful scrutiny, I think, leaves no doubt about the moral of the story.

The "Almost-Deduction" Theory

Although the preceding chapters have focused on the theorem of Bayes, we should recognize that it is not the only source of useful information concerning inductive reasoning provided by the calculus of probability—though we shall be brought back to prior probabilities in the end. For example, the theorem on total probability,

$$P(C|A) = P(B|A)\ P(C|A \cdot B) + P(\sim B|A)\ P(C|\sim B \cdot A) \tag{1}$$

has important consequences regarding transitivity (see Salmon 1965a); indeed, it is *the* fundamental transitivity relation among probabilities. It establishes a probabilistic relationship between A and C via an intermediary B. In the case of probability, however, in contrast to deductive implication, we must take account of the negation of B.

1. Transitivity

Let us use the simple arrow ($\rightarrow$) to represent deductive implication and the double arrow ($\Rightarrow$) for the relation of high probability. It is well known that deductive implication is transitive, that is,

$$(A \rightarrow) \ \& \ (B \rightarrow C) \rightarrow (A \rightarrow C) \tag{2}$$

What about the relation of high probability? The example is

If X smokes crack then very probably X previously smoked marijuana.
If X smokes marijuana then very probably X previously smoked cigarettes.

Therefore, if X smokes crack, probably X previously smoked cigarettes.

This example may look reasonable enough, but consider the following case:

> If X is a scientist then very probably X is alive today.[1]
> If X is alive today then very probably X is a microorganism.

From these two premises it is obviously improper to conclude

> If X is a scientist then probably X is a microorganism.

We therefore see that

$$(A \Rightarrow B) \ \& \ (B \Rightarrow C) \text{ does not imply } (A \Rightarrow C) \tag{3}$$

It is essential to understand that in examples of this sort, even if the probabilities are extremely high, there is no approximation to logical validity. Such arguments are completely fallacious.[2]

In order to see this situation from a formal point of view, we need to look back at equation (1). Given that $P(B|A)$ is close to 1, it follows that $P(\sim B|A)$ is close to 0. Obviously, nothing in the premises of (3) gives any information about $P(C|\sim B \cdot A)$; hence it can be arbitrarily small, even 0. Therefore, the second term on the right in equation (1) cannot be guaranteed to contribute to the sum. We need to examine the first term. We know that the first factor, $P(B|A)$, is high and that $P(C|B)$ is high. However, a high value of $P(C|B)$ does not imply that $P(C|A \cdot B)$ is high; for example, given that the vast majority of people are shorter than 6 feet, it does not follow that the great majority of professional basketball players are less than 6 feet in height. In fact, if B is an infinite class, it is possible to have

$$P(C|B) = 1 \quad \text{and} \quad P(C|A \cdot B) = 0 \tag{4}$$

This fact reveals a fundamental disanalogy with deductive logic, in which

$$(B \rightarrow C) \rightarrow (A \cdot B \rightarrow C) \tag{5}$$

is a basic theorem.

There are two other cases that we must consider, that is, two mixed cases containing one relation of deductive implication and one relation of high probability. In the first case we ask,

> If we have $(A \rightarrow B) \ \& \ (B \Rightarrow C)$ does it follow that $(A \Rightarrow C)$? (6)

The answer is negative, as is shown by a counterexample:

> If X is an even prime number, then X is a prime number.
> If X is a prime number, then the probability that X is odd is 1.
> _____
> If X is an even prime number, then probably X is odd.

This is obviously invalid. For readers who feel uncomfortable with arithmetic examples, another counterexample can be provided.

If X is human then very probably X does not cut other people.
If X is a surgeon then X is human.

Therefore, if X is a surgeon then probably X does not cut other people.

From a formal point of view, if $(A \to B)$ the theorem on total probability reduces to

$$P(C|A) = P(C|A \cdot B) \tag{7}$$

because $P(B|A) = 1$ and $P(\sim B|A) = 0$. As we have seen, the fact that $P(C|B)$ is high does not imply that $P(C|A \cdot B)$ is high. In this case transitivity is not valid.

In the other mixed case we have the formula

$$(A \Rightarrow B) \,\&\, (B \to C) \to (A \Rightarrow C) \tag{8}$$

Given that $(B \to C)$, we have $(A \cdot B \to C)$ and $P(C|A \cdot B) = 1$; thus, from (1) we have

$$P(C|A) = P(B|A) + P(\sim B|A)\, P(C|\sim B \cdot A). \tag{9}$$

Since the second term on the right cannot be negative, this implies that

$$P(C|A) \geq P(B|A) \tag{10}$$

The result is that in one of the mixed cases transitivity holds; in the other it does not.[3]

There is a fundamental reason for this difference. If A constitutes our total evidence and A→B, then B will generally not constitute total evidence. Therefore, an inductive inference on the basis of A may be correct, but an inductive inference on the basis of B will not be because it violates the requirement of total evidence for inductive logic.

2. Contraposition and Modus Tollens

Leaving transitivity aside, let us consider the probabilistic analogue of contraposition. If one writes $P(B|A)$ and $P(\sim A|\sim B)$, there is no temptation to suppose that these two probabilities are equal. An example will reinforce this point. From the statement that the large majority of men are not physicists it does not follow that the vast majority of physicists are not men. As Reichenbach (1954; [1954] 1976, 131) has noted, if we consider the famous paradox of the ravens from a probabilistic point of view, we see that the paradox vanishes immediately.

Notwithstanding that occasions to use probabilistic contraposition do not often present themselves, there is a closely related type of inference that is extremely tempting. Let us call it "probabilistic modus tollens." For example, suppose that someone who is worried about having a particular disease consults a physician.

According to the physician there is a very effective test to determine whether one has that disease or not. The test is conducted and the result is negative. The physician reports the result and says that, if the patient had had the disease, the result would very probably have been positive. In this situation one would conclude with high probability that the patient does not have disease. The inference seems straightforwardly valid, but the situation is not so simple. Consider another example:

> If X is a woman, very probably she is not the mother of quintuplets.
> Mrs. Dionne was the mother of quintuplets.
> ───────────────────────────────
> Therefore [probably], Mrs. Dionne was not a woman.

Even though both of these inferences have the same form, the latter is clearly invalid.

This form of inference is extremely important because it *seems* to represent a common form of inference in statistics. For example, suppose one conducts a controlled experiment to check on the efficacy of a particular drug. The subjects in the experimental group receive the drug; the subjects in the control group receive a placebo. Suppose, furthermore, that there is a statistically significant difference between the results in the two groups. One would then say that probably the drug is efficacious, for if it were not, these results would have been extremely improbable. Ronald N. Giere (1979), in the first edition of his introductory textbook, *Understanding Scientific Reasoning*, characterized this form as *the* standard form for the elimination of scientific hypotheses.[4]

How are we to interpret this situation? A final example, already employed in chapter 4, may help. Suppose we have a coin that might be standard and fair or that might have two heads; for purposes of the example there are no other possibilities. We cannot examine the coin to learn to which type it belongs. Later we are reliably informed that in ten tosses it has come up heads every time. For a standard coin the probability of this outcome is less than 1 in 1000; therefore, we say that this coin has two heads. But what should we say if we found that this coin was selected at random from a bag containing 100,000 coins, all of which, with only one exception, are standard? In this case, it would be more probable that the coin is standard, as a simple calculation, using Bayes's theorem, will demonstrate.

What attitude should we take? Clearly it is necessary to take the prior probabilities into consideration. Reichenbach (1954; [1954] 1976, 30–31) drew the same conclusion from his discussion of probabilistic contraposition. He offered a necessary and sufficient condition for the correctness of the inference from $P(B|A) \approx 1$ to $P(\sim A|\sim B) \approx 1$. Suppose that

$$P(B|A) = 1 - \delta \quad \text{and} \quad P(\sim A|\sim B) = 1 - \varepsilon \tag{11}$$

Then

$$\varepsilon \leq \delta \quad \text{if and only if} \quad P(A) + P(B) \leq 1 \tag{12}$$

That is, the probability of the probabilistic contrapositive is greater than or equal to the probability of the original if and only if $P(A) + P(B) \leq 1$. Moreover,

$$\varepsilon = \delta \quad \text{if and only if} \quad P(A) + P(B) = 1^5 \tag{13}$$

We see, therefore, that probabilistic contraposition is a valid argument if the classes A and B are not too large. This condition reveals a method for the analysis of counterexamples to probabilistic modus tollens. In the example of the coin that may have two heads, the probability of obtaining ten heads in ten tosses with a coin selected at random from the bag is slightly greater than 1/1024 and the probability of drawing a normal coin from the bag is 99,999/100,000. The sum of these two probabilities is obviously greater than 1. Even though I have given precise numerical values in this example, we must recognize a qualitative lesson. In any situation in which we wish to use probabilistic modus tollens we must consider the size of the classes—that is, we must consider the prior probabilities. It is easy to see how this consideration reveals the fallacy in the example of Mrs. Dionne.

3. Incremental vs. Absolute Confirmation

In several of the foregoing chapters, especially chapter 7, I have tried to show how Bayes's theorem can provide a useful guide to informal inductive reasoning. In this chapter I have enlarged on the point by showing how other formal relations in the probability calculus could serve the same function. There remains, however, an ambiguity in the term "confirmation" that has sometimes been the source of serious problems. In the first meaning of the term we are concerned with high probability, which is the type of inductive support discussed in the present chapter. In the second meaning we are dealing with increase in probability; in cases of the application of Bayes's theorem this is the kind of relation that is usually involved. It is essential not to confuse them. In the context of scientific confirmation I call the first "absolute confirmation" and the second "incremental confirmation." In the discussion of the relations of transitivity, contraposition, and probabilistic modus tollens, we have dealt with the cases in which the aim is to establish a conclusion with high probability. In contrast, in the discussion of the hypothetico-deductive method and Bayesian conditionalization, the aim was to produce an increment in the probability; obviously, in general a new piece of evidence does not produce a high probability for the hypothesis under consideration. Carnap (1950b, chap. 6) has presented this distinction clearly, and in addition, he studied the basic formal characteristics of the relation of probabilistic relevance.[6]

Carnap (1950b) discovered many surprising facts concerning probabilistic relevance. For example, there are situations in which, given two new items of evidence, each one taken separately may confirm the hypothesis, but taken in conjunction they do not confirm the hypothesis but rather diminish its probability. In addition, Carnap has demonstrated how a new item of evidence can confirm a hypothesis and at the same time diminish the probability of another hypothesis that is a logical consequence of the first. I have shown how it is possible for a piece of evidence to refute deductively the conjunction of two hypotheses but to confirm each one of them taken separately. This possibility has interesting implications for the problem of Duhem. It is beyond the scope of this article to discuss probabilistic relevance in detail, for that is the main topic of chapter 12, where we can find many further lessons regarding informal reasoning that can be drawn from the calculus of probability.[7]

4. Conclusion

The main purpose of this chapter has been to demonstrate how a careful examination of formal and numerical probability relations can furnish useful information when applied to informal reasoning. I have not attempted to construct a formal calculus of qualitative probability; I have tried, instead, to show how the quantitative calculus can be extremely useful even in many cases in which precise numerical values of probabilities are not available. These indications can help us to avoid such fallacious arguments as probabilistic modus tollens, among others. It is my long-standing conviction that, even for relatively sophisticated people, intuitions about inductive and probabilistic reasoning are quite unreliable. Some of these intuitions arise, I think, from a conscious or unconscious feeling that inductive logic is strongly analogous to deductive logic — a belief that might be called "the almost-deduction view of induction." I have tried to overcome this notion by showing how deductive relations are in many cases completely destroyed by the replacement of logical implication by relations of high probability. Great care must be taken in drawing parallels between deduction and induction. This theme will be further reinforced in chapter 12.

The "Partial-Entailment" Theory

1. Introduction

Ever since the famous H_2O papers of the 1940s (Hempel 1943, 1945; Helmer and Oppenheim 1945; Hempel and Oppenheim 1945), Carl G. Hempel has been closely associated with the concepts of confirmation and logical probability. His intriguing "paradoxes of confirmation" have brought forth an avalanche of responses, and more keep coming.[1] Some of the conditions of adequacy he enunciated, especially the "equivalence condition," are central to current controversies (see, e.g., Wallace 1966; Hanen 1967; Smokler 1967). Moreover, he has, in the meantime, continued to make significant contributions to inductive logic (Hempel 1960, 1966), as well as to such closely allied topics as the nature of theories and inductive (statistical) explanation (Hempel 1962a; 1965b, pts. III and IV).

In the early days, Helmer-Hempel-Oppenheim maintained that the qualitative concept of confirmation, as well as the quantitative concept of degree of confirmation, could be characterized in a purely syntactical manner. Hempel, at least, has subsequently been convinced by Nelson Goodman's (1951) "grue-bleen paradox" that a purely syntactical definition is impossible (Hempel "Postscript" 1965b, 50). Although he points out that Goodman's own solution via "entrenchment" is obviously pragmatic, Hempel does not commit himself to the thesis that confirmation and degree of confirmation are inescapably pragmatic. Personally, I think is possible to provide a straightforwardly semantical solution to Goodman's paradox by imposing certain conditions upon the interpretation of primitive descriptive predicates.[2] In this way we can avoid making confirmation a pragmatic relation. If such an approach to Goodman's paradox is viable, it leaves open the possibility that confirmation and degree of confirmation are logical relations, in the broad sense of "logical" that encompasses both syntactical and semantical considerations. This is the sense of "logical" that will be employed in the following discussion.

The purpose of this chapter is to examine the concept of logical probability with a view to considering its suitability as a fundamental logical relation upon which to ground inductive logic. A good deal of attention will be focused upon the relation of partial entailment that has sometimes been used to explain the concept of logical probability. The basic thesis I wish to examine is that the logical status of induction can be understood in terms of a strong analogy with deduction, according to which inductive logic rests upon a relation of partial entailment, just as deductive logic rests upon a relation of full logical entailment. Taking deductive entailment with its correlative logical independence as a point of departure, I shall show how a notion of partial entailment arises naturally within the context of duductive logic, and I shall introduce a relation of complete independence, quite distinct from logical independence, which obtains when there is neither full (deductive) entailment nor partial entailment.[3] As a matter of fact, when it comes to providing formal definitions, it will be easier to define complete independence first. This relation can then be used to help clarify partial entailment. In the course of the discussion, the concept of (semantic) content, which at first blush might seem to hold promise of providing a simple basis for the definition of complete independence, will be considered. I shall claim that the concept of content has no particular utility along these lines, but our discussion of complete independence will lead to a useful concept of independence of content.

In the end I shall be arguing that full clarity on the foregoing concepts, and the relations among them, militates against taking logical probability (or any other logical relation) as the basic relation for inductive logic. The main difficulty is that the notion of partial entailment arising naturally out of an examination of deductive relations is radically incongruous with the relation of partial entailment needed to serve as the fundamental relation of inductive logic. (This point was set out sketchily in Salmon 1967b, 69ff.) This is not to say that it is impossible to define an alternative concept of partial entailment for the purposes of inductive logic—clearly it is possible to do so—but the incongruity removes all plausibility from the analogy between deduction and induction as logical systems based upon the relations of full and partial entailment, respectively. If there is serious question about the status of the logical relation upon which inductive logic is supposed to rest—and how can there fail to be?—it is not answered by this analogy. I shall be arguing, as it develops, that the incongruity between the two distinct relations of partial entailment—that arising naturally from deductive logic and that needed for inductive logic—rests upon the extremely fundamental and sometimes unnoticed distinction between degree of confirmation and degree of relevance (Carnap's 1962b "concepts of firmness" and "concepts of increase of firmness").

This chapter is, I must hasten to emphasize, largely an exercise in heuristics. The concept of partial entailment has not, I believe, been used directly in the explication of the relation of logical probability upon which inductive logic is to be erected but rather in the course of attempts to clarify the explicandum. In this role it has borne a large part of the burden of lending plausibility to the whole enterprise of attempting to ground inductive logic on a logical relation. It is my conviction that a close examination of the salient concepts removes the plausibility of this undertaking, and it lends much greater plausibility to the view that the only relations upon which inductive logic can be said to rest are factual. The outcome, I believe, is entirely Humean: The unobserved is logically independent of the observed, and this means that neither deductive nor inductive logical relations provide the bridge from the observed to the unobserved needed to ground inductive logic.

2. The Languages L_N

Inasmuch as Carnap (1950b, 1962b) has provided precise foundations for both deductive and inductive logic, I shall utilize his terminology and his apparatus.[4] In fact, I shall confine my attention to his finite languages L_N. These have the structure of first-order applied functional calculi with identity, and each has a finite number of primitive predicates and a finite number N of individual constants. Such languages are interpreted over finite domains of N individuals. The restriction to such simple languages is a severe curtailment of generality, but it will still be possible to elicit the fundamental issues.

In the languages L_N, an *atomic statement* is one that consists of a primitive predicate followed by an appropriate number of individual constants. Any truthfunctional compound of atomic statements is a *molecular statement*. A *basic statement* is an atomic statement or the negation of an atomic statement. An atomic statement together with its negation is a *basic pair*. A *state description* is a conjunction that contains as conjuncts one and only one member of each basic pair. For the languages L_N the notion of a statement *holding in* a state description can be defined precisely (Carnap 1950b; 1962b, sec. D 18–4, 78ff.). The *range $R(p)$* of a statement p in L_N is the set of all state descriptions of L_N in which p holds. Since such relations as entailment, consistency, and incompatibility are defined in terms of ranges, a flagrant circularity would arise if they were used to *define* "holding in," but it is an immediate consequence of the definitions that a statement holds in any state description with which it is compatible, and it does not hold in any state description with which it is inconsistent. The range of a selfcontradictory statement is null, but all consistent statements have nonempty ranges. The basic logical concepts, such as L-truth, L-implication, L-equivalence, L-exclusiveness, and L-disjunctiveness, are defined in terms of ranges; for example, p is L-true if $R(p)$ is the set of all state descriptions, p L-implies q if $R(p)$ is contained in $R(q)$, and p and q are L-exclusive if $R(p \cdot q)$ is null (1962b, sec. 20). Any consistent statement can be expressed as (is L-equivalent to) the disjunction of all of the state description in its range (where a state description standing alone constitutes a disjunction with only one disjunct). The standard concept of entailment, I take it, coincides with L-implication; at any rate, I shall construe "entailment" in that way.

The languages L_N have essentially only a truth-functional structure, for every statement can be expressed as a truth function of atomic statements. Quantifiers can always be eliminated in favor of conjunctions or disjunctions.

3. Entailment and Independence

Since the main purpose of this chapter is to consider the possibility of grounding inductive logic in a logical relation between evidence and hypothesis, it seems reasonable to begin by considering the logical relations to be found in deductive logic, about which we have fairly clear conceptions. The central relation is that of entailment, in terms of which deductive validity can be understood. This relation provides the basis for correct deductive inference. If two statements do not have entailment relations between them, we say that they are *logically*—that is, deductively—*independent* of one another.

If inductive logic is to be grounded in a logical relation, it is natural to look for a relation that is somehow analogous to entailment to serve as a basis for correct inductive inference. Inductive logicians have felt that such a relation exists, and it has been called "partial entailment" or "partial implication" (Carnap 1950b; 1962b, sec. 55B). This relation is weaker than full logical entailment, but it does not constitute *complete* logical independence.

As a matter of fact, examination of the logical relations among statements does reveal a relation that looks like a plausible candidate for the title of "partial entailment." In order to clarify the notion of partial entailment, we shall need to introduce a concept of complete independence. There seem to be two extremes, logical entailment and complete independence, and between them lies the relation of partial entailment. Complete independence is a narrower concept than logical independence; logical independence obtains when there are no full entailment relations, whereas complete independence demands that there be no full or partial entailments. Definitions of "complete independence" and "partial entailment" arise very naturally within the context of deductive logic, and it is easy to find paradigms of these relations.

The trouble that beclouds this apparently happy situation is that the relation of partial entailment that arises naturally out of the general consideration of deductive logical relations is radically different from the concept of partial entailment that inductive logicians have offered. In fact, they are basically incompatible with one another. For one thing, they differ on the very admissibility of the relation of complete independence. The relation that arises from deductive considerations demands that some statements stand in the relation of complete independence; there are clear examples of partial entailments and clear examples of complete independence. The relation inductive logicians have supplied, by contrast, excludes the very possibility of the relation of complete independence ever obtaining.

The relation of complete independence will play a central role in the subsequent discussion for several reasons. First, as we have just seen, it is a focal point of difference between the two concepts of partial entailment. Second, it will actually prove convenient to define "complete independence" before "partial entailment." Third, and most important, in seeking a logical relation to underlie inductive logic, we are looking for a relation that embodies inductive dependence. By way of contrast, we shall need to be clear on what constitutes complete independence—that is, lack of any inductive or deductive dependency relation. As we shall see, the concept of logical independence can be characterized quite straightforwardly; the concept of complete independence is somewhat trickier. However, it is essential that we make a clear distinction between the two types of independence.

Let me now attempt to give some substance to these remarks by clarifying the main concepts with which we are concerned. The basic notion is logical entailment, which I take to be adequately explicated for present purposes by Carnap's concept of L-implication. Then we may say,

p and q are logically independent $=_{df}$ None of the following relations holds: p entails q, p entails $\sim q$, $\sim p$ entails q, $\sim p$ entails $\sim q$

In other words, two statements are logically independent unless a truth value of one of them determines a truth value for the other.

Suppose that p, q, r are three distinct atomic statements. In that case there will be no entailment relations between $p \cdot q$ and $q \cdot r$, but it is reasonable to say that there is some sort of logical dependency between them. Since they share a conjunct it seems appropriate to say that the relation is one of partial entailment. Furthermore, even in certain cases in which we do not have logical independence, it still seems natural to say that we have partial entailment, for we may have full entailment in one direction and partial entailment in the other. If p entails q but q does not entail p, we are inclined to say that q partially entails p, for q asserts part of what p asserts. One exception should perhaps be noted; if q is L-true it might be denied that q asserts anything at all, so we might be reluctant to say that it asserts part of what p does.

Since inductive logicians have often maintained that the truth of deductive consequences can lend inductive support to a hypothesis, the temptation is particularly strong to claim that q partially entails p whenever (but not necessarily only whenever) q is not L-true and p entails q. For instance, in L_3 let p be $(x)Fx$, which is equivalent to $Fa \cdot Fb \cdot Fc$, and let q be Fa. In this case also, q entails a conjunct of a statement L-equivalent to p.[5] In these cases we have found a deductive relation of dependency that falls short of full entailment, but which does not represent complete independence either. It is *one* plausible candidate for partial entailment.

Entailment relations can, as previously mentioned, be explicated in terms of ranges, that is, p entails q if and only if $R(p)$ is included within $R(q)$. If $R(p)$ overlaps $R(q)$ without being included in it, or if $R(q)$ is included within $R(p)$ but not conversely, it is sometimes said that p partially entails q. This is another concept of partial entailment that has seemed attractive to inductive logicians. Such relations among ranges are often presented diagrammatically by a device such as Euler circles, where the inclusion, exclusion, and overlap relations are shown pictorially (Carnap 1950b, 1962b, sec. 55B). At this point we must guard against an elementary confusion. When the ranges of p and q are mutually exclusive we do not have a relation of logical independence; rather, p entails $\sim q$ and q entails $\sim p$, so a full entailment holds in both directions. Thus, if we construe partial entailment as overlap of ranges, it follows immediately that between any two statements, going either way, there is either a full entailment or a partial entailment. The relation of complete independence, which is to consist in the absence of both full and partial entailments, never obtains.

These two incompatible concepts of partial entailment seem to arise because we have one picture of logical relations running from full entailment, through partial entailment, to complete independence, and another picture of relations among ranges extending from complete inclusion, through partial overlap, to

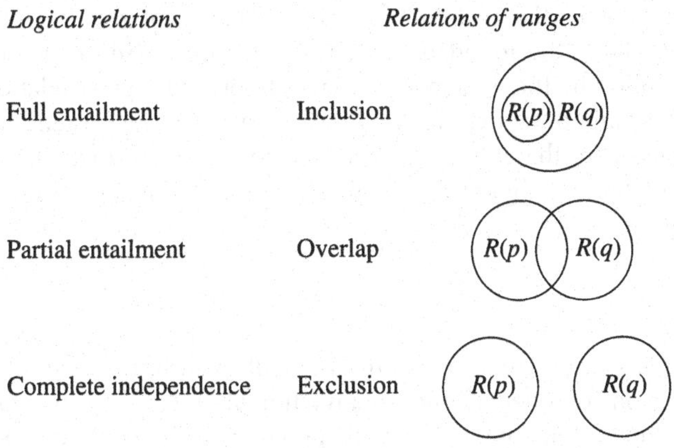

Logical relations   Relations of ranges

Full entailment   Inclusion

Partial entailment   Overlap

Complete independence   Exclusion

FIGURE 11.1 Two incongruous scales.

complete exclusion. Each of these pictures is appealing, but they are mutually incongruous (see figure 11.1). In particular, if we insist upon the possibility of complete independence, then we shall have to abandon the inductive logicians' approach that identifies partial entailment with overlap of ranges. Complete independence will have to be found in cases in which the ranges overlap, not in cases of mutually exclusive ranges. At this point it is not tempting to give up the concept of complete independence, for our paradigms point to three distinct relations: if p, q, and r are three distinct atomic statements, (i) $p \cdot q$ entails p, (ii) $p \cdot q$ partially entails $q \cdot r$, and (iii) p is completely independent of q. Hence, it is not clear that we should abandon the concept of partial entailment that allows room for complete independence.

In the sequel, I shall attempt to show how definitions of "complete independence" and "partial entailment" arise very naturally within the context of deductive logic. The trouble is that they are of no use in founding an inductive logic. Partial entailments of this sort fail to obtain just where we most need them in inductive logic. An inductive logic founded upon this concept of partial entailment would be transparently inadequate. I shall not argue, however, that this is the only possible definition of "partial entailment." As we have already seen, inductive logicians seem to have another concept in mind; indeed, this alternative becomes a vast set of alternatives when we consider the variety of ways in which overlap of ranges can be measured. Some of these alternatives appear to be far more suitable for inductive logic than is the concept that arises out of our consideration of deductive relations. However, if the concept of partial entailment is to provide a rationale for holding that inductive logic is based upon a logical relation, it must be a concept that arises naturally out of a consideration of logical relations. Otherwise the burden is reversed. If a relation of partial entailment suitable for inductive logic does not emerge from an examination of logical relations, but rather is extracted from a characteristically inductive concept such as degree of confirmation, then inductive logic must bear the burden of establishing the significance of the relation of partial entailment. Partial entailment can bear none of the burden of clarifying the foundations of inductive logic. But above all, we must avoid being lulled by the existence of a straightforward logical relation of partial entailment into supposing that we are thereby assured of the existence of a relation of partial entailment that can serve as a basis for inductive logic, for two distinct relations of partial entailment are involved, and they differ from one another in fundamental ways.

4. Content

As we have seen, the relation between statements with mutually exclusive ranges, far from being one of independence, is a relation of mutual logical entailment. It is initially tempting to suppose, however, that the concept of content will prove more useful than the notion of range in getting at the distinctions that concern us. For it certainly seems obvious, in the light of our simple paradigms of full and partial entailment, that the pairs of statements that sustain such relations share common content. Indeed, it is easy to suppose that these full or partial entailment relations obtain *by virtue of* shared content, so that lack of common content would characterize the relation of complete independence we have been seeking. I shall argue, nevertheless, that the concept of common content is actually of no use at all for this purpose, but in order to do so I shall, of course, need a reasonably precise account of the content of statements. Fortunately, one is available.

There are two notions of content that have strong preanalytic appeal.[6] First, it seems reasonable to think of the content of a statement p in terms of the statements p entails. In order to elucidate the content of a statement, one can exhibit its implications; indeed, we often characterize the very process of deductive inference as one that makes explicit the implicit content of the premises. Thinking along these lines, it is plausible to identify the content of a statement with the set of all of the statements it entails. Second, it is natural to think of the content of a statement in terms of the possible states of affairs it excludes. For instance, we feel that tautologies have no content precisely because they are compatible with any possible state of affairs; they rule nothing out. Statements with factual content, in contrast, are incompatible with some states of affairs, and the stronger the factual statement, the greater the number of possible states of affairs with which it is incompatible. According to this second conception of content, the greater the content of a statement the more completely it narrows down the range of possible states of affairs that could be actual if that statement were true.

The foregoing notions of content, despite possible appearances to the contrary, are not in conflict with one another; they fit together harmoniously in terms of an account of semantic content by Bar-Hillel and Carnap (1953–1954). This explication is provided for the simple finite languages L_N with which we are working. In such languages, the state descriptions, which are *conjunctions* containing one member from each basic pair of statements, are the *strongest* consistent statements in the language. Any consistent statement is equivalent to the disjunction of the state descriptions in its range, and it is entailed by any such state description. Since every consistent statement has a nonempty range, every consistent statement is entailed by some state description. Analogously, it is possible to define the set of *weakest* factual statements in such a language. Such statements are simply *disjunctions* containing as disjuncts one statement from each basic pair. These are weakest statements in the sense that every non-L-true statement in the language entails at least one such. Any factual statement is equivalent to the conjunction of all of these weakest statements it entails. If we think of content in the first of our two ways, in terms of what a statement entails, then it is natural to characterize the content of statements in terms of these weakest statements, just as the range of a statement is characterized in terms of the strongest statements (state descriptions). Indeed, since these disjunctions are weakest statements, we may think of them as having minimal content—as being content atoms, so to speak—or "content elements," as Bar-Hillel and Carnap call them. According to this conception of content, then, the content of a factual statement of L_N is the set of content elements it entails.

Each content element is, of course, the negation of a state description. In L_3, with one predicate F, we have the following:

State descriptions	Content elements
1. $Fa \cdot Fb \cdot Fc$	1. $\sim Fa \vee \sim Fb \vee \sim Fc$
2. $Fa \cdot Fb \cdot \sim Fc$	2. $\sim Fa \vee \sim Fb \vee Fc$
3. $Fa \cdot \sim Fb \cdot Fc$	3. $\sim Fa \vee Fb \vee \sim Fc$
4. $\sim Fa \cdot Fb \cdot Fc$	4. $Fa \vee \sim Fb \vee \sim Fc$
5. $Fa \cdot \sim Fb \cdot \sim Fc$	5. $\sim Fa \vee Fb \vee Fc$
6. $\sim Fa \cdot Fb \cdot \sim Fc$	6. $Fa \vee \sim Fb \vee Fc$
7. $\sim Fa \cdot \sim Fb \cdot Fc$	7. $Fa \vee Fb \vee \sim Fc$
8. $\sim Fa \cdot \sim Fb \cdot \sim Fc$	8. $Fa \vee Fb \vee Fc$

Identification of the content of p with the content elements that it entails is precisely equivalent to identifying that content with the set of state descriptions with which p is incompatible, that is, with the range of $\sim p$. Since the state descriptions are intended to be complete descriptions of possible states of the miniature universe describable by our language, the identification of the content of p with the statements that p entails comes to exactly the same thing as identification with the content of p with the possible states of affairs incompatible with p.

Note, incidentally, that any state description is incompatible with all other state descriptions; one reason, therefore, for saying that state descriptions have maximal content is that each one specifies the state of the universe completely by excluding all other possible states. (The only statement stronger than a state description is an L-false statement; such a statement is incompatible with every state description and entails every content element. The only statement weaker than a content element is a tautology; it entails no content elements and is incompatible with no state descriptions.) It is evident then, that the two notions of content with which we began merge into one and are neatly captured by the account given by Bar-Hillel and Carnap (1953–1954).

The foregoing approach immediately yields a quantitative measure of content or, more accurately, a large class of such measures. In his explication of "degree of confirmation" Carnap has shown how various confirmation functions could be defined in terms of measures assigned to state descriptions. The measure $m(p)$ of any statement p is then taken to be the sum of the measures of the state descriptions in $R(p)$ (Carnap 1950b; 1962b, sec. 55A). If each content element is given a content measure equal to the measure of the state description it negates, then the content measure $cont(p)$ will be the measure of the range of $\sim p$. If the measure of p is taken as the initial probability (null confirmation) $c_0(p)$, then we have a suitable sort of inverse relation between initial probability and content measure (see Popper [1935] 1959, secs. 31–35; Carnap 1966b):

$$cont(p) = m(\sim p) = 1 - m(p) = 1 - c_0(p)$$

There can be little doubt that the foregoing analysis of content is useful and illuminating, but it cannot be used directly as a basis for an analysis of complete independence of statements.[7] The reason is obvious. Given any two statements p and q, each of them entails $p \vee q$, so if we think of content in terms of entailments, any two statements—with one exception to be noted immediately—will have common content by virtue of a common consequence. The one exception arises when the statements p and q are mutually exhaustive, for in that case $p \vee q$ is L-true and has no content, so it can hardly represent shared content for p and q. For example, consider the statements $(\exists x)Fx$ and $(\exists x) \sim Fx$, which are mutually exhaustive but not mutually exclusive. Content element 8 is the entire content of the former, and content element 1 is the entire content of the latter. This is a case of a pair of statements with mutually exclusive content, that is, no content in common. These two statements are, of course, far from independent, for there is a full entailment relation between them: The falsity of either entails the truth of the other. Thus, non-overlapping content is no more indicative of complete independence than is non-overlapping of range. This is no surprise, for the concepts of content and range are essentially dual to one another, so there is not much hope of getting our relation of complete independence out of one if it does not come out of the other.

There is some danger, I fear, that these results may seem so highly counterintuitive as to suggest that the whole analysis of content has been fundamentally misguided. A little reflection can, I believe, dispel this initial reaction (if in fact it occurs). If there is a relation of complete independence—that is, if there are statements between which neither full nor partial entailment holds—then consider a paradigm pair such as Fa and Fc. Obviously they have overlapping content, just as they have overlapping ranges; they share content elements 6 ($Fa \vee \sim Fb \vee Fc$) and 8 ($Fa \vee Fb \vee Fc$). The fact that these independent statements have common content elements should occasion no surprise; in fact, this is precisely what we should have expected. If each of two independent statements rules out sets of possibilities, some of the possibilities ruled out by one will be excluded by the other as well, and some not. That they are independent is attested by the fact that neither all nor none of the possibilities incompatible with the one is excluded by the other as well. If it were otherwise, the statements would surely be far from independent. For example, Fa and Fc each rule out state description 8 ($\sim Fa \cdot \sim Fb \cdot \sim Fc$), but Fa rules out state descriptions containing $\sim Fc$ indiscriminately, for it also rules out state description 7 (∼*Fa*·∼*Fb*·*Fc*), which is just like state description 8, except for containing *Fc* where 8 contains ∼*Fc*. The state description *Fc* does not rule out state descriptions containing ∼*Fc* in this indiscriminate manner, though it does rule out state descriptions containing ∼*Fa* indiscriminately. This situation does not seem counterintuitive after all. It strongly suggests that what we began by regarding as a relation of *having no common content* needs more properly to be explicated as a relation of *independence of content*. Indeed, when we succeed in characterizing the relation of complete independence, in terms of which we shall be able to define partial entailment, a suitable concept of independence of content will emerge automatically.

The discussion of content was undertaken to see whether it would be feasible to explicate partial entailment in terms of possession of common content, and complete independence in terms of lack of common content. The result is essentially the same as was obtained in considering relations between ranges: If partial entailment is defined in terms of overlapping, a full entailment relation holds. I do not mean to maintain now, any more than I did in the discussion of overlap of ranges, that it is impossible to provide a stipulative definition of "partial entailment" according to which it obtains whenever there is overlapping of content. Obviously this can be done. However, the consequence is the same as before. The relation of partial entailment that results from such a definition is radically incongruous with that which arises naturally out of the consideration of the relations to be found in deductive logic. Any heuristic value of the analogy between full entailment and partial entailment vis-à-vis the foundations of inductive logic is entirely lost.

5. Complete Independence

I have taken *Fa* and *Fc* as a paradigm of complete independence, but any two distinct atomic statements would serve equally well. Likewise, I have taken *Fa*·*Fb* and *Fa*·*Fc* as a paradigm of partial entailment, and thus, of the absence of complete independence. The difference, as already noted, seems to be the presence of common conjuncts in the case of nonindependence and the absence of common conjuncts in the case of complete independence. However, this cannot be used as a criterion for distinguishing between the two relations. If *p* and *q* are L-equivalent, respectively, to *p'* and *q'*, then surely *p* must be completely independent of *q* if *p'* is completely independent of *q'*. Now *Fa* and *Fb* are L-equivalent, respectively, to (*Fa*∨*Fb*)·(*Fa*∨∼*Fb*) and (*Fa*∨*Fb*)·(∼*Fa*∨*Fb*), where the latter pair share a common conjunct and the former pair are, by hypothesis, completely independent. As a matter of fact any statement is L-equivalent to the conjunction of its content elements, so any two statements that have common content are L-equivalent, respectively, to a pair of statements that share a common conjunct. As we have seen, only L-disjoint statements fail to have common content.

If we impose a condition that statements be reduced to some sort of minimal form before we apply the criterion of shared conjuncts, a different difficulty will arise, *Fa*∨*Fb* and *Fb*∨*Fc* are not completely independent of each other, but they do not share a common conjunct. They do, however, share a common atomic part.

Thus, we might make the absence of common nonvacuous atomic parts the defining condition for complete atomic independence. More precisely,

> *p* is *completely atomically independent* of *q* = $_{df}$ *p* and *q* are L-equivalent, respectively, to two molecular (quantifier free) statements *p'* and *q'*, which share no common atomic parts.

This definition will be more useful if we can provide an effective procedure for finding out whether the statements *p'* and *q'* exist. This can be done through the use of normal forms. By application of a set of rules, which I shall sketch, any statement of L_N can be reduced to minimal disjunctive normal form (see Carnap 1950b, 1962b, sec. 21C for a full and precise statement of the rules). To reduce a statement *p* to minimal form, one begins by replacing universal and existential quantifiers by conjunctions and disjunctions, respectively, thereby reducing the statement to a molecular form. Truth-functional connectives other than negation, conjunction, and disjunction are eliminated according to standard definitions. Rules are provided for the elimination of redundant parts of disjunctions and conjunctions. DeMorgan's rules can be used to break up negations of large scope, and double negation is used to eliminate excess negations. The distribution rule breaks up conjunctions that contain disjunctions as conjuncts. The result will be a statement *p'* that is L-equivalent to *p* and that contains no vacuous atomic parts. It will have the form of a basic statement standing alone, a conjunction of basic statements in which no atomic statement occurs more than once, or a disjunction of such conjunctions of basic statements. Two statements *p* and *q* will be completely atomically independent if their minimal disjunctive normal forms have no atomic parts in common.

I am confident that complete atomic independence provides at least a sufficient condition of the kind of complete independence we seek. I am less sure that it is an appropriate necessary condition, for another definition of complete independence has a degree of plausibility. For languages of the type we are considering, complete independence would seem to be complete truth-functional irrelevance. We want to find a way of saying that the statement *p* makes no difference to the truth-functional character of *q*; that is, the truth or falsity of *p* has no bearing on the truth-value pattern of *q*. This condition obtains, I would suggest, if, in a truth table adequate to represent the truth patterns of the statements involved, the proportion of true cases for *q*, given that *p* is true, is equal to the proportion of true cases for *q*, given that *p* is false. This condition is equivalent to the condition, cited by Kemeny and Oppenheim (1952), that there be a zero correlation between the state descriptions in which *p* holds and the state descriptions in which *q* holds.[8] Let *p* and *q* be neither L-true nor L-false, and let *n*(*p*) be the number of cases in which the truth value of *p* is *T*. Then,

> *p* is *completely truty-functionally independent* of *q* =$_{df}$ $n(p \cdot q)/n(p)$ = $n(\sim p \cdot q)/n(\sim p)$

TABLE 11.1   Complete Truth-Functional Independence and Its Absence

p	q	r	p·q	q·r	p∨q	q∨r	p	p≡q
T	T	T	T	T	T	T	T	T
T	T	F	T	F	T	T	T	T
T	F	T	F	F	T	T	T	F
T	F	F	F	F	T	F	T	F
F	T	T	F	T	T	T	F	F
F	T	F	F	F	T	T	F	F
F	F	T	F	F	F	T	F	T
F	F	F	F	F	F	F	F	T

It follows immediately that, in case of complete truth-functional independence, the proportion of cases of q true, given that p is true, equals the proportion of cases of q true regardless of the truth value of p:

$$n(p·q)/n(p) = n(q)/n(t)$$

where t is a tautology.[9] Moreover, complete truth-functional independence is, as is to be expected, a symmetrical relation, for from the definiens we get the result that q is completely truth-functionally independent of p; that is,

$$n(q·p)/n(q) = n(\sim q·p)/n(\sim q)^{10}$$

Complete truth-functional independence and its absence are illustrated in table 11.1. This definition of complete truth-functional independence translates immediately into a definition in terms of ranges of statements. Simply read $n(p)$ as the number of state descriptions in the range of p, and the definition stands unaltered. I shall return below to the obvious relation between this definition and the confirmation function $c^\dagger$.

Let $t =$ tautology.

Examples of nonindependence

$$\frac{n[(p·q)·(q·r)]}{n(p·q)} = 1/2 \neq \frac{n[(\sim(p·q)·(q·r)]}{n[(\sim(p·q)]} = 1/6 \neq \frac{n(q·r)}{n(t)} = 1/4$$

$$\frac{n[(p \lor q)·(q \lor r)]}{n(p \lor q)} = 5/6 \neq \frac{n[\sim(p \lor q)·(q \lor r)]}{n[\sim(p \lor q)]} = 1/2 \neq \frac{n(q \lor r)}{n(t)} \neq 3/4$$

Examples of independence

$$\frac{n(p·q)}{n(p)} = \frac{n(\sim p·q)}{n(\sim p)} = \frac{n(q)}{n(t)} = 1/2$$

$$\frac{n[p·(p \equiv q)]}{n(p)} = \frac{n[\sim p·(p \equiv q)]}{n(\sim p)} = \frac{n(p \equiv q)}{n(t)} = 1/2$$

Complete truth-functional independence is not the same as complete atomic independence. Complete atomic independence implies complete truth-functional independence—this is a direct consequence of our methods for construction of truth tables. If you make a truth table for a formula with $n - 1$ atomic parts, but with all of the cases for n atomic parts, you simply repeat the truth-value pattern for that formula, producing the truth-value pattern for the formula, producing the truth-value pattern once for the value T of the extraneous atom, and reproducing it for the value F of the extraneous atom. If the truth table contains more than one extraneous atomic formula, the truth pattern for the original formula will be reproduced for each combination of truth values for the extraneous atoms. Such repetitions cannot change the ratios involved in the definition of complete truth-functional independence, for the net effect is simply to multiply numerator and denominator by the same integer. Thus, given two formulas p and q with no atomic parts in common, the proportion of true cases of either is not a function of the truth value of the other. To show that complete truth-functional independence does not imply complete atomic independence, we can provide a counterexample. It is seen immediately, by counting cases, that Fa and $Fa \equiv Fb$ are completely truth-functionally independent (see table 11.1). It is evident, however, that $Fa \equiv Fb$ is not L-equivalent to any statement not containing Fa. The minimal disjunctive normal form of $Fa \equiv Fb$ is $Fa \cdot Fb \lor \sim Fa \cdot \sim Fb$, whereas the minimal disjunctive normal form of Fa is itself.

Do we want to say that Fa partially entails $Fa \equiv Fb$? The reasons seem to me to favor denying it. Without regard to the truth value of Fa, $Fa \equiv Fb$ is true in half of the possible cases; given that Fa is true, it is still true in half of the remaining possible cases. To be sure, $Fa \equiv Fb$ is L-equivalent to $(Fa \supset Fb) \cdot (\sim Fa \supset \sim Fb)$, neither of whose conjuncts is completely independent of Fa. This does not pose any great obstacle to independence, however, for we are basically concerned with inductive relevance, and it is well known that evidence can be relevant to each of the two statements without being relevant to their conjunction (see Carnap 1950b; 1962b, T69–1, table 1b, 378).

6. Partial Entailment

The main reason for concern over complete independence is to use it in clarifying the concept of partial entailment. Since partial entailment, as we shall construe it, is a relevance relation, we shall have to make allowance for positive and negative relevance. In case p is positively relevant to q, we shall say the p partially entails q; if p is negatively relevant to q, we shall say that p partially entails $\sim q$. The general idea is that a partial entailment (positive or negative) shall obtain from p to q iff neither a full (deductive) entailment nor complete truth-functional independence obtains. In this discussion we shall continue to restrict p and q to statements that are neither L-true nor L-false.[11] The following definitions seem reasonable:

p partially entails $q =_{df} n(p \cdot q)/n(p) > n(\sim p \cdot q)/n(\sim p)$, and p does not entail q

p partially entails $\sim q =_{df} n(p \cdot q)/n(p) < n(\sim p \cdot q)/n(\sim p)$, and p does not entail $\sim q$

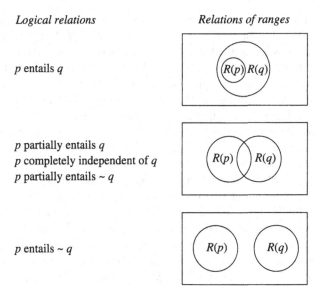

Logical relations   *Relations of ranges*

p entails q

p partially entails q
p completely independent of q
p partially entails ~ q

p entails ~ q

FIGURE 11.2   Two congruous scales.

We shall say informally that there is a partial entailment from p to q if either of the foregoing relations holds. Given these definitions, it will be useful to look once more at the comparison between *logical relations* and *relations of ranges* given in a preliminary way in figure 11.1; figure 11.2 now provides a more accurate representation. This diagram shows clearly how inclusion and exclusion of ranges constitute entailment relations, whereas overlap of ranges includes both types of partial entailment, as well as complete independence. It should be explicitly noted that complete independence is not necessarily represented diagrammatically by the picture in which $R(p)$ covers half of $R(q)$ and conversely. Complete independence is more complicated; it obtains when the proportion of $R(p)$ overlapped by $R(q)$ equals the proportion of the universe $R(t)$ occupied by $R(q)$. The universe is shown in the diagrams of figure 11.2 because of this relation.

Now that the relations of partial entailment and complete independence have been defined, let us look at some of the typical kinds of problems that concern inductive logic. The difficulty that immediately arises is that there are pairs of statements that are completely independent on our definition of complete truth-functional independence (or on the definition of complete atomic independence for that matter) between which inductive logicians have claimed that a relation of partial entailment holds. Consider, for example, the classic setup for drawing balls from an urn without replacement. Each draw reveals the color of a different ball. Suppose we draw two balls and find that each is red, and we wonder whether the next ball to be drawn will also be red. Letting F be the property red, our "evidence" is $Fa \cdot Fb$. We are entertaining the "hypothesis" Fc. These statements are completely independent on either definition: They have no nonvacuous atomic parts in common, and they are truth-functionally irrelevant to one another. Or, consider the same situation except that each ball is replaced after it is drawn. Different draws need not involve distinct balls. If our individual constants denote events, such as draws from the urn, everything is the same as before. This is the way it would be handled in the Carnapian framework. But even if individual constants must designate physical things so that the "evidence" and "hypothesis" may be about the same things, they will still be completely independent.

These two situations give us typical inductive examples. As a matter of fact, the fundamental inductive problem concerns inference from the past to the future or, more generally, from the observed to the unobserved. Past events are distinct from future events; observed events are distinct from unobserved events. Therefore, statements exclusively about past events are completely atomically (and a fortiori truth-functionally) independent of statements exclusively about future events, and statements strictly about observed events are completely independent of statements strictly about unobserved events. Hume ([1739–1740] 1888) made this point quite explicitly, and we see that he was right on either definition of complete independence.

There are, of course, inductive examples in which evidence and hypthesis are not completely independent. For instance, in the example of draws without replacement, the hypothesis that all of the balls in the urn are red is not completely independent of the evidence that the two balls so far drawn are red. There is, as we noted at the outset, a very reasonable sense in which this evidence partially entails this hypothesis. Peirce (1931) and Carnap (1950b, 1962b) both saw, however, that one can separate "tested parts" from "untested parts" of the hypothesis; for instance, if the urn contains only three balls, $Fa \cdot Fb$ is a tested part of $(x)Fx$, whereas Fc is an untested part. An untested part is a part that is completely independent of the evidence; indeed, it seems natural to define "untested part" in terms of complete independence:

> h' is an untested part (with respect to e) of $h =_{df} h$ entails h', and h' is completely independent of e

You may choose either complete atomic independence or complete truth-functional independence; the problem that remains for inductive logic will be equally severe either way.

7. Inductive Relevance

Both Peirce (1931, 2.744–746) and Carnap (1950b; 1962b, sec. 110A) recognized the nature of the problem involved in the relation between the evidence and an untested part of the hypothesis. If we construe "partial entailment" and "complete independence" in accord with the foregoing definitions, and if we identify inductive irrelevance with the absence of full or partial entailment relations (that is, with complete independence), this is tantamount to the choice of $c^{\dagger}$ as a confirmation function. It would preclude "learning from experience," for it makes the past inductively irrelevant to the future, the observed inductively irrelevant to the unobserved, and the "evidence" inductively irrelevant to the untested part of the hypothesis. Peirce reacted to this problem by rejecting the notion that inductive relevance is any kind of logical relation, but Carnap proposed different confirmation functions that do not rule out learning from experience. Carnap does not explicitly recognize the relation of complete truth-functional independence, and he does not see partial entailment as a relation standing somewhere between complete independence and full entailment. Rather, he identifies partial entailment with overlapping ranges. On his conception, full or partial entailment always obtains, so inductive irrelevance, if there is to be such a relation, cannot be identified with the absence of full and partial entailments. Carnap does, however, introduce a relation of inductive relevance based upon his confirmation functions,[12] and in the case of $c^{\dagger}$, inductive irrelevance coincides with our complete truth-functional independence. For other confirmation functions, however, the extension of the relation of inductive irrelevance is different and no longer coincides with complete truth-functional independence. Thus, for example, e may be irrelevant to h on $c^{\dagger}$ but relevant to h on c^{*}. Confirmation functions other than $c^{\dagger}$ provide for the possibility of learning from experience. However, a price is paid for the benefit. On $c^{\dagger}$, inductive irrelevance is identified straightforwardly with a logical irrelevance relation, but other confirmation functions yield inductive irrelevance relations that have no simple coordination with any logical irrelevance relation. In order to achieve fuller clarity on these matters, it will be useful to examine the concept of inductive relevance in some detail.

One of Carnap's chief contributions to inductive logic, in my opinion, is his careful attention to the concept of inductive relevance and his emphasis upon the importance of that relation. Not only does he devote a major chapter of his 1950b and 1962b book to the analysis of relevance, but also, in his preface to the second edition, he emphasizes the importance of distinguishing clearly between two sets of concepts, namely, "concepts of *firmness*" and "concepts of *increase of firmness*" (1962b, xv–xx). The concepts of firmness have to do with the degree of confirmation of a hypothesis on given evidence; the concepts of increase of firmness have to do with the *change in* degree of confirmation as a result of additional evidence. Change in degree of confirmation is, of course, an *inductive relevance* concept. If h has a certain degree of confirmation $c(h,i)$ on prior evidence i (in case there is no prior evidence, let i be the tautological evidence t), then e is positively relevant to h provided $c(h,i\cdot e\ )>c(h,i)$, negatively relevant if $c(h,i\cdot e)<c(h,i)$, and irrelevant if $c(h,i\cdot e)=c(h,i)$. We can define a relevance measure, degree of increase of degree of confirmation, as follows:

$$d(h,i\cdot e)=c(h,i\cdot e)-c(h,i)$$

This quantity is obviously negative when e is negatively relevant and zero when e is irrelevant. When i is the tautological evidence, we may write

$$d(h,e)=c(h,e)-c(h,t)$$

The point on which confusion is most apt to arise is in a failure to distinguish carefully between zero degree of confirmation and zero degree of increase of confirmation, especially if the prior evidence is tautological, that is, between $c(h,e)=0$ and $d(h,e)=0$. These are clearly not the same. In the finite languages

L_N, h has zero confirmation on e iff e is not L-false and e entails $\sim h$. Moreover, h has zero confirmation on t iff h is L-false. Therefore, $c(h,e) = d(h,e) = 0$ iff e is not L-false and h is L-false. In all of the interesting cases, namely, those in which neither e nor h is L-false, $c(h,e) = 0$ and $d(h,e) = 0$ are mutually incompatible.

Since Carnap (1950b, 1962b) has identified partial entailment with degree of confirmation, e partially entails h if e is not L-false, e and h are not L-exclusive, and e does not fully entail h (in the finite languages L_N). As emphasized above, it follows that e partially entails h whenever e and h are logically independent. Thus, Carnap makes partial entailment a "concept of firmness" and not a "concept of increase of firmness." In general, e partially entails h even if e is inductively irrelevant to h. The concept of "learning from experience," in sharp contrast, is a relevance concept or a "concept of increase of firmness"; it concerns the possibility of evidence raising or lowering probabilities. Therefore, if we take partial entailment as Carnap does, there is no guarantee that learning from experience is possible whenever we have partial entailments; quite the contrary. The confirmation function $c^\dagger$ provides an "inductive logic" with all the partial entailments Carnap demands, but in which learning from experience is impossible.

One of the many ways of looking at the fundamental problem of inductive logic is to view it as the problem of finding some relation of relevance between past and future, between observed and unobserved, which enables learning from experience to occur.[13] Inductive logic without appropriate relevance relations between past and future is useless. We can, to be sure, list the logical possibilities that may be realized in the future, and we may even assign measures to these possibilities on some a priori basis. If, however, experience and observational evidence cannot *alter* the status of these abstract logical possibilities by providing inductive grounds for expecting some more and others less, then inductive logic provides no means by which to overcome our complete ignorance of the future. Regardless of what we have experienced and observed, we would always face the future as nothing more than a set of abstract logical possibilities.

If one maintains that a logical relation of partial entailment holds between evidence and hypothesis and that this is the foundation of inductive logic, it seems reasonable to expect this relation to be a relevance relation. Carnap's (1950b, 1962b) relation of partial entailment simply is not a relevance relation, and it does not pretend to be. As we have seen, Carnap introduces a relevance relation that, in the case of $c^\dagger$, coincides with our relation of complete truth-functional independence but fails to coincide for other confirmation functions. Carnap's approach takes the nonrelevance relation of overlapping ranges as the basis for the concept of partial entailment and degree of confirmation. The particular confirmation function depends upon the way in which measures of overlapping are assigned. Inductive relevance is defined on the basis of degree of confirmation, so relevance relations can be changed by changing the measures assigned to the overlap of ranges. This means that inductive irrelevance, on Carnap's approach, is not tied directly to any logical irrelevance relation that is easily recognizable as such. The fact that inductive relevance and irrelevance are defined in terms of measures of overlap of ranges, and not on the basis of a well-defined logical relation like our complete truth-functional independence, fosters the idea that adjustments in the weights assigned to state descriptions constitutes a reasonable way to change the extension of the inductive relevance relation, thus providing for inductive relevance and "leaning from experience" in those cases in which we feel we should have it in inductive logic.

I am arguing, however, that the search for a *logical* relation upon which to ground inductive logic should proceed from an examination of the familiar relations of deductive logic—at least it should if the analogy between the logical relations of inductive and deductive logic is to have any value for our understanding of the foundations of inductive logic. Our examination of these relations has revealed a relation of maximal relevance (full entailment), a relation of irrelevance (complete truth-functional independence), and an intermediate relation of some sort of logical relevance (partial entailment). If we identify inductive relevance with partial entailment of this type (as opposed to the nonrelevance relation of overlapping ranges), and inductive irrelevance with complete truth-functional independence, then we have a strong analogy between the logical relations of deduction and those of induction. Unfortunately, these identifications make inductive logic entirely sterile by yielding irrelevance in just those cases in which relevance is needed if learning from experience is to be possible. If, on the other hand, we tinker with measures of overlapping, so as to introduce relevance relations that can serve as a basis for learning from experience, then we destroy the coincidence between inductive irrelevance and its straightforward deductive counterpart (complete truth-functional independence). Thus, for confirmation functions other than $c^\dagger$ no benefit accrues from the analogy between inductive irrelevance and logical irrelevance, and similarly, none accrues from any analogy between inductive relevance and any logical relevance relation.

We have already seen that two distinct concepts of partial entailment seem to be involved in the doctrine that inductive logic is grounded in a logical relation, and these two concepts are radically incongruous with one another. They are incongruous because one is a relevance relation and the other is not. The relevance relation yields a strong analogy between inductive and deductive logic, but it fails to provide the kind of inductive relevance that is needed for an inductive logic. When we use the nonrelevance concept, it is possible to endow inductive logic with more satisfying relevance relations, but we lose the heuristic value of the analogy. This is not to say we cannot define inductive relevance relations that allow for learning from experience; Carnap has done it. But the burden has been shifted; no longer can we argue that a strong analogy between relations of full and partial entailment makes the fundamental character of inductive logical relations intelligible; instead we must argue that the existence of intelligible inductive logical relations demonstrates an analogy between inductive and deductive logical relations. The heuristic value of the analogy is similarly destroyed when one departs from the clear concept of complete truth-functional independence in order to arrive at a more satisfactory relation of inductive irrelevance. There seems to be no unequivocal set of concepts that preserves the analogy and provides the relevance relations needed for inductive logic.

8. Logical vs. Factual Relations

The foregoing analysis of the relevance and irrelevance relations to be found in deductive logic and their comparison with those needed in inductive logic seem to me to strongly suggest that the conception of inductive logic as based upon a logical relation is fundamentally misconceived. There are two distinct kinds of relevance relations, logical and factual, and I am inclined to believe that although deductive logic requires and exploits logical relevance relations, induction is involved with factual relevance relations instead.

When I suggested above that any two distinct atomic statements constitute a paradigm pair of completely independent statements, I was, of course, referring to complete independence of some logical sort. I meant to make no suggestion about their factual independence. To take the most extreme case, let F and G be two distinct primitive predicates, and let a and b designate two distinct individuals. Fa and Gb are completely independent in any logical sense, but there may be a factual relation between them. It may be, for example, that $(x)(y)[(Fx{\cdot}Rxy) \supset Gy]$ is a law of nature and that Rab is true as a matter of contingent fact. Under these circumstances there is a strong dependency between Fa and Gb—that is, Fa materially implies Gb—but this is not based upon any logical relation between them. Similarly, some strong statistical relation might obtain between F's and G's that would make Fa constitute excellent inductive grounds for asserting or betting on Gb, but again, it need not be any sort of logical relation.

Let me make it entirely clear that I am not asserting that the entailment, partial entailment, and independence relations I have discussed are the only syntactical or semantical relations that can hold among the sentences of our languages L_N. For example, the primitive symbols of a language can be given an alphabetic order in terms of which the well-formed formulas of the language can be alphabetized. Sentences of the language can then stand in syntactical relations of before or after in the alphabetic order. Likewise, we could establish equivalence classes of L-equivalent sentences (such an equivalence class seems to represent what has traditionally been known as a proposition), and these equivalence classes could be ordered in terms of the length of their alphabetically first members. Moreover, it is possible to define "degree of confirmation" in such a way that true degree of confirmation statements are analytic; in this sense, degree of confirmation does constitute a logical relation. I do not intend to claim that there are no such logical relations, but only that they are not germane to inductive logic.

Consider the term "entailment." If we take it as an uncommitted symbol, we can assign it any meaning we wish, for example, parenthood. If we decide to let it stand for some logical—that is, syntactical or semantical—relation, we can define it so the p is said to entail q iff p is alphabetically prior to q. If we want it to represent a full-fledged semantical relation, we could stipulate that p entails q iff q L-implies p. However, all such definitions are obviously grossly inadequate as explications of the notion of entailment with which we are normally concerned. Entailment is a semantical relation that has to do with truth values. To use the word "entailment" for some completely nonlinguistic relationship, such as the relation of parenthood, is blatantly inadequate as an explication. It is likewise unsatisfactory to use this term to refer to a purely syntactical relation that has nothing to do with truth values, for example, the relation of alphabetic precedence. Moreover, it is unsound to use the word to designate the semantic relation that is the converse of L-implication. "Entailment" must refer to a relation between statements that is necessarily truth preserving, that is, a relation such that it is impossible to have p be true and q false if p entails q. Happily, we find it possible to define "entailment" in terms of L-implication in such a way that the semantic rules of the language guarantee that the relation so designated will have the desired characteristics.[14]

What I have said about entailment seems obvious enough, but I have spelled it out in laborious detail in order to have a basis for comparison with degree of confirmation. "Degree of confirmation," as well as a host of other related terms such as "inductive support," "inductive relevance," and even "partial entailment," are not completely uncommitted symbols. They do not call for stipulative definitions; instead, they demand explication. The aspect of the explication that requires closest attention, it seems to me, is its involvement with the truth values of statements. To define "degree of confirmation" in such a way that it has nothing to do with the truth values of statements bearing the relation of confirmation misses the point in just the same way as would an explication of "entailment" that referred only to the alphabetic ordering of sentences. In logic, whether it be deductive or inductive, we are concerned with inference—that is, we might say, with the truth values of some statements, given truth values of others. In deductive logic we have a logical relation of entailment that guarantees a truth value of one statement, given a truth value of another, and this guarantee depends only upon the semantical and syntactical rules of the language in which the statements occur. Another relation that has a similar character, even though it is a factual relation rather than a logical relation, is material implication. If p materially implies q but does not entail q, then q is true if p is, but we cannot certify that this relation obtains between two statements on the basis of semantical and syntactical relations alone.

In addition to the logical relation of entailment, we have found a logical relation of partial entailment. Whether this relation holds between two statements p and q depends only upon their syntactical and semantical features. When it does obtain, there is, as we have seen, an alteration of the truth-value pattern of q depending upon whether p is true or false. The truth-value pattern of q is a function of the truth value of p.

Our main question is whether a logical relation of confirmation can provide a foundation for inductive logic, that is, a basis for inductive inference. Without inductive inference we remain totally ignorant of the future. We can view it only as a set of abstract logical possibilities about which we might guess blindly or simply confess ignorance. The function of inductive logic is to provide a basis for inference that is some improvement over total ignorance. If we introduce the confirmation function $c^\dagger$ we find no improvement in our inductive situation at all. Whatever evidence we have is irrelevant to the future. With this confirmation function, inductive irrelevance coincides with complete truth-functional independence. Degree of confirmation, using $c^\dagger$, does have some direct relation to truth values, but it is no good as a foundation for inductive inference. Our real need, as we have seen, is not so much for a logical confirmation function as for a logical inductive relevance relation. In order to provide a suitable foundation for inductive logic, we need an inductive relevance relation that yields relevance even for statements that are completely truth-functionally independent. And this relevance relation must have some bearing upon inference.

The desideratum for inductive logic is to enable us to make inferences about the future that are some improvement upon the sheer blind guessing for which we would have to settle in the absence of inductive logic. This suggests—to me, irresistibly—that we are looking for a relation upon which to base inductive inference that will enable us to be right more often than could be expected by blind guessing. In other words, the relation upon which inductive inference is to be based is designed to yield true results more frequently when we take account of relevant evidence than we get by ignoring it. For example, we should be right more often in inferring or betting on Fc in the presence of $Fa \cdot Fb$ than we would be in inferring or betting on Fc either in the absence of relevant evidence or in the presence of negative evidence such as $\sim Fa \cdot \sim Fb$.

Unfortunately there is no known logical relation for which any such claim can be substantiated. Although it is possible to define a logical (that is, syntactical or semantical) relation between evidence and hypothesis that can be called "degree of confirmation," this relation has no useful bearing upon truth values of hypotheses in relation to evidence. If there are logical dependencies among truth values, any satisfactory concept of degree of confirmation will reflect them. For example if p entails q, the degree of confirmation of q on p must be 1, and if p entails $\sim q$, the degree of confirmation of q on p must be zero. Moreover, if there is a positive correlation between the truth cases for p and the truth cases for q—that is, a positive correlation between the state descriptions in which p holds and those in which q holds—we may expect this to be reflected as positive inductive relevance.[15] When, however, there is no functional relation between the truth-value pattern of p and the truth-value pattern of q, the procedure usually followed is to assign different weights to different total states of the universe (state descriptions). For example, Carnap's well-known confirmation function c^* results for L_3 when we assign the 1/4 to state descriptions 1 and 8, and the value 1/12 to the remaining ones.[16] If we find, by observation, that Fa holds, this eliminates low-valued state descriptions in which Fc holds and leaves the relatively higher valued one, whereas $\sim Fa$ eliminates the relatively high-valued one and leaves the low-valued state descriptions in which Fc holds. This is the source of the positive relevance of Fa to Fc. As we have noted repeatedly, if we assign equal weights to all of the state descriptions, namely, 1/8 to each, Fa is irrelevant to Fc. Of course, it is also possible to give state descriptions 1 and 8 a weight below that of the other; for instance, we can assign a weight of 1/14 to each of them and let 2–7 each have the value 1/7. The net result will be to make Fa negatively relevant to Fc, while making $\sim Fa$ positively relevant to Fc. For any two statements that are completely truth-functionally independent of each other, positive inductive relevance, negative inductive relevance, and inductive irrelevance are *determined entirely* by the a priori assignment of weights to the possible states of the universe.

If we knew in advance what state of the universe is the actual one—that is, if we knew which state description correctly characterizes the observed and unobserved portions of the universe—we would have no need for inductive logic. Our problem arises because we have no such knowledge. If, to borrow Peirce's (1931) apt phrase, universes were as plentiful as blackberries, we might try counting to see how often a particular state description turns out to be correct. As a matter of fact, universes are available in large numbers, for when we speak of a universe it can be a small thing such as an urn with small balls in it or a few tosses of a die. It need not be the entire physical world. A universe is whatever can be completely described by the particular language we are using; the language we have used as an example can describe three things with respect to one property. If universes containing three things characterized by the presence or absence of one property exemplify state descriptions 1 and 8 more frequently than the other state description, the kind of relevance relations established by c^* will yield increased truth frequency for inductive inferences. If the situation is reversed, with 1 and 8 being realized relatively infrequently, inference based on c^* will lead to decreased truth frequency. The essential point is simply that no syntactical or semantical features of state descriptions can be shown to have any relation whatever to the frequency with which they are realized in actual universes. Thus, there is no logical relation among state descriptions, and consequently no logical relation between evidence and hypothesis, that has any bearing upon the truth frequencies that will result from the choice of any particular confirmation function. We can define an unlimited supply of confirmation functions, each of which gives us different relevance relations, but we cannot establish anything about the truth frequencies any of them will yield if used as a basis for inductive inference. They are all as irrelevant to the aims of inductive inference as is the definition of "entailment" in terms of alphabetic precedence to the aims of deductive logic.

Given a particular definition of "degree of confirmation," it is impossible to show that one will be right more often by inferring or betting on highly confirmed hypotheses than on those with lower degrees of confirmation. If certain factual relations obtain—for example, if F is a property that usually occurs uniformly in universes of the type we are considering—then utilization of degrees of confirmation based upon a confirmation function like c^* will have some benefit in inductive inference. The benefit derives from a factual relation, however, not from a logical one. Similarly, if p materially implies q without entailing q, inferring q from p will be successful, but not because of any logical relation.

I am fully aware that strenuous objection will be raised against the view I have been proposing. Given the truth of e, we will be told, the logical relation of confirmation cannot guarantee either a truth value of h or a truth frequency for hypotheses that, like h, have a certain degree of confirmation with respect to e, but it can tell us what is reasonable to believe about the truth value of h. I must confess that I find it impossible to understand the force of the concept of rationality employed in this way. If adoption of an inductive logic does not give promise of greater success than sheer guessing, what is reasonable about it? If a particular inductive logic, based upon some particular definition of "degree of confirmation," is supposed to give greater promise of success, in what does that promise lie?[17] There are, as remarked above, two kinds of relevance relations, deductive and inductive. Deductive relevance is to be found in relations of entailment and partial entailment; these are logical (syntactical and semantical) relations. Inductive relevance, in contrast, seems to be fundamentally a relation of statistical relevance. Statistical correlation is an example of an inductive relevance relation, and so is material implication as a special limiting case of statistical correlation. Unlike deductive relevance relations, these relations cannot be established on the basis of semantical and syntactical considerations alone. They are factual relations. I am convinced that inductive logicians have been interested in confirmation because they believe that the logical relation of degree of confirmation mirrors the factual statistical relevance relations that actually obtain in the world. Such an assumption is entirely gratuitous; there is no way of establishing the correspondence of the logical with the factual relations. The search for logical surrogates for these factual relations appears to be futile.

9. Conclusion

The idea that inductive logic is based upon a *logical* relation between evidence and hypothesis—a relation identified as partial entailment—seems to me to rest upon an illusion. The illusion is fostered by a failure to make several important distinctions. First, we must distinguish the logical relation between evidence and hypothesis from the logical relation between evidence and an *untested part* of a hypothesis. Sometimes there is, in a plausible sense, a logical relation of partial entailment between evidence and hypothesis. This does not help as a foundation of inductive logic, however, for the relation does not hold between the evidence and any untested part of the hypothesis.

Second, we must distinguish statements whose content is mutually exclusive from those whose content is mutually *independent*. It is tempting to think preanalytically that "*p* partially entails *q*" means "*p* entails part of what *q* asserts," and that "*p* does not (at least) partially entail *q*" means "*p* does not entail part of what *q* asserts." Analysis shows, however, that this way of speaking can be misleading unless we understand that not entailing part of what *q* asserts means that the content is independent of the content of *q*, *not* that it is exclusive of the content of *q*.

Third, and most important, we must distinguish clearly between the relation of nonzero confirmation and that of inductive relevance. We must be careful which of these is made to correspond to partial entailment. It is all too easy to fasten upon the relation, *e does not entail ~h*, and then to slip unnoticed into the supposition that it is a relation of positive relevance of *e* to *h*.

Although all of these distinctions are elementary and obvious, I am strongly inclined to suspect that they have not always been kept clear. For this reason I believe that the concept of partial entailment has been misleading in inductive logic. No doubt concepts like partial entailment, content, and independence can be defined clearly, and *can* be used in ways that lead to no confusion, but I do not think this has happened in the tradition of the logical theory of probability. Indeed, I suspect that the notion of partial entailment has been a source of confusion so profound that it has led inductive logicians to mistake a factual relation for a logical one.

The spirit of this chapter is essentially Humean. I have tried to take seriously, and follow out in detail, Hume's dictum that the future is logically independent of the past. If there are any inductive relevance relations, I have maintained, they must be factual rather than logical. There are no inductive or deductive *logical* relations by which the Humean independence can be circumvented. The arguments I have offered, if sound, point to the futility of approaching inductive logic via the logical concept of probability or degree of confirmation.

My sole aim has been to argue against one approach without attempting to suggest or support any alternative. If I am correct in claiming that inductive inference depends upon factual, not logical, relevance relations, the obvious alternative is not attractive. Indeed, we find ourselves immediately confronting another of Hume's major claims, for the factual relations upon which inductive inference seems to be founded apparently cannot be established except by induction. A vicious circle threatens. The escape from this dilemma still constitutes a most serious philosophical challenge.

The foregoing attack upon the logical concept of probability may seem like an inappropriate way to pay homage to an admired friend who is one of the ablest champions of the logical conception, but I am not so immodest as to suppose that my arguments are faultless and definitive. Moreover, Hempel has always prized philosophical clarity above adherence to doctrine. If my discussion can add any clarity at all to the important concepts that provide the bases of inductive logic, I shall have paid him the highest tribute of which I am capable.

Appendix: The Popper-Miller Argument

Sir Karl Popper and David Miller (1983, 687) have offered "A proof of the impossibility of inductive probability" that, they modestly confess, "strikes both of us as pretty." Their "proof" goes as follows:

> Consider a hypothesis h, which in conjunction with unproblematic background knowledge b, entails evidence e. If $0 < p(e, b) < 1$, they point out, it follows that $p(h, e\cdot b) > p(h, b)$. "It is the support of h by e, the increase of the probability of h generated by e, that looks like a probabilistic inductive effect. We shall now show that this is an illusion." (687)

In order to demonstrate this "illusion," Popper and Miller "factorize" h as follows:

$$h = (h \leftarrow e)(h \lor e)$$

which is, of course,

$$h = (h \lor \sim e)(h \lor e)$$

"Thus," they continue, "we have split h into two factors, the second of which, $h \lor e$, deductively follows from the evidence e: it is the logically strongest part of h (or of all the content of h) that follows from e. The first factor, $h \leftarrow e$, contains all of h that goes beyond e" (678).

Popper and Miller (1983) clearly seek to distinguish the tested part of h from the untested part of h, but since they have no definition of the latter concept, they have not succeeded in their search. This chapter, however, contains a plausible definition:

h' is an untested part (with respect to e) of $h =_{df} h$ entails h' and h' is completely (truth-functionally) independent of e

Let us construct the relevant truth table:

h	e	$e \leftarrow h$	$h \vee e$
T	T	T	T
T	F	F	T
F	T	T	T
F	F	T	F

Obviously, $e \leftarrow h$ is not an untested part of h because, to qualify as such, it must be completely truth-functionally independent of e. However, if e is true, $e \leftarrow h$ is true in all of the cases, whereas if e is false, $e \leftarrow h$ is true in only half of the cases. Therefore, $e \leftarrow h$ and e are not completely truth-functionally independent of one another. Moreover, h does not entail $e \leftarrow h$, so it fails on both scores.

If h is to be "factorized" into a tested part and an untested part, we need an appropriate relation between the content of the two factors. As we saw in this chapter, this relationship is not one of mutually exclusive content but rather of mutually independent content. If $e \leftarrow h$ and $h \vee e$ are to represent the untested part of h and the tested part, respectively, they should be completely truth-functionally independent of one another. However, a look at the foregoing truth table shows that they are not; $e \leftarrow h$ is true in two-thirds of the cases in which $h \vee e$ is true, but it is true in all of the cases in which it is false, and conversely. The Popper-Miller factorization does not achieve its appropriate end.

Let us look more generally at all possible truth-functional combinations of h and e, letting the first column represent the truth pattern of h and the second that of e:

1	2	3	4	5	6	7	8	9	10	11	12	13	14	15	16
T	T	T	T	T	T	F	T	F	F	F	T	F	F	F	F
T	F	T	T	T	F	T	F	T	F	T	F	T	F	F	F
F	T	T	T	F	T	T	F	T	T	F	F	F	T	F	F
F	F	T	F	T	T	T	T	F	T	T	F	F	F	T	F

It is immediately evident that any tautology (col. 3) or self-contradiction (col. 16) is completely truth-functionally independent of e. The contradiction is not an untested part of h because it is not entailed by h. The tautology is an untested part of h because it is entailed by h, but it is of no interest in this context because it is vacuous. We have seen above that $e \leftarrow h$ (col. 6) is not completely truth functionally independent of e. By similar reasoning, it is obvious that no proposition with an odd number of T's and F's is completely truth-functionally independent of e. This eliminates columns 4–7 and 12–15; let us look at the table that results when we remove all of them:

h	e	8	9	10	11
T	T	T	F	F	F
T	F	F	T	F	T
F	T	F	T	T	F
F	F	T	F	T	T

Since h does not entail any of the remaining truth functions (cols. 8–11), none of them can qualify as untested parts of h. The moral to be drawn is that *the only truth-functional combination of* h *and* e *that can qualify as the untested part of* h *is the tautology,* which has no content. There is a simple reason for this fact. According to the truth table, e and h are truth-functionally independent of each other, but Popper and Miller (1983) stipulate that (in the presence of background knowledge b) h entails e. This can only be true by virtue of the internal logical structure of h and e. Thus to ascertain the untested parts of h with respect to e, we must also look at their internal structure.

It is obvious, I trust, that my purpose in this appendix is not to establish the existence of inductive proabilification. I am concerned only to show that the Popper-Miller (1983) disproof is unsound.

Confirmation and Relevance

I tem: One of the earliest surprises to emerge from Carnap's (1950b, 1962b, sec. 110A) precise and systematic study of confirmation was that the initially plausible Wittgenstein confirmation function $c^\dagger$ was untenable. Carnap's objection rested on the fact that $c^\dagger$ precludes "learning from experience" because it fails to incorporate suitable relevance relations. Carnap's alternative confirmation function c^* was offered as a distinct improvement because it does sustain the desired relevance relations.

Item: On somewhat similar grounds, it has been argued that the heuristic appeal of the concept of partial entailment, which is often used to explain the basic idea behind inductive logic, rests upon a confusion of relevance with nonrelevance relations. Once this confusion is cleared up, it seems, the apparent value of the analogy between full and partial entailment vanishes (see chapter 11; Salmon 1967a).

Item: In a careful discussion, based upon his detailed analysis of relevance, Carnap (1950b, 1962b, secs. 86–88) showed convincingly that Hempel's ([1945] 1965b) classic conditions of adequacy for any explication of the concept of confirmation are vitiated by another confusion of relevance with nonrelevance relations.

Item: A famous controversy, in which Popper ([1935] 1959) charges that Carnap's theory of confirmation contains a logical inconsistency, revolves around the same issue. As a result of this controversy, Carnap acknowledged in the preface to the second edition of *Logical Foundations of Probability* (1962b) that the first edition (1950b) had been unclear with regard to this very distinction between relevance and nonrelevance concepts.

Item: A new account of statistical explanation, based upon relations of relevance, has recently been proposed as an improvement over Hempel's (1945) well-known account, which is based upon relations of high degree of confirmation (Salmon 1971).

Item: The problem of whether inductive logic can embody rules of acceptance—that is, whether there are such things as inductive inferences in the usual sense—has been a source of deep concern to inductive logicians since the publication of Carnap's (1950b) work. Hilpinen (1968) has proposed a rule of inductive inference, which, he claims, avoids the "lottery paradox," thus overcoming the chief obstacle to the admission of rules of acceptance. Hilpinen's rule achieves this feat by incorporating a suitable combination of relevance and high confirmation requirements.

The foregoing enumeration of issues is designed to show the crucial importance of the relations between confirmation and relevance. Yet, in spite of the fact that many important technical results have been available at least since the publication of Carnap's (1950b) work, it seems to me that their consequences for the concept of confirmation have not been widely acknowledged or appreciated. In the first three sections of this chapter, I summarize some of the most important facts, mainly to make them available in a single concise and relatively nontechnical presentation. All the material of these sections is taken from the published literature, much of it from the latter chapters of Carnap's book. In section 4, in response to questions raised by Adolf Grünbaum in private conversation,[1] I apply some of these considerations to the Duhemian problems, where, to the best of my knowledge, they have not previously been brought to bear. In section 5, I attempt to pinpoint the source of some apparent difficulties with relevance concepts, and in the final section, I try to draw some morals from these results. All in all, I believe that many of the facts are shocking and counterintuitive and that they have considerable bearing upon current ideas about confirmation.

1. Carnap and Hempel

As Carnap (1950b, 1962b) pointed out, the concept of confirmation is radically ambiguous. If we say, for example, that the special theory of relativity has been confirmed by experimental evidence, we might have either of two quite distinct meanings in mind. On the one hand, we may intend to say that the special theory has become an accepted part of scientific knowledge and that it is very nearly certain in the light of its supporting evidence. If we admit that scientific hypotheses can have numerical degrees of confirmation, the sentence, on the construal, says that the degree of confirmation of the special theory on the available evidence is high. On the other hand, the same sentence might be used to make a very different statement. It may be taken to mean that some particular evidence—for example, observations on the lifetimes of mesons—renders the special theory more acceptable or better founded than it was in the absence of this evidence. If numerical degrees of confirmation are again admitted, this latter construal of the sentence amounts to the claim that the special theory has a higher degree of confirmation on the basis of the new evidence than it had on the basis of the previous evidence alone.

The discrepancy between these two meanings is made obvious by the fact that a hypothesis h, which has a rather low degree of confirmation on prior evidence e, may have its degree of confirmation raised by an item of positive evidence i without attaining a high degree of confirmation on the augmented body of evidence $e \cdot i$. In other words, a hypothesis may be confirmed (in the second sense) without being confirmed (in the first sense). Of course, we may believe that hypotheses can achieve high degrees of confirmation by the accumulation of many positive instances, but that is an entirely different matter. It is initially conceivable that a hypothesis with a low degree of confirmation might have its degree of confirmation increased repeatedly by positive instances, but in such a way that the confirmation approaches 1/4 (say) rather than 1. Thus, it may be possible for hypotheses to be repeatedly confirmed (in the second sense) without ever getting confirmed (in the first sense). It can also work the other way: a hypothesis h that already has a high degree of confirmation on evidence $e \cdot i$., even though the addition of evidence i does not raise the degree of confirmation of h. In this case, h is confirmed (in the first sense) without being confirmed (in the second sense) on the basis of additional evidence i.

If we continue to speak in terms of numerical degrees of confirmation, as I shall do throughout this chapter, we can formulate the distinction between these two senses of the term "confirm" clearly and concisely. For uniformity of formulation, let us assume some background evidence e (that may, upon occasion, be the tautological evidence t), as well as some additional evidence i on the basis of which degrees of confirmation are to be assessed. We can define "confirm" in the first (absolute, nonrelevance) sense as follows:

> D1. Hypothesis h is confirmed (in the absolute sense) by evidence i in the presence of background evidence $e =_{df} c(h, e \cdot i) > b$, where b is some chosen number, presumably close to 1.

This concept is absolute in that it makes no reference to the degree of confirmation of h on any other body of evidence.[2] The second (relevance) sense of "confirm" can be defined as follows:

> D2. Hypothesis h is confirmed (in the relevance sense) by evidence i in the presence of background evidence $e =_{df} c(h, e \cdot i) > c(h, e)$.

This is a relevance concept because it embodies a relation of change in degree of confirmation. Indeed, Carnap's main discussion of this distinction follows his technical treatment of relevance relations, and the second concept of confirmation is explicitly recognized as being identical with the concept of positive relevance (Carnap 1950b; 1962b, sec. 86).

It is this context that Carnap offers a detailed critical discussion of Hempel's criteria of adequacy for an explication of confirmation.[3] Carnap shows conclusively, I believe, that Hempel has conflated the two concepts of confirmation, with the result that he has adopted an indefensible set of conditions of adequacy. As Carnap says, he is dealing with two explicanda, not with a single, unambiguous one. The incipient confusion is signaled by Hempel's (1965b) characterization of

> the quest for general objective criteria determining (A) whether, and — if possible — even (B) to what degree, a hypothesis h may be said to be corroborated by a given body of evidence e. . . . The two parts of this . . . problem can be related in somewhat more precise terms as follows:
>
> (A) To give precise definitions of the two nonquantitative relational concepts of confirmation and disconfirmation; i.e. to define the meaning of the phrases 'e

confirms *h'* and '*e* disconfirms *h*'. (When *e* neither confirms nor disconfirms *h*, we shall say that *e* is neutral, or irrelevant, with respect to *h*.)
(B) (1) To lay down criteria defining a metrical concept "degree of confirmation of *h* with respect to *e*, "whose values are real numbers.... (6).[4]

The parenthetical remark under (A) makes it particularly clear that a relevance concept of confirmation is involved there, whereas a nonrelevance concept of confirmation is obviously involved in (B).

The difficulties start to show up when Hempel begins laying down conditions of adequacy for the concept of confirmation (A)—as opposed to degree of confirmation (B). According to the very first condition, "entailment is a special case of confirmation." This condition states the following:

H-8.1 Entailment Condition. Any sentence which is entailed by an observation report is confirmed by it.[5]

If we are concerned with the absolute concept of confirmation, this condition is impeccable, for $c(h) = 1$ if *e* entails *h*. It is not acceptable, however, as a criterion of adequacy for a relevance concept of confirmation. For suppose our hypothesis *h* is $(\exists x)Fx$, evidence *e* is *Fa*, and evidence *i* is *Fb*. In this case, *i* entails *h*, but *i* does not confirm *h* in the relevance sense, for $c(h,e \cdot i) = 1 = c(h,e)$.

Carnap offers further arguments to show that the following condition has a similar defect:

H-8.3 Consistency Condition. Every logically consistent observation report is logically compatible with the class of all the hypotheses which it confirms. (Hempel 1965b, 33)

This condition, like the entailment condition, is suitable for the absolute concept of confirmation, but not for the relevance concept, for, although no two incompatible hypotheses can have high degrees of confirmation on the same body of evidence, an observation report can be positively relevant to a number of different and incompatible hypotheses, provided that none of them has too high a prior degree of confirmation on the background evidence *e*. This happens typically when a given observation is compatible with a number of incompatible hypotheses—when, for example, a given bit of quantitative data fits several possible curves.

The remaining condition Hempel (1965b) wanted to adopt is as follows:

H-8.2 Consequence Condition. If an observation report confirms every one of a class *k* of sentences, then it also confirms any sentence which is a logical consequence of *k*. (31)

It will suffice to look at two conditions that follow from it.

H-8.21 Special consequence Condition. If an observation report confirms a hypothesis *h*, then it also confirms every consequence of *h*.
H-8.22 Equivalence Condition. If an observation report confirms a hypothesis *h*, then it also confirms every hypothesis that is logically equivalent with *h*. (31.)

The equivalence condition must hold for both concepts of confirmation. Within the formal calculus of probability (which Carnap's concept of degree of confirmation satisfies) we show that, if h is equivalent to h', then $c(h,e) = c(h',e)$, for any evidence e whatever. Thus, if h has a high degree of confirmation on $e \cdot i$, h' does also. Likewise, if i increases the degree of confirmation of h, it will also increase the degree of confirmation of h'.

The special consequence condition is easily shown to be valid for the non-relevance concept of confirmation:

If h entails k, then $c(k,e) \geq c(h,e)$; hence, if $c(h,e \cdot i) > b$, then $c(k,e \cdot i) > b$.

But here, I think, our intuitions mislead us most seductively. It turns out, as Carnap has shown with great clarity, that the special consequence condition fails for the relevance concept of confirmation. It is entirely possible for i to be positively relevant to h without being positively relevant to some logical consequence k. We shall return in section 3 to a more detailed discussion of this fact.

The net result of the confusion of the two different concepts is that obviously correct statements about confirmation relations of one type are laid down as conditions of adequacy for explications of concepts of the other type, where, upon examination, they turn out to be clearly unsatisfactory. Carnap (1950b; 1962b, 473) showed how the entailment condition could be modified to make it acceptable as a condition of adequacy. As long as we are dealing with the relevance concept of confirmation, it looks as if the consistency condition should simply be abandoned. The equivalence condition appears to be acceptable as it stands. The special consequence condition, surprisingly enough, cannot be retained.

Hempel tried to lay down conditions for a nonquantitative concept of confirmation, and we have seen some of the trouble he encountered. After careful investigation of this problem, Carnap (1950b; 1962b, 467) came to the conclusion that it is best to establish a quantitative concept of degree of confirmation and then to make the definition of the two nonquantitative concepts dependent upon it, as we have done in D1 and D2 above. Given a quantitative concept and the two definitions, there is no need for conditions of adequacy like those advanced by Hempel. The nonquantitative concepts of confirmation are fully determined by those definitions, but we may, if we wish, see what general conditions, such as H-8.1, H-8.2, H-8.21, H-8.22, and H-8.3, are satisfied. In a 1964 postscript to the earlier article, Hempel expresses general agreement with this approach of Carnap (Hempel 1965b, 50). Yet, he does so with such equanimity that I wonder whether he, as well as many others, recognizes the profound and far-reaching consequences of the fact that the relevance concept of confirmation fails to satisfy the special consequence condition and other closely related conditions (which will be discussed in section 3).

2. Carnap and Popper

Once the fundamental ambiguity of the term "confirm" has been pointed out, we might suppose that reasonably well-informed authors could easily avoid confusing the two senses. Ironically, even Carnap himself did not remain entirely free from this fault. In the preface to the second edition of *Logical Foundations of Probability* (1962a), he acknowledges that the first edition (1950) was not completely unambiguous. In the new preface, he attempts to straighten out the difficulty.

In the first edition, Carnap (1950b, sec.8) had distinguished a triplet of confirmation concepts.

1. Classificatory—*e* confirms *h*.
2. Comparative—*e* confirms *h* more than *e'* confirms *h'*.
3. Quantitative—the degree of confirmation of *h* on *e* is *u*.

In the second edition, Carnap (1962b, xv–xvi) sees the need for two such triplets of concepts. For this purpose, he begins by distinguishing what he calls "concepts of firmness" and "concepts of increase of firmness." The concept of confirmation we defined above in D1, which was called an "absolute" concept, falls under the heading "concepts of firmness." The concept of confirmation we defined in D2, and called a "relevance" concept, falls under the heading "concepts of increase of firmness." Under each of these headings, Carnap sets out a triplet of classificatory, comparative, and quantitative concepts:

I. Three Concepts of Firmness

I-1. *h* is firm on the basis of *e*. $c(h,e) > b$, where b is some fixed number.

I-2. *h* is firmer on *e* than is *h'* on *e'*. $c(h,e) > c(h',e')$.

I-3. The degree of firmness of *h* on *e* is *u*. $c(h,e) = u$.

To deal with the concepts of increase of firmness, Carnap introduces a simple relevance measure $D(h,i) =_{df} c(h,i) - c(h,t)$. This is a measure of what might be called "initial relevance," where the tautological evidence *t* serves as background evidence. The second triplet of concepts is given as follows:

II. Three Concepts of Increase of Firmness

II-1. *h* is made firmer by *i*. $D(h,i) > 0$.

II-2. The increase in firmness of *h* due to *i* is greater than the increase of firmness of *h'* due to *i'*. $D(h,i) > D(h',i')$.

II-3. The amount of increase of firmness of *h* due to *i* is *w*. $D(h,i) = w$.

Given the foregoing arrays of concepts, any temptation we might have had to identify the absolute (nonrelevance) concept of confirmation with the original classificatory concept, and to identify the relevance concept of confirmation with the original comparative concept, while distinguishing both from the original quantitative concept of degree of confirmation, can be seen to be quite mistaken. What we defined above in D1 as the absolute concept of confirmation is clearly seen to coincide with Carnap's (1962b) new classificatory concept I-1, and our relevance concept of confirmation defined in D2 obviously coincides with Carnap's other new classificatory concept II-1. Carnap's familiar concept of degree of confirmation (probability₁) is obviously his quantitative concept of firmness I-3, and his new quantitative concept II-3 coincides with the concept of degree of relevance. Although we shall not have much occasion to deal with the comparative concepts, it is perhaps worth mentioning that the new comparative concept II-2 has an important use. When we compare the strengths of different tests of the same hypothesis we frequently have occasion to say that a test that yields evidence i is better than a test that yields evidence i'; sometimes, at least, this means that i is more relevant to h than is i'—that is, the finding i would increase the degree of confirmation of h by a greater amount than would the finding i'.

It is useful to have a more general measure of relevance than the measure D of initial relevance. We therefore define the relevance of evidence i to hypothesis h on (in the presence of) background evidence e as follows:[6]

$$\text{D3. R } (i,h,e) = c(h,e \cdot i) - c(h,e)$$

Then we can say:

D4. i is positively relevant to h on $e =_{df} R(i,h,e) > 0$
      i is negatively relevant to h on $e =_{df} R(i,h,e) < 0$
      i is irrelevant to h on $e =_{df} R(i,h,e) = 0$

Hence, the classificatory concept of confirmation in the relevance sense can be defined by the condition $R(i,h,e) > 0$. Using these relevance concepts, we can define a set of concepts of confirmation in the relevance sense as follows:

D5. i confirms h given $e =_{df} i$ is positively relevant to h on e
      i disconfirms h given $e =_{df} i$ is negatively relevant to h on e
      i neither confirms nor disconfirms h given $e =_{df} i$ is irrelevant to h on e

Having delineated his two triplets of concepts, Carnap (1962b, xvii–xix) then acknowledges that his original triplet in the first edition was a mixture of concepts of the two types; in particular, he had adopted the classificatory concept of increase of firmness II-1, along with the comparative and quantitative concepts of firmness I-2 and I-3. This fact, plus some misleading informal remarks about these concepts, let to serious confusion.

As Carnap (1962b, xix, fn.) further acknowledges, Popper called attention to these difficulties, but he fell victim to them as well. By equivocating on the admittedly ambiguous concept of confirmation, he claimed to have derived a contradiction within Carnap's formal theory of probability₁. Popper (1959) offers the following example:

> Consider the next throw with a homogeneous die. Let x be the statement 'six will turn up'; let y be its negation, that is to say, let $y = \bar{x}$; and let z be the information 'an even number will turn up'.

We have the following absolute probabilities:

$$p(x) = 1/6; \; p(y) = 5/6; \; p(z) = 1/2.$$

Moreover, we have the following relative probabilities:

$$p(x,z) = 1/3; \ p(y,z) = 2/3.$$

We see that x is supported by the information z, for z raises the probability of x from 1/6 to 2/6 = 1/3. We also see that y is undermined by z, for z lowers the probability of y by the same amount from 5/6 to 4/6 = 2/3. Nevertheless, we have $p(x,z) < p(y,z)$. (390)

From this example, Popper quite correctly draws the conclusion that there are statements x, y, and z such that z confirms x, z disconfirms y, and y has a higher degree of confirmation on z than x has. As Popper points out quite clearly, this result would be logically inconsistent if we were to take the term "confirm" in its nonrelevance sense. It would be self-contradictory to say,

> The degree of confirmation of x on z is high, the degree of confirmation of y on z is not high, and the degree of confirmation of y on z is higher than the degree of confirmation of x on z.

The example, of course, justifies no such statement; there is no way to pick the number b employed by Carnap in his definition of confirmation in the (firmness) sense I-1 according to which we could say that the degree of confirmation of x on z is greater than b and the degree of confirmation of y on z is not greater than b. The proper formulation of the situation is this:

> The evidence z is positively relevant to x, the evidence z is negatively relevant to y, and the degree of confirmation of x on z is less than the degree of confirmation of y on z.

There is nothing even slightly paradoxical about this statement.

Popper's (1959) example shows clearly the danger of equivocating between the two concepts of confirmation, but it certainly does not show any inherent defect in Carnap's system of inductive logic, for this system contains both degree of confirmation $c(h,e)$ and degree of relevance $R(i,h,e)$. The latter is clearly and unambiguously defined in terms of the former, and there are no grounds for confusing them (Carnap 1950b; 1962b, sec. 67). The example shows, however, the importance of exercising great care in our use of English-language expressions in talking about these exact concepts.

3. The Vagaries of Relevance

It can be soundly urged, I believe, that the verb "to confirm" is used more frequently in its relevance sense than in the absolute sense. When we say that a test confirmed a hypothesis, we would normally be taken to mean that the result was positively relevant to the hypothesis. When we say that positive instances are confirming instances, it seems that we are characterizing confirming evidence as evidence that is positively relevant to the hypothesis in question. If we say that several investigators independently confirmed some hypothesis, it would seem sensible to understand that each of them had found positively relevant evidence. There is no need to belabor this point. Let us simply assert that the term "confirm" is often used in its relevance sense, and we wish to investigate some of the properties of this concept. In other words, let us agree for now to use the term "confirm" solely in its relevance sense (unless some explicit qualifications indicates otherwise) and see what we will be committed to.

It would be easy to suppose that, once we are clear on the two senses of the term "confirm," and once we resolve to use it in only one of these senses in a given context, it would be a simple matter to tidy up our usage enough to avoid such nasty equivocations as we have already discussed. This is not the case, I fear. For, as Carnap has shown by means of simple examples and elegant arguments, the relevance concept of confirmation has some highly counterintuitive properties.

Suppose, for instance, that two scientists are interested in the same hypothesis *h*, and they go off to their separate laboratories to perform tests of that hypothesis. The tests yield two positive results, *i* and *j*. Each of the evidence statements is positively relevant to *h* in the presence of common background information *e*. Each scientist happily reports his positive finding to the other. Can they now safely conclude that the net result of both tests is a confirmation of *h*? The answer, amazingly, is no. As Carnap (1950b, 1962b) has shown, two separate items of evidence can each be positively relevant to a hypothesis, whereas their conjunction is negative to that very same hypothesis. He offers the following example (383–384; paraphrased for brevity).

Example 1. Let the prior evidence *e* contain the following information. Ten chess players participate in a chess tournament in New York City; some of them are local people, some from out of town; some are junior players, some are seniors; some are men (M), some are women (W). Their distribution is known to be as follows:

TABLE 12.1

	Local players	Out-of-towners
Juniors	M, W, W	M, M
Seniors	M, M	W, W, W

Furthermore, the evidence *e* is supposed to be such that on its basis each of the ten players has an equal chance of becoming the winner, hence 1/10. It is assumed that in each case of evidence that certain players have been eliminated, the remaining players have equal chances of winning.

Let *h* be the hypothesis that a man wins. Let *i* be the evidence that a local player wins; let *j* be the evidence that a junior wins. Using the equiprobability information embodied in the background evidence *e*, we can read the following values directly from table 12.1:

$$c(h,e) = 1/2 \quad c(h,e \cdot i) = 3/5 \quad R(i,h,e) = 1/10$$
$$c(h,e \cdot j) = 3/5 \quad R(j,h,e) = 1/10$$
$$c(h,e \cdot i \cdot j) = 1/3 \quad R(i \cdot j,h,e) = -1/6$$

Thus, i and j are each positively relevant to h, and the conjunction $i \cdot j$ is negatively relevant to h. In other words, i confirms h and j confirms h but $i \cdot j$ disconfirms h.

The setup of example 1 can be used to show that a given piece of evidence may confirm each of two hypotheses individually, while that same evidence disconfirms their conjunction Carnap (1950b; 1962b, 394–395).

Example 2. Let the evidence e be the same as in example 1. Let h be the hypothesis that a local player wins; let k be the hypothesis that a junior wins. Let i be evidence stating that a man wins. The following values can be read directly from table 12.1:

$$c(h,e) = 1/2 \qquad c(h,e \cdot i) = 3/5 \qquad R(i,h,e) = 1/10$$
$$c(k,e) = 1/2 \qquad c(k,e \cdot i) = 3/5 \qquad R(i,k,e) = 1/10$$
$$c(h \cdot k,e) = 3/10 \qquad c(h \cdot k,e \cdot i) = 1/5 \qquad R(i,h \cdot k,e) = -1/10$$

Thus, i confirms h and I confirms k, but I disconfirms $h \cdot k$.

In the light of this possibility it might happen that a scientist has evidence that supports the hypothesis that there is gravitational attraction between any pair of bodies when at least one is of astronomical dimensions and the hypothesis of gravitational attraction between bodies when both are of terrestrial dimensions, but which disconfirms the law of universal gravitation. In the next section we shall see that this possibility has interesting philosophical consequences. A further use of the same situation enables us to show that evidence can be positive to each of two hypotheses, but nevertheless negative to their disjunction Carnap (1950b; 1962b, 384).

Example 3. Let the evidence e be the same as in example 1. Let h be the hypothesis that an out-of-towner wins; let k be the hypothesis that a senior wins. Let i be evidence stating that a woman wins. The following values can be read directly from table 12.1:

$$c(h,e) = 1/2 \qquad c(h,e \cdot i) = 3/5 \qquad R(i,h,e) = 1/10$$
$$c(k,e) = 1/2 \qquad c(k,e \cdot i) = 3/5 \qquad R(i,k,e) = 1/10$$
$$c(h \lor k,e) = 7/10 \qquad c(h \lor k,e \cdot i) = 3/5 \qquad R(i,h \lor k,e) = -1/10$$

Thus, i confirms h and i confirms k, but i nevertheless disconfirms $h \lor k$. Imagine the following situation (example adapted from Carnap (1950b; 1962b, 367): a medical researcher finds evidence confirming the hypothesis that Jones is suffering from viral pneumonia and also confirming the hypothesis that Jones is suffering from bacterial pneumonia—yet this very same evidence disconfirms the hypothesis that Jones has pneumonia. It is difficult to entertain such a state of affairs, even as an abstract possibility.

These three examples illustrate a few members of a large series of severely counterintuitive situations that can be realized.

i. Each of two evidence statements may confirm a hypothesis, while their conjunction disconfirms it (Example 1).

ii. Each of two evidence statements may confirm a hypothesis, while their disjunction disconfirms it (Example 2a, Carnap 1950b; 1962a, 384).

iii. A piece of evidence may confirm each of two hypotheses, while it disconfirms their conjunction (Example 2).

iv. A piece of evidence may confirm each of two hypotheses, while it disconfirms their disjunction (Example 3).

This list may be continued by systematically interchanging positive relevance (confirmation) and negative relevance (disconfirmation) throughout the preceding statements. Moreover, a large number of similar possibilities obtain if irrelevance is combined with positive or negative relevance. Carnap presents a systematic inventory of all of these possible relevance situations (secs. 69–71).

In section 1, we mentioned that Hempel's special consequence condition does not hold for the relevance concept of confirmation. This fact immediately becomes apparent upon examination of statement iv above. Since h entails $h \lor k$, and since i may confirm h while disconfirming $h \lor k$, we have an instance in which evidence confirms a statement but fails to confirm one of its logical consequences. Statement ii, incidentally, shows that the converse consequence condition, which Hempel (Hempel 1965b, 32–33) discusses but does not adopt, also fails for the relevance concept of confirmation. Since $h \cdot k$ entails h, and since i may confirm h without confirming $h \cdot k$, we have an instance in which evidence confirms a hypothesis without confirming at least one statement from which that hypothesis follows. The failure of the special consequence condition and the converse consequence condition appears very mild when compared with the much stronger results i–iv, as well as analogous ones. Although one might, without feeling too queasy, give up the special consequence condition—the converse consequence condition being unsatisfactory on the basis of much more immediate and obvious considerations—it is much harder to swallow possibilities like i–iv without severe indigestion.

4. Duhem and Relevance

According to a crude account of scientific method, the testing of a hypothesis consists in deducing observational consequences and seeing whether or not the facts are as predicted. If the prediction is fulfilled we have a positive instance; if the prediction is false the result is a negative instance. There is a basic asymmetry between verification and falsification. If, from hypothesis h, an observational consequence o can be deduced, then the occurrence of a fact o' that is incompatible with o (o' entails $\sim o$) enables us to infer the falsity of h by good old modus tollens. If, however, we find the derived observational prediction fulfilled, we still cannot deduce the truth of h, for to do so would involve the fallacy of affirming the consequent.

There are many grounds for charging that the foregoing account is a gross oversimplification. One of the most familiar, which was emphasized by Duhem ([1906] 1954), points out that hypotheses are seldom, if ever, tested in isolation; instead, auxiliary hypotheses a are normally required as additional premises to make it possible to deduce observational consequences from the hypothesis h that is being tested. Hence, evidence o' (incompatible with o) does not entail the falsity of h, but only the falsity of the conjunction $h \cdot a$. There are no deductive grounds on which we can say that h rather than a is the member of the conjunction that has been falsified by the negative outcome. To whatever extent auxiliary hypotheses are invoked in the deduction of the observational consequence, to that extent the alleged deductive asymmetry of verification and falsification is untenable.

At this point, a clear warning should be flashed, for we recall the strange things that happen when conjunctions of hypotheses are considered. Example 2 of the previous section shows that evidence that disconfirms a conjunction $h \cdot a$ can nevertheless separately confirm each of the conjuncts. Is it possible that a negative result of a test of the hypothesis h, in which auxiliary hypotheses a were also involved, could result in the confirmation of the hypothesis of interest h and in a confirmation of the auxiliary hypotheses a as well?

It might be objected, at this point, that in the Duhemian situation o' is not merely negatively relevant to $h \cdot a$; rather,

$$o' \text{ entails} \sim (h \cdot a) \tag{1}$$

This objection, though not quite accurate, raises a crucial question. Its inaccuracy lies in the fact that h and a together do not normally entail o'; in the usual situation some initial conditions are required along with the main hypothesis and the auxiliary hypotheses. If this is the case, condition (1) does not obtain. We can deal with this trivial objection to (1), however, by saying that, since the initial conditions are established by observation, they are to be taken as part of the background evidence e that figures in all of our previous investigations of relevance relations. Thus, we can assert that, in the presence of background evidence, e, o can be derived from $h \cdot a$. This allows us to reinstate condition (1). Unfortunately, condition (1) is of no help to us. Consider the following situation:

Example 4. The evidence e contains the same equiprobability assumptions as the evidence in example 1, except for the fact that the distribution of players is as indicated in the following table:

TABLE 12.2

	Local players	Out-of-towners
Juniors	W	M, M
Seniors	M, M	M, W, W, W, W

Let h be the hypothesis that a local player wins; let k be the hypothesis that a junior wins. Let i be evidence stating that a man wins. In this case, condition (1) is satisfied; the evidence i is logically incompatible with the conjunction $h \cdot k$. The following values can be read directly from the table:

$$c(h, e) = 0.3 \quad c(h, e \cdot i) = 0.4 \quad R(i, h, e) = 0.1$$
$$c(k, e) = 0.3 \quad c(k, e \cdot i) = 0.4 \quad R(i, k, e) = 0.1$$
$$c(h \cdot k, e) = 0.1 \quad c(h \cdot k, e \cdot i) = 0 \quad R(i, h \cdot k, e) = -0.1$$

This example shows that evidence i, even though it conclusively refutes the conjunction $h \cdot k$, nevertheless confirms both h and k taken individually.

Here is the situation:

> Scientist Smith comes home late at night after a hard day at the lab. "How did your work go today, dear?" asks his wife.
>
> "Well, you know the Smith hypothesis, h_8, on which I have staked my entire scientific reputation? And you know the experiment I was running today to test my hypothesis? Well, the result was negative."
>
> "Oh, dear, what a shame! Now you have to give up your hypothesis and watch your entire reputation go down the drain!"
>
> "Not at all. In order to carry out the test, I had to make use of some auxiliary hypotheses."
>
> "Oh, what a relief—saved by Duhem! Your hypothesis wasn't refuted after all," says the philosophical Mrs. Smith.
>
> "Better than that," Smith continues, "I actually confirmed the Smith hypothesis."
>
> "Why that's wonderful, dear," replies Mrs. Smith, "you found you could reject an auxiliary hypothesis, and that in so doing, you could establish the fact that the test actually confirmed your hypothesis? How ingenious!"
>
> "No," Smith continues, "it's even better. I found I had confirmed the auxiliary hypotheses as well!"

This is the Duhemian thesis reinforced with a vengeance. Not only does a negative test result fail to refute the test hypothesis conclusively—it may end up confirming both the test hypothesis and the auxiliary hypotheses as well.

It is very tempting to suppose that much of the difficulty might be averted if only we could have sufficient confidence in our auxiliary hypotheses. If a medical researcher has a hypothesis about a disease that entails the presence of a certain microorganism in the blood of our favorite victim Jones, it would be outrageous for him to call into question the laws of optics as applied to microscopes as a way of explaining failure to find the bacterium. If the auxiliary hypotheses are well enough established beforehand, we seem to know where to lay the blame when our observational predictions go wrong. The question is how to establish the auxiliary hypotheses in the first place, for if the Duhemian is right, no hypotheses are ever tested in isolation. To test any hypothesis, according to this view, it is necessary to utilize auxiliary hypotheses; consequently, to establish our auxiliary hypotheses a for use in the tests of h, we would need some other auxiliary hypotheses a' to carry out the tests of a. A vicious regress threatens.

A more contextual approach might be tried (see, for example, Grünbaum 1976b). Suppose that a has been used repeatedly as an auxiliary hypothesis in the successful testing of other hypotheses j, k, l, and so on. Suppose, that is, that the conjunctions $j \cdot a$, $k \cdot a$, $l \cdot a$, and so forth have been tested and repeatedly confirmed—that is, all test results have been positively relevant instances. Can we say that a has been highly confirmed as a result of all of these successes? Initially, we might have been tempted to draw the affirmative conclusion, but by now we know better. Examples similar to those of the previous section can easily be found to show that evidence positively relevant to a conjunction of two hypotheses can nevertheless be negative to each conjunct (see Carnap 1950b; 1962b, 394–395, 3b). It is therefore logically possible for each confirmation of a conjunction containing *a* to constitute a disconfirmation of *a* — and indeed a disconfirmation of the other conjunct as well in each such case.

There is one important restriction that applies to the case in which new observational evidence refutes a conjunction of two hypotheses, namely, hypotheses that are incompatible on evidence *e·i* can have, at most, probabilities that add to 1. If *e·i* entails $\sim(h\cdot k)$

$$c(h,e\cdot i) + c(k,e\cdot i) \leq 1$$

Since we are interested in the case in which *i* is positively relevant to both *h* and *k*, these hypotheses must also satisfy the condition

$$c(h,e) + c(k,e) < 1$$

We have here, incidentally, what remains of the asymmetry between confirmation and refutation. If evidence *i* refutes the conjunction *h·k*, that fact places an upper limit on the sum of the probabilities of *h* and *k* relative to *e·i*. If, however, evidence *i* confirms a conjunction *h·k* while disconfirming each of the conjuncts, there is no lower bound (other than zero) on the sum of their degrees of confirmation on *i*.

In this connection, let us recall our ingenious scientist Smith, who turned a refuting test result into a positive confirmation of both his pet hypothesis h_8 and his auxiliary hypotheses *a*. We see that he must have been working with a test hypothesis or auxiliaries (or both) that had rather low probabilities. We might well question the legitimacy of using hypotheses with degrees of confirmation appreciably less than 1 as auxiliary hypotheses. If Smith's auxiliaries *a* had decent degrees of confirmation, his own hypothesis h_8 must have been quite improbable. His clever wife might have made some choice remarks about his staking an entire reputation on so improbable a hypothesis. But I should not get carried away with dramatic license. If we eliminate all the unnecessary remarks about staking his reputation on h_8 and regard it rather as a hypothesis he finds interesting, then its initial improbability may be no ground for objection. Perhaps every interesting general scientific hypothesis starts its career with a very low prior probability. Knowing, as we do, that a positively relevant instance may disconfirm both our test hypothesis and our auxiliaries, whereas a negative instance my confirm them both, we see that there remains a serious, and as yet unanswered, question: How can any hypothesis ever become either reasonably well confirmed or reasonably conclusively disconfirmed (in the absolute sense). It obviously is still an open question of how we could ever get any well-confirmed hypotheses to serve as auxiliaries for the purpose of testing other hypotheses.

Suppose, nevertheless, that we have a hypothesis *h* to test and some auxiliaries *a* that will be employed in conducting the test and that somehow we have ascertained that *a* has a higher prior confirmation than *h* on the initial evidence *e*:

$$c(a,e) > c(h,e)$$

Suppose, further, that as the result of the test we obtain negative evidence o' which refutes the conjunction $h \cdot a$ but which confirms both h and a. Thus, o' entails $\sim(h \cdot a)$ and

$$c(h, e \cdot o') > c(h, e) \qquad c(a, e \cdot o') > c(a, e)$$

We have already seen that this can happen (in example 4), but now we ask the further question, is it possible that the posterior confirmation of h is greater than the posterior confirmation of a? In other words, can the negative evidence o' confirm both conjuncts and do so in a way that reverses the relation between h and a? A simple example will show that the answer is affirmative.

Example 5. The Department of History and Philosophy of Science at Polly Tech had two openings, one in history of science and the other in philosophy of science. Among the 1000 applicants for the position in history, 100 were women. Among the 2000 applicants for the position in philosophy, 100 were women. Let h be the hypothesis that the history job was filled by a woman; let k be the hypothesis that the philosophy job was filled by a woman. Since both selections were made by the use of a fair lottery device belonging to the inductive logician in the department,

$$c(h, e) = 0.1$$

$$c(k, e) = 0.05$$

$$c(h, e) > c(k, e)$$

Let i be the evidence that the two new appointees were discovered engaging in heterosexual intercourse with each other in the office of the historian. It follows at once that

$$c(h \cdot k, e \cdot i) = 0$$

$$c(h, e \cdot i) + c(k, e \cdot i) = 1$$

that is, one appointee was a woman and the other a man, but we do not know which is which. Since it is considerably more probable, let us assume, that the office used was that of the male celebrant, we assign the values

$$c(h, e \cdot i) = 0.2 \qquad c(k, e \cdot i) = 0.8$$

with the result that

$$c(h, e \cdot i) < c(k, e \cdot i)$$

This illustrates the possibility of a reversal of the comparative relation between the test hypothesis and auxiliaries as a result of refuting evidence. It shows that a's initial superiority to h is no assurance that it will still be so subsequent to the refuting evidence. If, prior to the negative test result, we had to choose between h and a, we would have preferred a, but after the negative outcome h is preferable to a.

There is one significant constraint that must be fulfilled if this reversal is to occur in the stated circumstances. If our auxiliary hypotheses a are initially better confirmed than our test hypothesis h, and if the conjunction $h \cdot a$ is refuted by evidence o' that is positively relevant to both h and a, and if the posterior confirmation of h is greater than the posterior confirmation of a, then the prior confirmation of a must have been less than 1/2. For,

$$c(h,e \cdot o') + c(a,e \cdot o') \leq 1$$

and

$$c(h,e \cdot o') > c(a,e \cdot o')$$

Hence,

$$c(a,e \cdot o') < 1/2$$

Moreover,

$$c(a,e) < c(a,e \cdot o')$$

Therefore,

$$c(a,e) < 1/2$$

It follows that if a is initially more probable than h and also initially more probable than its own negation $\sim a$, then it is impossible for a refuting instance o' that confirms both h and a to render h more probable than a. Perhaps that is some comfort. If our auxiliaries are more probable than not, and if they are better established before the test than our test hypothesis h, then a refuting test outcome that confirms both h and a cannot make h preferable to a.

But this is not really the tough case. The most serious problem is whether a refutation of the conjunction $h \cdot a$ can render h more probable than a by being positively relevant to h and negatively relevant to a, even when a is initially much more highly confirmed than h. You will readily conclude that this is possible; after all of the weird outcomes we have discussed, this situation seems quite prosaic. Consider the following example:

Example 6. Let

$e =$ Brown is an adult American male
$h =$ Brown is a Roman Catholic
$k =$ Brown is married

and suppose the following degrees of confirmation to obtain

$$c(h,e) = 0.3$$

$$c(k,e) = 0.8$$

$$c(h \cdot k, e) = 0.2$$

Let i be the information that Brown is a priest—that is, an ordained clergyman of the Roman Catholic, Episcopal, or Eastern Orthodox Church. Clearly, i refutes the conjunction $h \cdot k$, so

$$c(h \cdot k, e \cdot i) = 0$$

Since the overwhelming majority of priests in America are Roman Catholic, let us assume that

$$c(h, e \cdot i) = 0.9$$

and since some, but not all, non–Roman Catholic priests marry, let

$$c(k, e \cdot i) = 0.05$$

We see that i is strongly relevant to both h and k; in particular, it is positively relevant to h and negatively relevant to k. Moreover, whereas k has a much higher degree of confirmation than h relative to the prior evidence e, h has a much higher degree of confirmation than k on the posterior evidence $e \cdot i$. Thus, the refuting evidence serves to reverse the preferability relation between h and k.

It might be helpful to look at this situation diagrammatically and to think of it in terms of class ratios or frequencies. Since class ratios satisfy the mathematical calculus of probabilities, they provide a useful device for establishing the possibility of certain probability relations. With our background evidence e let us associate a reference class A, and with our hypotheses h and k let us associate two overlapping subclasses B and C, respectively. With our additional evidence i let us associate a further subclass D of A. More precisely, let

$$e = x \in A, \ h = x \in B, \ k = x \in C, \ i = x \in D$$

Since we are interested in the case in which the prior confirmation of k is high and the prior confirmation of h is somewhat lower, we want most of A to be included in C and somewhat less of A to be included in B. Moreover, since our hypotheses h and k should not be mutually exclusive on the prior evidence e alone, B and C must overlap. However, neither B nor C can be a subset of the other; they must be mutually exclusive within D since h and k are mutually incompatible on additional evidence i. Moreover, because we are not considering cases in which $e \cdot i$

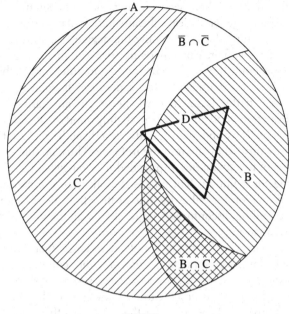

FIGURE 12.1

entails either *h* or *k* alone, the intersections of *D* with *B* and with *C* cannot be null. We incorporate all of these features in figure 12.1.

In order to achieve the desired result—that is, to have the posterior confirmation of *h* greater than the posterior confirmation of *k*—it is only necessary to draw *D* so that its intersection with *B* is larger than its intersection with *C*. This is obviously an unproblematic condition to fulfill. Indeed, there is no difficulty in arranging it so that the proportion of *D* occupied by its intersection with *B* is larger than the proportion of *A* occupied by its intersection with *C*. When this condition obtains, we can say not only that the evidence *i* has made the posterior confirmation of *h* greater than the posterior confirmation of *k* (thus reversing their preferability standing) but also that the posterior confirmation of *h* is greater than the prior confirmation of *k*. Translated into the Duhemian situation, this means that not only can the refuting evidence *o′* disconfirm the initially highly probable auxiliary hypotheses *a* but it can also confirm the test hypothesis *h* to the extent that its posterior confirmation makes it more certain than were the auxiliaries before the evidence *o′* turned up. This set of relationships is easily seen to be satisfied by example 6 if we let

$$A = \text{American men} \quad B = \text{Roman Catholics}$$
$$C = \text{married men} \quad D = \text{priests}$$

It is evident from figure 12.1 that *C* can be extremely probable relative to *A* without presenting any serious obstacle to the foregoing outcome, provided *D* is very much smaller than A. There seems little ground for assurance that a refuting experimental result will generally leave the auxiliaries intact, rather than drastically disconfirming them and radically elevating the degree of confirmation of the test hypothesis. This state of affairs can apparently arise with distressing ease, and there are no evident constraints that need to be fulfilled in order for it to happen. There seems to be no basis for confidence that it does not happen frequently in real scientific situations. If this is so, than whenever our auxiliaries have an initial degree of confirmation that falls ever so slightly short of certainty, we cannot legitimately infer with confidence that the test hypothesis h, rather than the auxiliaries, is disconfirmed by the refuting instance. Thus, no matter how high the initial probability of the auxiliaries a (provided it is somewhat less than certainty), it is still possible for a finding that entails the falsity of the conjunction $h \cdot a$ to constitute a confirmation for either the one or the other. We certainly cannot say that the negative finding disconfirms h rather than a on the ground that a is more probable initially than h.

A parallel remark can be made about an instance that confirms the conjunction $h \cdot a$. Such an instance might disconfirm either conjunct, and we have no way of saying which. In view of these dismal facts, we may well repeat, with even greater emphasis, the question posed earlier: How can any hypothesis (including those we need as auxiliary hypotheses for the testing of other hypotheses) ever be reasonably well confirmed or disconfirmed (in the absolute sense)?

The (to my mind) shocking possibilities that we have been surveying arise as consequences of the fact that we have been construing "confirm" in the relevance sense. What happens, I believe, is that we acknowledge with Carnap and Hempel that the classificatory concept of confirmation is best understood, in most contexts, as a concept of positive relevance defined on the basis of some quantitative degree of confirmation function. Then, in contexts such as the discussion of the Duhemian thesis, we proceed to talk casually about confirmation, forgetting that our intuitive notions are apt to be very seriously misleading. The old ambiguity of the absolute and the relevance sense of confirmation infect our intuitions, with the result that all kinds of unwarranted suppositions insinuate themselves. We find it extraordinarily difficult to keep a firm mental grasp upon such strange possibilities as we have seen in this section and the previous one.

5. Analysis of the Anomalies

There is, of course, as striking a contrast between the "hypotheses" and "evidence" involved in our contrived examples, on the one hand, and the genuine hypotheses and evidence to be found in actual scientific practice, on the other. This observation might easily lead to the charge that the foregoing discussion is not pertinent to the logic of actual scientific confirmation, as opposed to the theory of confirmation constructed by Carnap (1950b, 1962b) on highly artificial and oversimplified languages. This irrelevance is demonstrated by the fact, so the objection might continue, that the kinds of problems and difficulties we have been discussing simply do not arise when real scientists test serious scientific hypotheses.

This objection, it seems to me, is wide of the mark. I am prepared to grant that such weird possibilities as we discussed in previous sections do not arise in scientific practice; at least, I have no concrete cases from the current or past history of science to offer as examples of them. This is, however, a statement of the problem rather than a solution. Carnap has provided a number of examples that, on the surface at least, seem to make a shambles of confirmation; why do they not also make a shambles of science itself? There can be no question that, for example, one statement can confirm each of two other statements separately while at the same time disconfirming their disjunction or conjunction. If that sort of phenomenon never occurs in actual scientific testing, it must be because we know something more about our evidence and hypotheses than merely that the evidence confirms the hypotheses. The problem is to determine the additional factors in the actual situation that block the strange results we can construct in the artificial case. In this section, I shall try to give some indications of what seems to be involved.

The crux of the situation seems to be the fact that we have said very little when we have stated merely that a hypothesis *h* has been confirmed by evidence *i*. This statement means, of course, that *i* raises the degree of confirmation of *h*, but that in itself provides very little information. It is by virtue of this paucity of content that we can go on and say that this same evidence *i* confirms hypothesis *k* as well, without being justified in saying anything about the effect of *i* upon the disjunction or the conjunction of *h* with *k*.

This state of affairs seems strange to intuitions that have been thoroughly conditioned on the extensional relations of truth-functional logic. Probabilities are not extensional in the same way; given the truth values of *h* and *k* we can immediately ascertain the truth values of the disjunction and the conjunction. The degrees of confirmation ("probability values") of *h* and *k* do not, however, determine the degree of confirmation of either the disjunction or the conjunction. This fact follows immediately from the addition and multiplication rules of the probability calculus:

$$c(h \vee k, e) = c(h,e) + c(k,e) - c(h \cdot k, e) \tag{2}$$

$$c(h \cdot k, e) = c(h,e) \times c(k, h \cdot e) = c(k,e) \times c(h, k \cdot e) \tag{3}$$

To determine the probability of the disjunction, we need, in addition to the values of the probabilities of the disjuncts, the probability of the conjunction. The disjunctive probability is the sum of the probabilities of the two disjuncts if they are mutually incompatible in the presence of evidence *e*, in which case $c(h \cdot k, e) = 0$.[7] The probability of the conjunction, in turn, depends upon the probability of one of the conjuncts alone and the conditional probability of the other conjunct, given the first.[8] If

$$c(k, h \cdot e) = c(k,e) \tag{4}$$

the multiplication rule assumes the special form

$$c(h \cdot k, e) = c(h,e) \times c(k,e) \tag{5}$$

in which case the probability of the conjunction is simply the product of the probabilities of the two conjuncts. When condition (4) is fulfilled, h and k are said to be independent of one another.[9] Independence, as thus defined, is obviously a relevance concept, for (4) is equivalent to the statement that h is irrelevant to k, that is, $R(h,k,e) = 0$.

We can now see why strange things happen with regard to confirmation in the relevance sense. If the hypotheses h and k are mutually exclusive in the presence of e (and a fortiori in the presence of $e \cdot i$), then

$$c(h \vee k, e) = c(h,e) + c(k,e) \tag{6}$$

$$c(h \vee k, e \cdot i) = c(h, e \cdot i) + c(k, e \cdot i) \tag{7}$$

so that if

$$c(h, e \cdot i) > c(h,e) \quad \text{and} \quad c(k, e \cdot i) > c(k,e) \tag{8}$$

it follows immediately that

$$c(h \vee k, e \cdot i) > c(h \vee k, e) \tag{9}$$

Hence, in this special case, if i confirms h and i confirms k, then i must confirm their disjunction. This results from the fact that the relation between h and k is the same in the presence of $e \cdot i$ as it is in the presence of e alone.[10]

If, however, h and k are not mutually exclusive on evidence e, we must use the general formulas

$$c(h \vee k, e) = c(h,e) + c(k,e) - c(h \cdot k, e) \tag{10}$$

$$c(h \vee k, e \cdot i) = c(h, e \cdot i) + c(k, e \cdot i) - c(h \cdot k, e \cdot i) \tag{11}$$

Now, if it should happen that the evidence i drastically alters the relevance of h to k in just the right way, our apparently anomalous results can arise. For then, as we shall see in a moment by way of a concrete (fictitious) example, even though condition (8) obtains—that is, i confirms h and i confirms k—condition (9) may fail. Thus, if

$$c(h \cdot k, e \cdot i) > c(h \cdot k, e) \tag{12}$$

it may happen that

$$c(h \vee k, e \cdot i) < c(h \vee k, e) \tag{13}$$

that is, i disconfirms $h \vee k$. Let us see how this works.

Example 7. Suppose h says that poor old Jones has bacterial pneumonia, and k says that he has viral pneumonia. I am assuming that these are the only varieties of pneumonia, so that $h \vee k$ says simply that he has pneumonia. Let evidence e contain the results of a superficial diagnosis, as well as standard medical background knowledge about the disease, on the basis of which we can establish degrees of confirmation for h, k, $h \cdot k$, and $h \vee k$. Suppose, moreover, that the probability on e that Jones has both viral and bacterial pneumonia is quite low, that is, that people do not often get then both simultaneously. For the sake of definiteness, let us introduce some numerical values. Suppose that on the basis of the superficial diagnosis it is 98% certain that Jones has one or the other form of pneumonia, but the diagnosis leaves it entirely uncertain which type he has. Suppose, moreover, that on the basis of e, there is only a 2% chance that he has both. We have the following values:

$$c(h,e) = 0.50 \qquad c(k,e) = 0.50$$
$$c(h \vee k,e) = 0.98 \qquad c(h.k,e) = 0.02$$

These values satisfy the addition formula (2). Suppose, now, that there is a certain test that indicates quite reliably those rare cases in which the subject has both forms of pneumonia. Let i be the statement that this test was administered to Jones with a positive result, and let this result make it 89% certain that Jones has both types. Assume, moreover, that the test rarely yields a positive result if the patient has only one form of pneumonia (that is, when the positive result occurs for a patient who does not have both types, he usually has neither type). In particular, let

$$c(h,e \cdot i) = 0.90, \quad c(k,e \cdot i) = 0.90, \quad c(h \cdot k,e \cdot i) = 0.89$$

from which it follows that

$$c(h \vee k,e \cdot i) = 0.91 < c(h \vee k,e) = 0.98$$

The test result i thus confirms the hypothesis that Jones has bacterial pneumonia and the hypothesis that Jones has viral pneumonia, but it disconfirms the hypothesis that Jones has pneumonia.

It achieves this feat by greatly increasing the probability that he has both. This increase brings about a sort of clustering together of cases of viral and bacterial pneumonia, concomitantly decreasing the proportion of people with only one of the two types.

The effect is easily seen diagrammatically in figure 12.2. Even though the rectangles in b_2 are larger than those in a_2, those in b_2 cover a smaller total area because of their very much greater degree of overlap. Taking the rectangles to represent the number of cases of each type, we see graphically how the probability of each type of pneumonia can increase simultaneously with a decrease in the overall probability of pneumonia. The evidence i has significantly altered the relevance relation between h and k. Using the multiplication formula (3), we can establish that

$$c(k,h \cdot e) = 0.04 \qquad c(k,h \cdot e \cdot i) = 0.99$$
$$R(h,k,e) = -0.46 \qquad R(h,k,e \cdot i) = 0.09$$

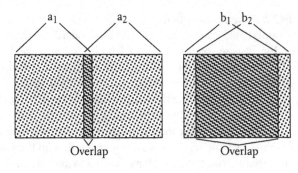

FIGURE 12.2

In the presence of *e* alone, *h* is negatively relevant to *k*; in the presence of *i* as well, *h* becomes positively relevant to *k*. There is nothing outlandish in such changes of relevance in the light of additional evidence. This case thus exemplifies condition (13) by satisfying condition (12).

A similar analysis enables us to understand how an item of evidence can confirm each of two hypotheses, while disconfirming—indeed, even while conclusively refuting—their conjunction. If hypothesis *h* and *k* are independent of each other in the presence of evidence *e·i* and also in the presence of *e* alone, the following relations obtain:

$$c(h \cdot k, e) = c(h, e) \times c(k, e) \tag{14}$$

$$c(h \cdot k, e \cdot i) = c(h, e \cdot i) \times c(k, e \cdot i) \tag{15}$$

so that if

$$c(h, e \cdot i) > c(h, e) \quad \text{and} \quad c(k, e \cdot i) > c(k, e) \tag{16}$$

it follows immediately that

$$c(h \cdot k, e \cdot i) > c(h \cdot k, e) \tag{17}$$

Hence, in this special case, if *i* confirms *h* and *i* confirms *k*, then *i* must confirm *h·k*.

A different situation obtains if *h* and *k* are not independent on both *e* and *e·i*; in that case we must use the general formulas

$$c(h \cdot k, e) = c(h, e) \times c(k, h \cdot e) \tag{18}$$

$$c(h \cdot k, e \cdot i) = c(h, e \cdot i) \times c(k, h \cdot e \cdot i) \tag{19}$$

Even given that condition (16) still obtains, so that

$$c(k,e \cdot i) > c(k,e) \tag{20}$$

it is still possible that

$$c(k,h \cdot e \cdot i) < c(k,h \cdot e)^{11} \tag{21}$$

which makes it possible, in turn, that

$$c(h \cdot k, e \cdot i) < c(h \cdot k, e) \tag{22}$$

Since, according to (20) and (21),

$$c(k,e \cdot i) - c(k,e) = R(i,k,e) > 0 \tag{23}$$

$$c(k,h \cdot e \cdot i) - c(k,h \cdot e) = R(i,k,h \cdot e) < 0 \tag{24}$$

the possibility of i confirming each of two hypotheses while disconfirming their conjunction depends upon the ability of h to make a difference in the relevance of i to k. We said above, however, that the occurrence of the strange confirmation phenomena depends upon the possibility of a change in the relevance of the hypotheses to one another in the light of new evidence. These characterizations are, however, equivalent to one another, for the change in relevance of i to k brought about by h is equal to the change in relevance of h to k brought about by i; that is,

$$R(h,k,e) - R(h,k,e \cdot i) = R(i,k,e) - R(i,k,h \cdot e)^{12} \tag{25}$$

We can therefore still maintain that the apparently anomalous confirmation situation arises from the ability of new evidence to change relations of relevance between the hypotheses, as was suggested by our initial examination of the general addition and multiplication rules (2) and (3).

Let us illustrate the conjunctive situation with another concrete (though fictitious) example.

Example 8. Suppose that the evidence e tells us that two radioactive atoms A and B decay, each ejecting a particle, and that the probability in each case is 0.7 that it is an alpha particle, 0.2 that it is a negative electron, and 0.1 that it is a positive electron (positron). Assume that the two emissions are independent of one another. Let h be the statement that atom A emits a negative electron; let k be the statement that atom B emits a negative electron. We have the following probabilities:

$$c(h,e) = 0.2, \quad c(k,e) = 0.2, \quad c(h \cdot k, e) = 0.04$$

Let i be the observation that the two particles approach one another and annihilate upon meeting. Since this occurrence requires the presence of one positive and one negative electron, i entails $\sim(h \cdot k)$. At the same time, since a negative electron must have been present, and since it is just as probable that it was emitted by atom A as atom B, we have

$$c(h,e \cdot i) = 0.5 \quad \text{and} \quad c(k,e \cdot i) = 0.5$$

Hence, evidence i, which refutes the conjunction of the two hypotheses, confirms each one of them.[13] This occurs because evidence i makes the hypotheses h and k, which were independent of one another on evidence e alone, into mutually exclusive and exhaustive alternatives; that is,

$$c(k,h \cdot e) - c(k,e) = R(h,k,e) = 0$$

$$c(k,h \cdot e \cdot i) - c(k,e \cdot i) = R(h,k,e \cdot i) = -0.5$$

Hypotheses that were totally irrelevant to each other in the absence of evidence i become very strongly relevant in the presence of i. Again, there is nothing especially astonishing about such a change in relevance as a result of new evidence.

Since, as we have seen, all the trouble seems to arise out of a change in the relevance of one hypothesis to the other as a result of new evidence, the most immediate suggestion might be to choose hypotheses h and k, whose mutual relevance relations will not change in the light of the new evidence i. We have noted that this constancy of relevance is guaranteed if we begin with hypotheses that are mutually exclusive on evidence e; they remain mutually exclusive on any augmented evidence $e \cdot i$. But when we use the conjunction of a test hypothesis h with auxiliary hypotheses a in order to attempt a test of h, we certainly do not want h and a to be mutually exclusive—that is, we do not want to be in the position of knowing that we must reject our test hypothesis h if we are prepared to accept the auxiliaries a, even without the addition of any new evidence i. It would be more reasonable to insist that the auxiliary hypotheses a should themselves be neutral (irrelevant, independent) to the test hypothesis h. If that condition is satisfied, we can accept the auxiliary hypotheses a and still keep an entirely open mind regarding h. We cannot, however, demand that h and a remain irrelevant to one another after the new evidence i has been obtained. The interesting test situation is that in which, given e, $h \cdot a$ entails some observational consequence o. If a result o' occurs that is incompatible with o, then our hypotheses h and a, which may have been independent in the presence of e alone, are mutually exclusive in the light of the new evidence o'. Thus, the very design of that kind of test requires hypotheses whose mutual relevance relations are bound to change in the face of new evidence. Several of our examples (5, positions at Polly Tech; 6, celibacy among priests; 8, electron-positron annihilation) show exactly what can happen when new evidence renders independent hypotheses mutually incompatible.

6. Conclusions

The crude hypothetico-deductive account of scientific inference, according to which hypotheses are confirmed by deducing observational consequences that are then verified by observation, is widely recognized nowadays as an oversimplification (even leaving aside the Duhemian objections). One can hardly improve upon Russell's (1939, 149) classic example. From the hypothesis, "Pigs have wings," in conjunction with the observed initial condition, "Pigs are good to eat," we can deduce the consequence, "Some winged things are good to eat." Upon observing that such winged creatures as ducks and turkeys are good to eat, we have a hypothetico-deductive confirmation of the hypothesis "Pigs have wings." I am inclined to agree with a wide variety of authors who hold that something akin to a Bayesian schema must be involved in the confirmation of scientific hypotheses. If this is correct, it is entirely possible to have positive hypothetico-deductive test results that do not confirm the hypothesis (that is, they do not add anything to its degree of confirmation on prior evidence). To emphasize this point, Reichenbach (1949, 96) aptly described the crude hypothetico-deductive inference as an instance of "the fallacy of incomplete schematization." Recognition of the basic inadequacy of the hypothetico-deductive schema does no violence to the logic of science; it only shows that the methods of science are more complex than this oversimplified schema.

Acknowledging the fact that positive hypothetico-deductive instances may not be confirming instances, I have been discussing the logic of confirmation—that is, I have been investigating the conclusions that can be drawn from the knowledge that this or that evidence confirms this or that hypothesis. By and large, we have found this logic to be poverty-ridden. Although evidence *i* confirms hypotheses *h* and *k*, we have found that we cannot infer that *i* confirms *h·k*. Evidence *i* may in fact confirm *h·k*, but to draw that conclusion from the given premises would be another instance of the fallacy of incomplete schematization. Indeed, our investigations have revealed exactly what is missing in the inference. In addition to knowing that *i* is positively relevant to *h* and positively relevant to *k*, we must know what bearing *i* has on the relevance of *h* to *k*. If this is known quantitatively, and if the degrees of relevance of *i* to *h* and to *k* are also known quantitatively, we can ascertain the relevance of *i* to *h·k* and to *h∨k*. Without this quantitative knowledge, we cannot say much of anything. The moral is simple: Even if we base our qualitative concept of confirmation (in the relevance sense) upon a quantitative concept of degree of confirmation, the resulting qualitative concept is not very serviceable. It is too crude a concept, and it doesn't carry enough information to be useful. In order to make any substantial headway in understanding the logic of evidential support of scientific hypotheses, we must be prepared to work with at least crude estimates of quantitative values of degree of confirmation and degree of relevance. Then, in contexts such as the discussion of the Duhemian problem, we must bring the more sophisticated concepts to bear if we hope to achieve greater clarity and avoid logical fallacies. In detailing the shortcomings of the qualitative concept of confirmation, we have, in a way, shown that this sort of confirmation theory is a shambles, but we have done no more violence to the logic of science than to show that it embodies more powerful concepts.

If we are willing, as Carnap has done, to regard degree of confirmation (in the nonrelevance sense) as a probability—that is, as a numerical functor that satisfies the probability calculus—then we can bring the structure of the quantitative probability concept to bear on problems of confirmation. With this apparatus, which gives us the power of Bayes's theorem, we can aspire to a much fuller understanding of relations of confirmation (in both the absolute and the relevance senses).

We can also provide an answer to many who have claimed that confirmation is not a probability concept. Confirmation in the relevance sense is admittedly not a probability; as we have insisted, it is not to be identified with high probability. A quantitative concept of degree of relevance can nevertheless be defined in terms of a concept of degree of confirmation. Degree of relevance, as thus defined, is not a probability; it obviously can take on negative values, which degree of probability cannot do. It is a probability concept, however, in the sense that it is explicitly defined in terms of degree of confirmation, which is construed as a probability concept. Thus, even though degree of confirmation in the relevance sense cannot be construed as degree of probability, this fact is not a basis for concluding that the concept of probability is an inadequate or inappropriate tool for studying the logic of evidential relations between scientific hypotheses and observational evidence. Moreover, it provides no basis whatever for rejecting the notion that high probabilities, as well as high content, are what we want our scientific hypotheses eventually to achieve on the basis of experimental testing.

Notes

Introduction to Part I: Reality

1. That does not mean, according to many, that the menace of logical positivism no longer exists. Logiposiphobia is rampant. Perhaps there still are positivists who simply refuse to admit what they are. Perhaps there are philosophers who are not, strictly speaking, logical positivists but whose views are nevertheless seriously contaminated by positivistic tendencies. I am vividly reminded of the deplorable era of Senator Joseph McCarthy, who found no actual communist spies but who believed that the United States and its institutions were thoroughly infested with communist sympathizers whose aim was to undermine the institutions of that nation.

2. See Rescher (1985). The preface (vii) incorrectly gives 1982 as the year of the conference.

3. For example, van Fraassen (1980) seems to presuppose a negative answer without explicitly raising the question.

4. By 1938 all three had emigrated to the United States.

5. In this year Perrin published his first results on the study of Brownian motion and the ascertainment of Avogadro's number.

6. Nye (1984) contains many of the original sources for the period 1860–1911.

Chapter 1

1. I now believe that Carnap never accepted scientific realism; see chapter 2, "Scientific Realism in the Empiricist Tradition" (note added in 1994).

2. Van Fraassen maintains that a viable, though imprecise, distinction can be made between what is observable and what is unobservable, and I shall go along with this assumption for purposes of the present discussion.

3. John Locke ([1690] 1924) characterizes as a secondary quality "the power that is in any body, by reason of its *insensible* primary qualities to operate after a peculiar manner on any of our senses, and thereby produce in *us* the different ideas of several colours, sounds, smells, tastes, &c." (bk. II, chap. 8, sec. 23; my emphasis). Locke further characterizes as a tertiary quality "the power that is in any body, by reason of the particular constitution of its primary qualities, to make such a change in the bulk, figure, texture, and motion of another

body, as to make it operate on our senses differently from what it did before. Thus the sun has a power to make wax white, and fire, to make lead fluid." Such qualities thus influence the insensible primary qualities that are responsible for the secondary qualities of objects. I am grateful to Adolf Grünbaum for reminding me of Locke's doctrine of tertiary qualities in connection with this chapter.

4. As Brian Skyrms pointed out in discussion, the fact that "hidden variable" theories appear to be untenable does not mean that common cause explanations are precluded. The question, rather, is whether whatever it is that produces quantum-mechanical correlations of the sort found in the Einstein-Podolsky-Rosen paradox can appropriately be considered a *causal* mechanism.

5. In a personal letter van Fraassen writes:

"Induction" is used in a broad and narrow sense. I believe in it in the broad sense that it is possible to form reasonable expectations of the future (and reasonable opinions about the past and elsewhere) and that it is possible "to attain the truth" (as William James put it), i.e., possible that these expectations/opinions should be correct.

But "induction" is also used in a narrow sense, to denote a putative theory or logic which systematically presents quasi-logical relationships between hypotheses and evidence or something like argument forms, going beyond considerations of consistency (even of the generalized consistency that Bayesians call coherence). I now think that no such theory or logic has ever really existed, because even the most ardent inductivists have presented putative canons of induction themselves as defeasible, rough, provisional, etc. (These canons have included numerical induction, statistical syllogism, various forms of inference to the best explanation, common cause principles, principles of analogical inference. Let me come back to this below.) So there is, as it were, nothing (as yet?) to believe in. I must say I am highly skeptical now of the program to produce such a theory or logic (and I think philosophers have jumped the gun in talking as if they were referring to an extant theory.).

Maybe you think this is a bit harsh, but I formed the opinion first of all about the false (also putative) theory of confirmation (for which a similar distinction can be drawn). In *Fact, Fiction, and Forecast* (1955), Chap. 2, Goodman tells his University of London audience that his theory is the twentieth century replacement for inductive logic, and that Hempel has already constructed its rudiments. He then proceeds to precipitate this fledgling theory into the projectability mess—i.e., he showed that the announced principles of the theory had to be regarded as defeasible, rough, provisional, . . . , etc. even by its adherents (or I should say, pursuants).

Having reached those conclusions about the theory of confirmation, I then began to think that the same also applied to induction. That theory of confirmation is one whose *esse* is *sperari*! How absolutely indispensable is free logic for a discipline which talks about its own theories before they exist!

In the light of this, let us look at your remarks about the common cause principle (pp. 16–17). In your view, common cause is a good principle, but not universally applicable—it needs to be qualified. As far as I can see, all proffered canons of induction have suffered this same fate, except those too vague to suffer any fate. As long as the nature of the qualification is still left open, however, I do not see how you can be sure of your *next* statement: that the appeal to the common cause principle is legitimate in connection with Hacking and Perrin.

I agree with van Fraassen's criticism, at least to the extent of endorsing the view that any empiricist who wants to be a realist must be prepared to exhibit the inductive logic or theory of confirmation in which the required inferences can be carried out. This is a large part of the motivation for the inquiries in part II of this volume.

Chapter 2

1. Chapter 1 of Carnap (1966a) deals with explanation, but it contains a blatant inconsistency. On p. 7 it is said that "all explanation" conforms to a deductive schema, whereas on the next page the existence of statistical explanations, which do not conform to that schema, are affirmed. The inconsistency was removed—by adding the word "deductive" between "all" and "explanation"—when this book was reissued in paperback with the new title, *Introduction to the Philosophy of Science*. Reichenbach's (1956) *The Direction of Time* contains many suggestive remarks about explanation but not an articulated theory.

2. For an account of philosophical discussions of scientific explanation following (Hempel and Oppenheim 1948), see Salmon ([1989] 1990).

3. Gereon Wolters (1989, 654–655) argues that Mach did eventually accept the atomic theory, but he did so as an instrumentalist and a phenomenalist. He thus continued to refuse to draw any ontological conclusions regarding the existence of atoms.

4. These remarks occur in the last paragraph of chap. 26 (256). He retracted them by modifying that paragraph in the subsequent edition, *An Introduction to the Philosophy of Science*; for full details see my "Carnap on Realism: Comments on Parrini" in (Salmon and Wolters 1994).

5. This question about the interpretation of Carnap is discussed in detail in Parrini (1994) and (Salmon 1994).

6. Reichenbach makes no reference to Einstein-Podolsky-Rosen (1935) in his 1938 book. Because he spent the years from 1933 to 1938 in Turkey, he may not have seen it before completing this book. He did discuss the Einstein-Podolsky-Rosen correlations in his 1946 book, where he took them very seriously indeed, invoking a nonstandard (three-valued) logic to deal with them.

7. I have discussed this issue in greater detail in Salmon (1999).

8. For purposes of the present discussion I am omitting explanations of general laws.

9. My remarks in this paragraph are based on my reading of the paragraph that begins on the bottom of p. 219 and the next two paragraphs on p. 220 of "The Theoretician's Dilemma" as reprinted in Hempel (1965c). I hope I have not distorted his meaning.

10. Although *The Philosophy of Rudolf Carnap* was published in 1963, it was more than a decade in the making. Hempel's contribution to this volume was apparently composed before "The Theoretician's Dilemma," but "The Theoretician's Dilemma" lists the contribution to the Schilpp volume as forthcoming.

11. My view of the matter was conditioned in part by Carnap's (1966a) misleading remarks on instrumentalism and realism in *Philosophical Foundations of Physics*, which was published shortly after the Schilpp volume.

12. Although Carnap is here referring to the debate over epistemological realism, it seems safe to say that the same considerations would apply to the debate over scientific realism.

13. There are, of course, dissenters, chief among whom is Bas van Fraassen (1980). He indicated that we may accept the atomic hypothesis as empirically adequate, without being committed to belief in the existence of atoms. His view seems to resemble, in some respects, the position of Mach (see note 3).

Chapter 3

1. This phrase is borrowed from a different context in Perrin (1923, 207).

2. Carnap's *Aufbau* (1928a; English trans 1967) contains the logical positivists' most heroic attempt to carry out the program of phenomenalism. See Goodman (1951) for a penetrating study of phenomenalism and an illuminating critique of Carnap.

3. Glymours's (1980) account of the history of atomism ends with the work of Cannizzaro. Glymour's intent was to show how his boot-strapping theory of confirmation is exemplified in the determination of the relative atomic weights, and he argues his case effectively. He does not claim, however, that Cannizzaro's work established the existence of atoms (263); in fact, it would appear to be contrary to the spirit of his remarks at the beginning of his discussion of atomism to accept that view without the ascertainment of such atomic quantities as (nonrelative) mass or Avogadro's number (227).

4. The brand of positivism here involved is Comptian; logical positivism is a twentieth-century phenomenon.

5. This is how the second law was seen at the time. Much more recently such philosophers as Reichenbach (1956) and Grünbaum (1963, 1973) have shown how the asymmetry can be attributed to the initial conditions rather than to the law.

6. Energy is not an invariant in relativity theory, but it is in classical mechanics.

7. This remark applies to Rutherford's planetary model. Thomson's "plum pudding" model does not suffer from this defect, but it was incapable of dealing with Rutherford's experimental findings on large-angle scattering of alpha particles by gold foil.

8. By this time, Perrin was, of course, thoroughly familiar with the Einstein-Smoluchowski theory.

9. Einstein ([1905] 1989, 124) states that "a dissolved molecule differs from a suspended body in size *alone* ..." (emphasis in original). Some of Perrin's early work dealt with colloids, that is, suspensions of particles larger than molecules but smaller than Brownian particles. This study was made possible by the invention of the ultramicroscope in 1903.

10. One method is described in Holton and Brush (1973, 364–365):

An especially interesting technique, developed by Otto Stern and others in the 1920s, requires that atoms of a metal be evaporated in all directions from a hot wire of known temperature. The wire runs down the center of a close-fitting hollow cylinder which has a thin slit that can be opened and shut like a fast shutter. When the [slit] is briefly uncovered, a short, thin, well-defined 'squirt' of atoms escapes into the evacuated space beyond the cylinder; as they continue undisturbed in a straight line, the fast atoms outdistance the slower ones, thus spreading into a procession what started out to be a thickly packed crowd of atoms.

If now all atoms are allowed to impinge on a screen or slide of glass, they will quickly condense on it and form a thin coating of metal at a well-defined spot. But in order to separate on the screen the fast atoms from the slow ones, the screen is pulled across the path of the atoms as they fall on it; thus the coating becomes a lengthy smear instead of one spot, with the fastest atoms falling at the head of the deposit, the slowest ones at the tail end.

This is rather like shooting off a load of buckshot against a moving train at the very moment when the first car passes by. The speediest pellets will indeed hit the first car, and the slowest might hit the last. From the distribution of hits along the side of the whole train, you could analyze the distributions of velocities of all pellets.

11. To the best of my knowledge, this relationship yields the first formula that links microquantities and macroquantities, making it possible to calculate the former from the latter, pressure and volume being macroscopically determinable.

12. In his 1908 work Einstein gives a less technical version of his argument for the benefit of chemists.

13. In this argument Einstein takes Avogadro's number N to be a universal constant whose value is given. For this purpose he adopts a value that he admits to be a rough

estimate derived from the kinetic theory of gases, $N = 6 \times 10^{23}$, which is, ironically, closer to the currently accepted value of 6.0235×10^{23} than are any of Perrin's experimental results. Though accurate, this estimate cannot be considered reliable.

14. Miller (1987, 479) goes on to say, "Given early twentieth-century ignorance about intermolecular forces, molecule theorists might, in principle, have accepted substantial departures from Avogadro's hypothesis." Avogadro's hypothesis, that at equal temperatures and pressures a given volume of *any* gas contains the same number of molecules, played an important role in the work on relative molecular weights that culminated in Cannizzaro's table of values. It is unclear what relevance this hypothesis has in the context of Miller's discussion.

15. At this point we should recall Ian Hacking's argument regarding microscopic observability, discussed in "Realism and Empiricism: The Key Question" (see chapter 1), and Hacking's thirteen methods for cross-checking the characteristics of the microscopic grid.

16. Since, for technical reasons that were not revealed to the jury, the trial was cancelled after opening statements by both attorneys and the testimony, I do not know the legal outcome in this case. However, it appeared to me to be a textbook example of the common cause principle. It is interesting (to me anyhow) that particularly clear real-life examples have excellent academic credentials.

17. At this point I cited Chaffee (1980) for his clear basic exposition in an article in the *Scientific American*. At the time I wrote the passage quoted here I was in close personal contact with its author, who was affiliated with the Harvard-Smithsonian Observatory on Mt. Hopkins, a few miles south of Tucson, Arizona, where the discovery of the "twin quasar" occurred, Chaffee being one of the discoverers. It was fascinating to share vicariously the excitement of this discovery and to learn of further developments, including the fact that it now seems that a cluster of galaxies rather than a single one acts as the lensing object.

18. The reader of Reichenbach must note that his probability notation is not standard; in his probability expressions the term representing the reference class stands in the first place and the term representing the attribute class stands in the second place. This is just opposite to current usage.

19. A, B, and C are not necessarily classes of events; this point has crucial importance in connection with the interpretation of Perrin's argument.

20. If events A and B have both a common cause C and a common effect E, formulas (3)–(6) may hold with E substituted for C throughout.

21. Since A, B, and C are classes, whereas causal processes are particulars, we should think in terms of triplets x, y, z of particulars belonging respectively to A, B, and C. We assume the instance $z \in C$ associated with a given $x \in A$ and $y \in B$ can be distinguished from C's connected with other members of A and B.

22. This passage is attributed to Einstein in Mermin (1985, 38).

23. As I recall, the context was a satire of paranormal phenomena.

24. See d'Espagnat (1979), where the common cause principle is explicitly invoked, for a clear exposition of this point.

25. Brian Skyrms has pointed out in oral discussion that only *local* common causes are precluded by the quantum mechanical considerations raised here.

26. This particular example concerning wings is not mentioned in Sober (1988); I have introduced it into the discussion.

27. For example, it is implausible to suppose that the mean values of the populations differ markedly and that the agreement of the means of the samples results only from the draw of unrepresentative samples that happen by chance to agree.

28. In Reichenbach's notation the order of the two arguments in the probability expression is reversed. To avoid unnecessary confusion I have not adopted this feature of his notation.

29. The following question might arise in connection with inequality (5). Suppose we have an experiment on alpha radiation in which the pertinent initial conditions are satisfied, but the initial conditions for the corresponding experiment on Brownian motion are not fulfilled. In this case the pair will not qualify for membership in *C*. What possible bearing can the failure of initial conditions on the Brownian motion experiment have on the outcome of the alpha-radiation experiment? The answer is that the class ∼*C* also contains pairs in which the initial conditions for alpha-radiation experiments fail to be fulfilled. The same problem arises, mutatis mutandis, for inequality (6).

30. If the expanded notation were not used, the probability P(*A·B*|*C*) would appear to be zero because the sets *A* and *B* are disjoint. In addition, if *C* were not carefully specified, it might seem to signify different classes in P(*A*|*C*) and P(*B*|*C*). I am grateful to Christopher Hitchcock for pointing out these difficulties to me.

31. This fact presents a special problem for Reichenbach because of his view that the frequency interpretation is the only legitimate interpretation of probability. However, Reichenbach (1994, appendix) proposes a way of assigning probabilities counterfactually in reference classes that are, in fact, empty.

32. Feynman's pagination is peculiar. The number before the dash refers to the chapter; the number after the dash refers to the page in the designated chapter.

33. In this remark I am referring to those mechanists who tried to construct mechanical models of gravitation.

Chapter 4

1. *Harper's Magazine* 202 (June 1951): 9–11.

2. In Salmon, (1967b), I have discussed these issues in greater detail and have argued this case at greater length.

3. As we shall see in chapter 12, "Confirmation and Relevance," under mild assumptions it can be shown that a positive instance of H-D confirmation does increase, to some degree, the probability of the hypothesis, but the size of the increment may be incredibly small. This consideration does not mitigate the basic problems that plague the H-D method.

4. In nonstandard analysis the infinitesimal has been reinstated on firm logical foundations, but this is strictly a twentieth-century development.

5. For a very readable account of recent developments and a comparison with the earlier situation in the foundations of geometry, see Cohen and Hersh (1967).

6. If P(*B*|*A*) ≠ 0, a positive H-D instance does increase the probability of the hypothesis somewhat, but the amount can be exceedingly small.

7. Editors of *Consumer Reports* (1961, chap. 2); the original version of this chapter was published before (Pauling 1970); see Salmon (1961, chap. 5) for a detailed discussion of the experimental evidence.

8. Gardner (1957, 326, n. 4) quotes a delightful remark of Bertrand Russell on his method for dealing with cranks.

9. An extended discussion of the distinction between discovery and justification is given in chapter 5 of this volume.

10. Since this chapter was first written the so-called "anthropic principle" has been introduced into scientific cosmology; see Gale (1981) for a popular account and Barrow and Tipler (1986) for the technical details.

11. In the years since the original version of this chapter was first published we have seen an alarming rise of the doctrine of special creation under the misleading title of "creation science." For details see Kitcher (1982).

12. A word of caution is required here. My criticisms of teleological theories do not apply to functional explanations in evolutionary biology, sociology, or anthropology. As Larry Wright (1976) has effectively argued, it is possible to construe functional ascriptions in such a way as to avoid any involvement with final causation. Moreover, he offers an analysis of teleological explanation that is similarly free from any element of final causation. Wright's account affords no support to theories of the type of Lecomte du Noüy.

13. Lucretius, *The Nature of the Universe*, was originally titled *De Rerum Natura* and has been translated with the title *On the Nature of Things* (1951). The Latham translation is far more intelligible than the older ones.

14. Although many earlier authors made general statements about the meaning of "probability," they were more like anticipations than well-articulated analyses.

15. Many philosophers today adopt a so-called *propensity interpretation* of probability and interpret physical probabilities in terms of it. As Paul Humphreys (1985) has shown, however, propensities do not constitute an admissible interpretation of the probability calculus, and, in particular, they do not furnish a suitable interpretation of Bayes's theorem.

16. The nature and rationale of probabilistic coherence is discussed in detail in chapter 8, "Dynamic Rationality."

17. In chapter 7, "Revisions of Scientific Convictions," and chapter 8, "Dynamic Rationality," I discuss in some detail *Bayesian conditionalization* as a method for revising degrees of belief.

Chapter 5

1. Robert McLaughlin (1982a) has argued persuasively that the terms "context of invention" and "context of appraisal" should replace "context of discovery" and "context of justification," respectively; since, however, the original version of the present chapter was first published in 1970, McLaughlin's improved nomenclature was not available.

2. Reichenbach (1938, sec. 1). I have offered an elementary discussion of the distinction in Salmon (1984a, secs. 1–3).

3. Justification includes negative, as well as positive, appraisal.

4. As I explain in chapter 3, "An Empiricist Argument for Realism," close examination of the arguments actually employed by scientists early in the twentieth century to establish the existence of atoms and molecules casts considerable light on the logical structure of arguments that validly support scientific realism.

5. For further elaboration of this point see Salmon (1967b, 5–17).

6. This case is discussed in chapter 4, "Plausibility Arguments in Science."

7. In Salmon (1984a, sec. 30), I tried to offer an introductory Bayesian account of the confirmation of hypotheses without introducing any technicalities of the probability calculus.

8. See Salmon (1967a) for a more detailed examination of Carnap's views.

9. An excellent exposition is found in Edwards, Lindman, and Savage (1963).

10. It is to be noted that, despite the title of his book, *The Logic of Scientific Discovery*, Popper accepts the distinction between discovery and justification and explicitly declares that he is concerned with the latter but not the former.

11. See Salmon (1981) for an examination of the validity of this claim.

12. At a conference held in Konstanz, Germany, in 1991, W. V. Quine stated publicly that he considered the H-D method, taken in conjunction with a principle of simplicity, as the best available account of scientific confirmation. This is the closest thing to an endorsement of the H-D method that I have heard from a prominent philosopher in recent decades.

13. Shimony (1970) argues that hypotheses seriously proposed by serious scientists should have nonnegligible prior probabilities because in the last four centuries they have had a nonnegligible frequency of success.

14. Watson (1969) provides a fascinating account of the discovery of an important scientific hypothesis, and it illustrates many of the points I have been making. If literary reviewers had had a clearer grasp of the distinction between the context of discovery and the context of justification, perhaps they would have been less shocked at the emotions reported in the narrative.

Chapter 6

1. "Postscript—1969," in Kuhn (1970) contains discussions of some of the major topics that are treated in the present chapter.

2. Such philosophers are often characterized by their opponents as *logical positivists*, but this is an egregious historical inaccuracy. Although some of them had been members of or closely associated with the Vienna Circle in their earlier years, none of them retained the early positivistic commitment to phenomenalism and/or instrumentalism in their more mature writings. Reichenbach and Feigl, for example, were outspoken realists, and Carnap regarded physicalism as a tenable philosophical framework. Reichenbach (1935) never associated himself with positivism; indeed, he regarded his book *Experience and Prediction* as a refutation of logical positivism. I could go on.

3. Ironically, perhaps, Kuhn (1962) was published in the *International Encyclopedia of Unified Science*, an ambitious project of the logical positivists and logical empiricists. Carnap was one of the editors. An illuminating commentary on the historical relationship between Kuhn's book and logical empiricism, citing Carnap's comments, can be found in "Did Kuhn Kill Logical Empiricism?" (Reisch 1991).

4. Hempel, Kuhn, and Salmon (1983).

5. The response is given in greater detail in Kuhn's article than in the "Postscript."

6. Throughout this chapter I use the terms "hypothesis" and "theory" more or less interchangeably. Kuhn tends to prefer "theory," whereas I tend to prefer "hypothesis," but nothing of importance hinges on this usage here.

7. This equation is identical to equation (3) in chapter 4, "Plausibility Arguments in Science," with B, T, and E substituted for A, B, and C, respectively.

8. As I remarked above, three probabilities are required to calculate the posterior probability—a prior probability and two likelihoods. Obviously, in view of the theorem on total probability, if we have a prior probability, one of the likelihoods, and the expectedness, we can compute the other likelihood; likewise, if we have one prior probability and both likelihoods, we can compute the expectedness.

9. A set of degrees of conviction is coherent provided that its members do not violate any of the conditions embodied in the mathematical calculus of probability. For further details see chapter 8, "Dynamic Rationality."

10. I reject the so-called propensity *interpretation* of probability because, as Humphreys (1985) pointed out, the probability calculus accomodates inverse probabilities of the type that occur in Bayes's theorem, but the corresponding inverse propensities do not exist.

11. A so-called Dutch book is a combination of bets such that, no matter what the outcome of the event upon which the wagers are made, the subject is bound to suffer a net loss. These will be discussed in greater detail in Chapter 8.

12. Grünbaum (1984, 202–204) criticizes Motley's account of Freud's theory; he considers Motley's version a distortion, and he points out that Freud's motivational explanations were explicitly confined to a very circumscribed set of slips. He defends Freud against

Motley's criticism on the ground that Freud's actual account has greater complexity than Motley gives it credit for.

13. As Duhem ([1906] 1954) has made abundantly clear, in such cases we may be led to reexamine our background knowledge B, which normally involves auxiliary hypotheses, to see whether it remains acceptable in the light of the negative outcome E. Consequently, refutation of T is not usually as automatic as it appears in the simplified account just given. Nevertheless, the probability relation just stated is correct.

14. Exchangeability is the personalist's surrogate for randomness; it means that the subject would draw the same conclusion regardless of the order in which the members of an observed sample occurred.

15. Note that, in order to get the posterior probability—the probability that the observed results were produced by a biased device—the prior probabilities have to be taken into account.

16. Indeed, stellar parallax was not detected until the nineteenth century.

17. This question was, in fact, raised by Adolf Grünbaum in a private communication.

18. If more than two theories are serious candidates, the pairwise comparison can be repeated as many times as necessary.

19. Weinberg (1972, 198). Note that this correction is smaller by an order of magnitude than the correction of 43 seconds of arc per century for Mercury.

20. Unfortunately, recent evidence strongly suggests that Pluto is not sufficiently massive to explain the perturbations of Neptune. It may turn out, however, that the alleged perturbations of Neptune do not even exist. We have not observed Neptune through an entire circuit of its orbit.

21. See chapter 8 for a more detailed discussion of various grades of rationality.

22. I should like to express my deepest gratitude to Adolf Grünbaum and Philip Kitcher for important criticism and valuable suggestions with respect to an earlier version of this chapter.

Chapter 7

1. It should be noted, incidentally, that Carnap (1962a) adopted in his article an essentially Bayesian account of revision of degrees of belief.

2. Both Carnap and Reichenbach insisted on probabilistic coherence; on this point they were in agreement with the Bayesians. According to Carnap (1950b), probability$_1$— degree of confirmation—is a degree of rational belief. Carnap imposed many conditions, but probabilistic coherence is the most fundamental. According to Reichenbach ([1935] 1949) probabilities are frequencies, and he demonstrated that frequencies necessarily conform to the calculus of probability (at least as he had axiomatized it). In his calculus, probabilities are finitely additive, and he demonstrated that they satisfy this condition. Countable additivity, however, can be violated by limiting frequencies, so he was unable to establish that condition. This technical point does not affect the arguments in this chapter.

3. As in the preceding chapter I shall use the terms "hypothesis" and "theory" interchangeably, and use T as the symbol in formulas.

4. Such might be my reaction to an experiment in extrasensory perception that yielded a posterior probability for the theory that I considered unacceptably high.

5. As Brian Skyrms (1985) explained in an elegant lecture at the University of Pittsburgh, it is the bookmaker—not the client—who is threatened by the perils of a Dutch book. The bookmaker posts odds on a set of alternative outcomes; if the posted odds are incoherent, a bettor or group of bettors can make a set of bets in which the net outcome for the bookie is necessarily negative. That is the static Dutch book situation. If, in addition, the

bookie posts payoffs for conditional bets, then the possibility of a kinematic Dutch book threatens unless the conditional probabilities satisfy Bayesian conditionalization.

6. This kinematic Dutch book argument is not altogether uncontroversial; Bacchus et al. (1990) and Patrick Maher (1993, chap. 5) have raised serious objections against it.

7. Only in a system of inductive logic as arbitrary and egregiously oversimple as that of Carnap (1950b) is this possibility precluded. According to Carnap's requirement of descriptive completeness, the language of the system must contain all of the predicates of which science will ever have need. It is necessary to list these predicates before formulating any inductive logic. In this system no new hypothesis or new concept can possibly arise. As Carnap clearly realized, such a system is completely unrealistic, and this requirement is not contained in his later, more complex systems.

8. During a recent visit to Anchorage, Alaska, I found artifacts, allegedly made from fossil mammoth ivory, for sale in gift shops.

9. As a matter of fact, it now appears that T_3 is the correct hypothesis—that the radiocarbon dating was in error.

10. Physicists were, of course, fully aware of this relationship, but the operators of steamboats and steam locomotives were not sufficiently cognizant of it.

11. The double line separating the premises from the conclusion signifies that the argument is inductive.

Chapter 8

1. Wyden (1984, 50–51) mentions this incident, but without throwing any light on the questions I am raising.

2. Eliot Marshall (27 June 1986, 1596). All of the following quotations from this article appear on the same page.

3. If the probability of failure of a given rocket is .01, the probability of at least 1 failure in 50 firings is .395; if the probability of failure of a given rocket is .02, the probability of at least 1 failure in 50 firings is .636.

4. Marshall offers the following interpretation of that figure: "It means NASA thinks it could launch the shuttle, as is, every day for the next 280 years and expect not one equipment-based disaster." One could hope that NASA statisticians would not follow Marshall in making such an inference. I presume that the number 280 was chosen because that many years contain a little over 100,000 days—102,268 to be precise (allowing for leap years). However, it must be recalled, as was pointed out in the article, that each launch involves two rockets, so the number of rocket firings in that period of time would be 204,536. If the probability of failure on any given rocket is 1 in 100,000, the probability of at least 1 failure in that many firings is about .870. In one century of daily launches, with that probability of failure, there is about a 50:50 chance of at least 1 failure. This latter estimate is, of course, absurd, but to show the absurdity of the NASA estimate we should try to get the arithmetic right.

5. Rudolf Carnap and David Lewis, among others, use the *credence* terminology.

6. I realize that a great deal of important work has been done on the handling of inconsistent systems, so that, for example, the presence of an inconsistency in a large body of data does not necessarily bring a scientific investigation to a grinding halt. See, for example, Rescher and Brandom (1980). However, inasmuch as our main concern is with probabilistic coherence, it will not be necessary to go into that issue in detail.

7. When the modification is drastic enough, Kuhn calls it a scientific revolution.

8. Another way of handling these betting considerations—one that I would prefer—is to remember that two factors bear upon our bets. The first is the degree of probability of the event that we are betting on; the second is our degree of assurance that that probability is

objectively correct. In his theory of probability, Carnap, in effect, eliminated this second consideration by making all such probability statements either logically true or logically false. This meant that the caution factor had to be built into the value of the probability; strict coherence does the job.

9. Hacking argues that this could not have occurred before the seventeenth century. I am not at all sure that his historical thesis about the time at which the concept of probability emerged is correct, but his conceptual analysis strikes me as sound.

10. Carnap continued work on the project of constructing an adequate inductive logic until his death in 1970. His later work on the subject is in Carnap and Jeffrey (1971) and Jeffrey (1980). Jeffrey is the sole editor of volume 2.

11. In order to discuss quotations from Ramsey's article I shall revert to his terminology and refer to *partial beliefs* and *degrees of belief*.

12. Ramsey clearly means that the subject behaves so as to maximize his or her expected utility.

13. In discussing Mellor's views I shall continue to use the traditional "partial belief" and "degree of belief" terminology. I have treated Mellor's book in considerable detail in Salmon (1979). My review also contains a general discussion of propensity theory with special emphasis on the work of Karl Popper, its originator.

14. In discussing Lewis's views, I use his terminology of "partial belief" and "credence."

15. In explaining this principle Lewis (1980, 266) makes excursions into nonstandard analysis and possible worlds semantics. I do not think either of these tools is required for our discussion.

16. One who buys a ticket each day for a year pays \$365 for a set of tickets whose expectation value is \$182.50. This is a poor investment indeed—especially for people at or near the poverty level—but regrettably it is not an uncommon occurrence.

17. Fortunately, this particular incident was discovered, and the perpetrators were brought to trial and punished. Presumably this sort of thing is not going on any more.

18. Consider different roulette wheels as different games. The same goes for slot machines, blackjack tables, and so forth. It is not necessary that every game have a probability distribution different from all others.

19. This assertion is based upon a report sponsored by the American Physical Society, *Science News* 131 (2 May, 1987): 276. The report refers to the need for "improvements of several orders of magnitude."

20. Oscar Wilde in "An ideal Husband" nicely conveyed the unrealistic nature of this assumption: "When the gods want to punish us they answer our prayers."

21. It may be that some probabilities are not generated by propensities as I am construing them. See Hacking (1982) for discussion and an interesting example in which this appears to be the case.

22. I am adopting the felicitous terminology, "context of invention" and "context of appraisal," proposed by Robert McLaughlin (1982b, 69–100), as a substitute for the traditional "context of discovery" and "context of justification."

23. This approach bears some strong resemblance to a suggestion offered by Ian Hacking (1968, sec. 9). I deplore the fact that it has taken me twenty years to appreciate the value of this suggestion.

24. The considerations advanced by Hacking in note 23 make me suspect that it is not.

Chapter 9

The paper on which this chapter is based was written for presentation at a conference held at the University of Arizona in 1976, the bicentennial of Hume's death. On that occasion it seemed appropriate to discuss his *Dialogues Concerning Natural Religion*, for this post-humously published work is a significant part of Hume's legacy. The composition of the dialogues was begun in 1751, with a first draft completed by 1756. Hume wanted to publish them at that time, but because of the controversial nature of this work, friends persuaded him to delay publication. He revised the *Dialogues* in 1761 and again in 1776. One of Hume's chief concerns during the last year of his life was that they be published shortly after his death. To this end he made elaborate provisions in his will, but the actual publication did not occur until 1779. The first edition did not carry the publisher's name or any editorial comment.

I am grateful to Professor Nancy Cartwright for stimulating and penetrating comments on the version of this paper that was presented at the Hume conference. I am grateful to Professor Frederick Ferré for important information concerning William Paley's construal of the design argument.

Page references in the text are to Hume's *Dialogues Concerning Natural Religion*, edited by Nelson Pike (Indianapolis: Bobbs-Merrill, 1970). In addition, for readers who do not have this edition at hand, I indicate which dialogue any quoted passage is taken from.

1. As Frederick Ferré kindly pointed out to me, Paley—unlike Cleanthes—rests his version of the design argument upon the existence of particular contrivances in nature; such as the human eye, rather than appealing to the alleged machinelike character of the whole universe.

2. By employing a slightly different form of Bayes's theorem, we can replace $P(C/A \cdot B)$ by $P(C/A)$, but in either case, three distinct probabilities are required.

3. Paley (1851) was much more careful than was Cleanthes to refrain from literally imputing omnipotence, omniscience, and perfect benevolence to God. Paley also held that natural theology must be supplemented by revealed theology in order to ascertain the attributes of God.

4. The term "deism" is often used to refer to the doctrine that there is a God who created the universe but after the creation does not interfere with it in any way.

5. Some of these historical developments are presented in elementary terms in Holton and Brush (1973, chaps. 17.7–10 and 18).

6. In the spirit of bicentennialism, which inspired the Hume conference at which this chapter was presented, it should be noted that the fundamental work on statistical mechanics by J. Willard Gibbs dates from 1876. See M. J. Klein (1990).

7. Many of these considerations are discussed with philosophical clarity in Reichenbach (1956, chaps. 3–4).

8. See Weinberg (1972) for an accurate and reliable popular presentation of current cosmological theory. I do not mean to assert that the history of the universe is completely understood or that there are no substantial controversies. My point is merely that such knowledge, incomplete and tentative as it may be, is not negligible. Bringing it to bear in the present context is not contrary to the spirit of Hume's discussion.

9. The one exception—hardly worth mentioning—was mine (Salmon 1951), published just 200 years after Hume began his first draft. That article was defective in many ways, and I am thankful for the obscurity it enjoyed.

10. If I recall correctly, Bertrand Russell was once asked if there were any conceivable evidence that could lead him to a belief in God. He offered something similar to Cleanthes's suggestion. He was then asked what he would say if, after dying, he were transported to the presence of God; how would he justify his failure on earth to be a believer? "I'd say, 'Not enough evidence, God, not enough evidence!'" Some years ago I gave a talk on the design argument at De Pauw University. After the lecture there was a party, and this story about Russell was told. The next day we learned that he had died that very evening.

11. Here I find myself in sharp disagreement with Pike's commentary in Hume ([1779] 1970).

Chapter 10

1. In the early 1960s it was estimated that 90% of all scientists who ever lived were alive at that time. I suspect the percentage is not very different today.

2. John Patrick Day (1961) took transitivity to be a legitimate inductive relation. Salmon (1963c) reviews Day's "ordinary language" approach to probability and induction.

3. Nicholas Rescher (1961) claims that both of the mixed cases are sound for plausible reasoning.

4. Giere, however, has informed me in private correspondence that this was a simplification introduced for didactic purposes.

5. Reichenbach's proof requires that $P(A) \neq 0$ and $P(A|\sim B) \neq 0$.

6. Carnap (1962b, xiii–xxii) adds further clarification of this distinction.

7. Salmon (1973a) offers a brief and popularized account of these issues.

Chapter 11

1. For citations to some of the more important articles see note 1 to the "Postscript" in Hempel (1965b, 47).

2. See Salmon (1963b). This article contains a resolution of Goodman's paradox that I still think to be substantially correct.

3. This study was initiated by consideration of Mary Shearer (1966). In her thesis, Shearer undertook to demonstrate Carnap's claim that $c^\dagger$ does not admit "learning from experience." In carrying out her proof she used the notion I have introduced below under the name of "complete atomic independence," but it turns out that essentially the same proposition can be demonstrated very much more simply by using what I call "complete truth-functional independence." This is a strong motive for introducing the latter as a distinct concept of complete independence.

4. A full account can be found in Carnap (1950b; 1962b, chap. 3). I shall sketch a few of the essential points, but the foregoing source should be consulted for greater precision and detail. In one deliberate departure from Carnap's usage, I shall use the term "statement" in almost all cases in which he would use the term "sentence." I do this to emphasize the fact that we are dealing with interpreted languages and not just uninterpreted strings of symbols. In order to avoid considerable complication of the notation, I frequently use italicized letters and formulas containing italicized letters as names of themselves. I believe the context makes clear in all cases whether such symbols are being mentioned or only used.

5. Throughout this discussion I shall make free use of substitution under L-equivalence, for all of the L-concepts are invariant with respect to such replacement, and the logical relations of inductive logic as discussed by Hempel and Carnap are likewise invariant as long as we do not switch from one language L_N to another.

6. I have expressed both of these notions in very elementary terms in Salmon (1963a, 14–15, 98–100).

7. Exception could be taken to the Bar-Hillel and Carnap (1953–1954) analysis of content on the ground that L-implication is unsatisfactory as an explanation of entailment. Anderson and Belnap (1962) for example, have worked extensively on alternative concepts of entailment. Interesting as these investigations are, they do not seem likely to enable us to salvage the idea that completely independent statements have no common content, for the chief obstacle is the logical principle of addition (p entails $p \lor q$), which holds for the

stronger entailment relations, as well as for Carnap's L-implication. My own feeble efforts to cook up an entailment relation that would not have this property produced nothing plausible. I am grateful to Alan Anderson for pointing out a severe difficulty in one proposal.

8. See especially CA11 and Theorem 2. If t is a tautology, $n(t)$ becomes the total number of state descriptions. Kemeny and Oppenheim show that zero correlation obtains iff

$$n(p)n(q) = n(p{\cdot}q)n(t)$$

which is shown in note 9 to be equivalent to our definition.

9. Given that

$$\frac{n(p{\cdot}q)}{n(p)} = \frac{n({\sim}p{\cdot}q)}{n({\sim}p)}$$

we have

$$n(p{\cdot}q)n({\sim}p) = n({\sim}p{\cdot}q)n(p)$$

Adding $n(p{\cdot}q)n(p)$ to both sides yields

$$n(p{\cdot}q)n(p) + n(p{\cdot}q)n({\sim}p) = n(p{\cdot}q)n(p) + n({\sim}p{\cdot}q)n(p)$$

or

$$n(p{\cdot}q)[n(p) + n({\sim}p)] = n(p)[n(p{\cdot}q) + n({\sim}p{\cdot}q)].$$

Since $n(t) = n(p) + n({\sim}p)$ and $n(q) = n(p{\cdot}q) + n({\sim}p{\cdot}q)$

$$n(p{\cdot}q)n(t) = n(p)n(q)$$

or, dividing,

$$\frac{n(p{\cdot}q)}{n(p)} = \frac{n(q)}{n(t)}$$

10. This fact is proved as follows. Assume

$$\frac{n(p{\cdot}q)}{n(p)} = \frac{n({\sim}p{\cdot}q)}{n({\sim}p)}$$

Since $n(p) = n(p{\cdot}q) + n(p{\cdot}{\sim}q)$ and $n({\sim}p) = n({\sim}p{\cdot}q) + n({\sim}p{\cdot}{\sim}q)$,

$$\frac{n(p{\cdot}q)}{n(p{\cdot}q) + n(p{\cdot}{\sim}q)} = \frac{n({\sim}p{\cdot}q)}{n({\sim}p{\cdot}q) + n({\sim}p{\cdot}{\sim}q)}$$

By cross-multiplication,

$$n(p{\cdot}q)[n({\sim}p{\cdot}q) + n({\sim}p{\cdot}{\sim}q)] = n({\sim}p{\cdot}q)[n(p{\cdot}q) + n(p{\cdot}{\sim}q)]$$

or

$$n(p \cdot q)n(\sim p \cdot q) + n(p \cdot q)n(\sim p \cdot \sim q) = n(\sim p \cdot q)n(p \cdot q) + n(\sim p \cdot q)n(p \cdot \sim q)$$

Subtracting $n(p \cdot q)n(\sim p \cdot q)$ from both sides yields

$$n(p \cdot q)n(\sim p \cdot \sim q) = n(\sim p \cdot q)n(p \cdot \sim q)$$

which is equivalent to

$$n(q \cdot p)n(\sim q \cdot \sim p) = n(\sim q \cdot p)n(q \cdot \sim p)$$

by commutation of conjunction. Adding $n(q \cdot p)n(\sim q \cdot p)$ to both sides, we have

$$n(q \cdot p)n(\sim q \cdot p) + n(q \cdot p)n(\sim q \cdot \sim p) = n(q \cdot p)n(\sim q \cdot p) + n(\sim q \cdot p)n(q \cdot \sim p)$$

or

$$n(q \cdot p)[n(\sim q \cdot p) + n(\sim q \cdot \sim p)] = n(\sim q \cdot p)[n(q \cdot p) + n(q \cdot \sim p)]$$

Since $n(\sim q) = n(\sim q \cdot p) + n(\sim q \cdot \sim p)$ and $n(q) = n(q \cdot p) + n(q \cdot \sim p)$,

$$n(q \cdot p)n(\sim q) = n(\sim q \cdot p)n(q)$$

Dividing by $n(q)$ and $n(\sim q)$, neither of which is zero, we get

$$\frac{n(q \cdot p)}{n(q)} = \frac{n(\sim q \cdot p)}{n(\sim q)}$$

11. We impose this restriction to prevent the denominators of the fractions in the definitions from being zero. This is no serious restriction since in inductive logic we are concerned with relations among factual statements.

12. Chapter 6 of Carnap (1950b, 1962b) is devoted to the subject of relevance, and it provides considerable illumination for this seriously neglected topic.

13. Though I shall often refer only to inference from past to future, it should be understood that the general problem concerns inference from the observed to the unobserved, regardless of the temporal locations of the events involved.

14. Some philosophers, for example, Anderson and Belnap (1962), would maintain that we still have not found a fully adequate explication of "entailment" and that a relation must fulfill additional conditions to deserve the name. That does not matter to the point I am making. They would surely agree that the relation must fulfill at least the conditions I have mentioned, and that is all I need to insist upon.

15. It would be a mistake, I believe, to conclude that this type of positive relevance must be reflected, in turn, in increased truth frequencies, but that is a separate issue. The issue at present concerns inductive relevance where no correlations of truth values or state descriptions obtain.

16. See Carnap (1950b; 1962b, 110A) for the definition of c^*. Although Carnap no longer regards c^* as fully adequate, he does maintain that it is a good approximation in many circumstances. In any case, it illustrates nicely the point about relevance under discussion here.

17. I have discussed the problem of the force of the term "rational" in inductive logic in several articles, especially in the lead article and reply to S. Barker and H. Kyburg in Salmon (1965b).

Chapter 12

Some of the ideas in this chapter were discussed nontechnically in Salmon (1973a).

1. At that time Grünbaum reported that much of his stimulus for raising these questions resulted from discussions with Professor Laurens Laudan. See also Grünbaum (1976b).

2. The term "absolute probability" is sometimes used to refer to probabilities that are not relative or conditional. For example, Carnap's null confirmation $c_0(h)$ is an absolute probability, as contrasted with $c(h,e)$, in which the degree of confirmation of h is relative to, or conditional upon, e. The distinction I am making between the concepts defined in D1 and D2 is quite different. It is a distinction between two different types of confirmation, where one is a conditional probability and the other is a relevance relation defined in terms of conditional probabilities. In this chapter, I shall not use the concept of absolute probability at all; in place of null confirmation I shall always use the confirmation $c(h,t)$ on tautological evidence, which is equivalent to the null confirmation but which is a conditional or relative probability.

3. These conditions of adequacy are presented in Hempel (1945; reprinted, with a 1964 postscript, in 1965b). Page references in succeeding notes will be to the reprinted version.

4. Hempel's capital letters "H" and "E" have been changed to lowercase for uniformity with Carnap's notation.

5. Hempel (1965b, 31). Following Carnap, an "H" is attached to the numbers of Hempel's conditions.

6. Carnap introduces the simple initial relevance measure D for temporary heuristic purposes in the preface to 1962a. In chapter 6 he discusses both our relevance measure R and Keynes's relevance quotient: $c(h,e\cdot i)/c(h,e)$, but for his own technical purposes he adopts a more complicated relevance measure, $r(i,h,e)$. For purposes of this chapter, I prefer the simpler and more intuitive measure R, which serves as well as Carnap's measure r in the present context. Since this measure differs from that used by Carnap (1950b; 1962b, chap. 6), I use a capital "R" to distinguish it from Carnap's lowercase symbol.

7. The condition $c(h\cdot k,e) = 0$ is obviously sufficient to make the probability of the disjunction equal to the sum of the probabilities of the disjuncts, and this is a weaker condition than e entails $\sim(h, k)$. Since the difference between these conditions has no particular import for this discussion, I shall, in effect, ignore it.

8. Because of the commutativity of conjunction, it does not matter whether the probability of h conditional only on e or the probability of k conditional only on e is taken. This is shown by the double equality in formula (3).

9. Independence is a symmetric relation; if h is independent of k then k will be independent of h.

10. To secure this result it is not necessary that $c(h\cdot k,e\cdot i) = 0$; it is sufficient to have $c(h\cdot k,e) = c(h\cdot k,e\cdot i)$, though obviously this condition is not necessary either.

11. To establish the compatibility of (20) and (21), perhaps a simple example, in addition to the one about to be given in the text, will be helpful. Let

$e = $ X is a man. $i = $ X is American.
$h = $ X is very wealthy. $k = $ X vacations in America.

Under this interpretation, relation (20) asserts: It is more probable that an American man vacations in America than it is that a man (regardless of nationality) vacations in

America. Under the same interpretation, relation (21) asserts: It is less probable that a very wealthy American man will vacation in America than it is that a very wealthy man (regardless of nationality) will vacation in America. The interpretation of formula (20) seems like an obviously true statement; the interpretation of (21) seems likely to be true owing to the apparent tendency of the very wealthy to vacation abroad. There is, in any case, no contradiction in assuming that every very wealthy American man vacations on the French Riviera, whereas every very wealthy man from any other country vacations in America.

12. This equality can easily be shown by writing out the relevance terms according to their definitions, as follows:

$$R(h,k,e) =_{df} c(k,h\cdot e) - c(k,e)$$
$$R(h,k,e\cdot i) =_{df} c(k\cdot h,e\cdot i) - c(k,e\cdot i)$$
$$R(h,k,e) - R(h,k,e\cdot i) = c(k,h\cdot e) - c(k,e) - c(k,h\cdot e\cdot i) + c(k,e\cdot i) \qquad (*).$$
$$R(i,k,e) =_{df} c(k,e\cdot i) - c(k,e)$$
$$R(i,k,h\cdot e) =_{df} c(k,h\cdot e\cdot i) - c(k,h\cdot e)$$
$$R(i,k,e) - R(i,k,h\cdot e) = c(k,e\cdot i) - c(k,e) - c(k,h\cdot e\cdot i) + c(k,h\cdot e) \qquad (**).$$

The right-hand sides of equations (*) and (**) obviously differ only in the arrangement of terms.

13. This case constitutes a counterexample to the Hempel consistency condition H-8.3, discussed in section 2 above.

References

American Physical Society. 1987. Report. *Science News* 131: 276.

Anderson, Alan, and Nuel Belnap. 1962. Tautological entailments. *Philosophical Studies* 13: 9–24.

Armstrong, David. 1989. Singular causes and laws of nature. *Center for Philosophy of Science, Annual Lecture Series*, September 22, University of Pittsburgh.

Ayer, A. J. 1946. *Language, Truth, and Logic.* 2nd ed. New York: Dover.

Bacchus, F., H. Kyburg, and M. Thalus. 1990. Against conditionalization. *Synthese* 85: 475–506.

Bacon, Francis. [1620] 1889. *Novum Organum.* 2nd ed. Thomas Fowler (ed.). Oxford: Oxford University Press.

Bar-Hillel, Yehoshua, and R. Carnap, 1953–1954. Semantic information. *British Journal for the Philosophy of Science* 4: 147–157.

Barrow, John D., and Frank J. Tipler. 1986. *The Anthropic Cosmological Principle.* Oxford: Clarendon Press.

Born, Max, and Emil Wolf. 1964. *Principles of Optics: Electromagnetic Theory of Propagation, Interference and Diffraction of Light.* Oxford: Pergamon Press.

Boyle, Robert. [1772] 1965–1966. *The Works of the Honourable Robert Boyle.* 3rd ed. Thomas Birch (ed.). Introduced by Douglas McKie. Reprint, Hildesheim: G. Olms.

Braithwaite, R. B. 1953. *Scientific Explanation: A Study of the Function of Theory Probability and Law in Science.* New York and London: Cambridge University Press.

Brewster, David. 1990. Report on the present state of physical optics. In John Worrall, Scientific revolutions and scientific rationality: The case of the elderly holdout. *Minnesota Studies in Philosophy of Science, Volume 14: Scientific Theories*, ed. C. Wade Savage, 319–354. Minneapolis: University of Minnesota Press.

Brush, Stephen. 1989. Prediction and theory evaluation: On the case of light bending. *Science* 246: 1124–1129.

Butler, Bishop Joseph. 1940. *The Analogy of Religion*, 3rd ed. London: John & Paul Knapfton.

Cairns-Smith, A. G. 1985. *Seven Clues to the Origin of Life.* Cambridge: Cambridge University Press.

Cannizzaro, S. 1858. Sketch of a course of chemical philosophy. *Il Nuovo Cimento*, 7: 321–66.

Carnap, Rudolf, 1928a. *Der Logische Aufbau der Welt*. Berlin-Schlachtensee: Weltkreis-Verlag.

——. 1950a. Empiricism, semantics, and ontology. *Revue Internationale de Philosophie* 4e *Année*, 11: 20–40.

——. 1950b. *Logical Foundations of Probability*. Chicago: University of Chicago Press.

——. 1962a. The aim of inductive logic. In *Logic, Methodology and Philosophy of Science*, ed. Ernest Nagel, Patrick Suppes, and Alfred Tarski, 303–318. Stanford: Stanford University Press.

——. 1962b. *Logical Foundations of Probability*. 2nd ed. Chicago: University of Chicago Press.

——. 1963. Replies and systematic expositions. In *The Philosophy of Rudolf Carnap*, ed. Paul Arthur Schilpp, 859–1013. La Salle, Ill.: Open Court.

——. 1966a. *Philosophical Foundations of Physics*. New York: Basic Books.

——. 1966b. Probability and content measure. In *Mind, Matter and Method: Essays in Honor of Herbert Feigl*, ed. P. Feyerband and G. Maxwell, 248–260. Minneapolis: University of Minnesota Press.

——. 1967. *The Logical Structure of the World and Pseudoproblems in Philosophy*. Rolf A. George (trans.). Berkeley and Los Angeles: University of California Press.

——. 1974. *An Introduction to the Philosophy of Science*. New York: Basic Books.

Carnap, Rudolf, and Richard Jeffrey, eds. 1971. *Studies in Inductive Logic and Probability*, vol. 1. Berkeley, Los Angeles, London: University of California Press.

Chaffee, Frederic H., Jr. 1980. The discovery of a gravitational lens. *Scientific American* 243 (5): 70–88.

Church, Alonzo. 1949. Review of Ayer (1946). *Journal of Symbolic Logic* 14: 52–53.

Cohen, Paul J., and Reuben Hersh. 1967. Non-Cantorian set theory. *Scientific American* 217 (6): 104–117.

Cohen, Robert W., ed. 1981. *Herbert Feigl: Inquiries and Provocations*. Dordrecht: Reidel.

Cooper, Leon. 1968. *An Introduction to the Meaning and Structure of Physics*. New York, Evanston, London: Harper and Row.

Dalton, John. 1808. *A New System of Chemical Philosophy*, Vol. 1, part 1. Manchester and London: Printed by S. Russell for R. Bickerstaff.

Day, John Patrick. 1961. *Inductive Probability*. New York: Humanities Press.

d'Espagnat, Bernard. 1979. The quantum theory and reality. *Scientific American* 241 (5): 158–181.

Duhem, Pierre. [1906] 1954. *The Aim and Structure of Physical Theory*. Philip P. Weiner (trans.). Princeton, N.J.: Princeton University Press.

du Noüy, Pierre Lecomte. 1947. *Human Destiny*. New York and London: Longmans, Green.

Earman, John. 1992. *Bayes or Bust: A Critical Examination of Bayesian Confirmation Theory*. Cambridge, Mass.: MIT Press.

Editors of *Consumer Reports*, 1961. *The Medicine Show*. New York: Simon and Schuster.

Edwards, Ward, Harold Lindman, and Leonard J. Savage. 1963. Bayesian statistical inference for psychological research. *Psychological Review* 70: 193–242.

Einstein, Albert. [1905] 1989. On the movement of small particles suspended in stationary liquids required by the molecular-kinetic theory of heat. In *The Collected Papers of Albert Einstein*, vol. 2 (doc. 16): 123–134. Princeton, N.J.: Princeton University Press.

——. 1908. Elementary theory of Brownian motion. In *The Collected Papers of Albert Einstein*, vol. 2 (doc. 50): 318–328. Princeton, N.J.: Princeton University Press.

——. 1949. Autobiographical Notes. In *Albert Einstein: Philosopher-Scientist*, ed. Paul Arthur Schilpp, 2–95. New York: Tudor.

Einstein, Albert, B. Podolsky, and N. Rosen. 1935. Can quantum-mechanical description of physical reality be considered complete? *Physical review* 47: 777–780.

Feigl, H., and G. Maxwell, eds. 1962. *Minnesota Studies in the Philosophy of Science, Vol. 3: Scientific Explanation, Space and Time*. Minneapolis: University of Minnesota Press.

Fetzer, James H. 1971. Dispositional probabilities. In *PSA 1970: In Memory of Rudolph Carnap: Proceedings of the 1970 Bienniel Meeting, Philosophy of Science Association*, ed. Roger C. Buck and Robert C. Cohen, 473–482. Dordrecht: Reidel. Reprinted in *Scientific Knowledge*. 1981. Dordrecht: Reidel.

Fetzer, James H., ed. 1988. *Probability and Causality: Essays in Honor of Wesley C. Salmon*. Dordrecht: Reidel.

Feynman, R. P., R. B. Leighton, and M. Sands. 1963. *The Feynman Lectures on Physics, vol. 1*. Reading, Mass: Addison-Wesley.

Fine, Arthur. 1984. The natural ontological attitude. In *Scientific Realism*, ed. Jarrett Leplin, 83–107. Berkeley and Los Angeles: University of California Press.

Gale, George. 1981. The anthropic principle. *Scientific American* 245 (6): 154–171.

Gantner, Michael. 1979. Realism and instrumentalism in 19th-century atomism. *Philosophy of Science* 46: 1–34.

Garber, Daniel. 1983. Old evidence and logical omniscience in Bayesian confirmation theory. In *Minnesota Studies in the Philosophy of Science, Vol. 10: Testing Scientific Theories*, ed. John Earman, 99–131. Minneapolis: University of Minnesota Press.

Gardner, Martin, 1957: *Fads and Fallacies in the Name of Science*. New York: Dover.

Gemes, Kenneth B. 1991. "Content and confirmation." Ph.D. diss., University of Pittsburgh.

Giere, Ronald N. 1979. *Understanding Scientific Reasoning*. New York: Holt, Rinehart and Winston.

———. 1991. *Understanding Scientific Reasoning*. 3rd ed. New York: Holt, Rinehart and Winston.

Glymour, Clark. 1980. *Theory and Evidence*. Princeton, N.J.: Princeton University Press.

Goodman, Nelson. 1951. *The Structure of Appearance*. Cambridge, Mass.: Harvard University Press.

———. 1955. *Fact, Fiction, and Forecast*. Cambridge, Mass.: Harvard University Press.

Grant, Edward. 1962. Late medieval thought, Copernicus, and the scientific revolution. *Journal of the History of Ideas* 23: 197–220.

Grünbaum, Adolf. 1963. *Philosophical Problems of Space and Time*. New York: Knopf.

———. 1973. Second enlarged edition. Dordrecht: Reidel.

———. 1976a. Can a theory answer more questions than one of its rivals? *British Journal for the Philosophy of Science* 27: 1–23.

———. 1976b. Is falsifiability the touchstone of scientific rationality? Karl Popper versus inductivism. In *Boston Studies in the Philosophy of Science Series, vol. 39: Essays in Memory of Imre Lakatos*, ed. R. S. Cohen, P. K. Feyerabend, and M. W. Wartofsky, 213–252. Dordrecht: Reidel.

———. 1984. *The Foundations of Psychoanalysis*. Berkeley: University of California Press.

Hacking, Ian. 1968. One problem about induction. In *The Problem of Inductive Logic*, ed. Imre Lakatos, 52–54. Amsterdam: North-Holland.

———. 1975. *The Emergence of Probability*. Cambridge: Cambridge University Press.

———. 1981a. Do we see through a microscope? *Pacific Philosophical Quarterly* 62: 305–322.

———. 1982. Grounding probabilities from below. In *PSA 1980*, ed. Peter Asquith and Ron Giere, 110–116. East Lansing, Mich.: Philosophy of Science Association.

———. 1983. *Representing and Intervening*. Cambridge: Cambridge University Press.

———. 1989. Extragalactic reality: The case of gravitational lensing. *Philosophy of Science* 56: 555–581.

Hacking, Ian, ed. 1981b. *Scientific Revolutions*. Oxford: Oxford University Press.

Hanen, Marsha P. 1967. Goodman, Wallace, and the equivalence condition. *The Journal of Philosophy* 64: 271–280.

Hanson, N. R. 1961. Is there a logic of discovery. In *Current Issues in the Philosophy of Science*, ed. Herbert Feigl and Grover Maxwell, 20–35. New York: Holt, Rinehart and Winston.

Harari, H. 1983. The structure of quarks and leptons. *Scientific American*, April, 56–68.

Hardy, G. H., P. V. Seshu Aiyar, and B. M. Wilson. 1927. *Collected Papers of Srinivasa Ramanujan*. Cambridge: Cambridge University Press.

Helmer, Olaf, and Paul Oppenheim. 1945. A syntactical definition of probability and degree of confirmation. *The Journal of Symbolic Logic* 10: 26–60.

Hempel, Carl G. 1942. The function of general laws in history. *Journal of Philosophy* 39: 35–48. Reprinted in Hempel (1965b, 231–243).

———. 1943. A purely syntactical definition of confirmation. *The Journal of Symbolic Logic* 8: 122–143.

———. 1945. Studies in the logic of confirmation. *Mind* 54: 1–26, 97–120. Reprinted with some changes and a "Postscript" in Hempel (1965b).

———. 1951. The concept of cognitive significance: A reconsideration. *Proceedings of the American Academy of Arts and Sciences* 80: 61–77.

———. 1958. The theoretician's dilemma. In *Minnesota Studies in the Philosophy of Science, Volume II: Concepts, Theories, and the Mind-Body Problem*, ed. H. Feigl, M. Scriven, and G. Maxwell, 37–98. Minneapolis: University of Minnesota Press. Reprinted in Hempel (1965b, 173–226). Page references in this chapter are to the reprinted version.

———. 1960. Inductive inconsistencies. *Synthese* 12: 439–469; reprinted with slight changes in Hempel (1965b).

———. 1962a. Deductive-nomological vs. statistical explanation. In *Minnesota Studies in the Philosophy of Science, Volume III: Scientific Explanation, Space, and Time*, ed. H. Feigl and G. Maxwell, 98–169. Minneapolis: University of Minnesota Press.

———. 1962b. Explanation in science and history. In *Frontiers of Science and Philosophy*, ed. Robert Colodny, 7–34. Pittsburgh: University of Pittsburgh Press.

———. 1963. Implications of Carnap's work for the philosophy of science. In *The Philosophy of Rudolf Carnap*, ed. Paul Arthur Schilpp, 685–709. La Salle, Ill.: Open Court.

———. 1965a. Aspects of scientific explanation. In *Aspects of Scientific Explanation and Other Essays in the Philosophy of Science*, 331–496. New York: Free Press.

———. 1965b. *Aspects of Scientific Explanation and Other Essays in the Philosophy of Science*. New York: Free Press.

———. 1965c. The theoretician's dilemma. In *Aspects of Scientific Explanation and Other Essays in the Philosophy of Science*, 173–226 (slightly revised version of Hempel 1958). New York: Free Press.

———. 1966. Recent problems of induction. In *Mind and Cosmos*, ed. Robert Colodny, 112–134. Pittsburgh: University of Pittsburgh Press.

Hempel, C. G., T. S. Kuhn, and W. C. Salmon. 1983. Symposium: The philosophy of Carl G. Hempel. *Journal of Philosophy* 80: 555–572.

Hempel, C. G., and P. Oppenheim. 1945. A definition of 'degree of confirmation'. *Philosophy of Science* 12: 98–115.

———. 1948. Studies in the logic of explanation. *Philosophy of Science* 15: 135–175. Reprinted in Hempel (1965b, 135–175).

Hilpinen, Risto. 1968. *Rules of Acceptance and Inductive Logic; Acta Philosophica Fennica*, vol. 22. Amsterdam: North-Holland.

Hintikka, Jaakko. 1966. A two-dimensional continuum of inductive methods. In *Aspects of Inductive Logic*, ed. Jaakko Hintikka and Patrick Suppes, 113–132. Amsterdam: North-Holland.

Holton, Gerald, and Stephen G. Brush. 1973. *Introduction to Concepts and Theories in Physical Science*. 2nd ed. Reading, Mass.: Addison-Wesley.

Hubbard, L. Ron. 1950. *Dianetics: The Modern Science of Mental Healing*. New York: Hermitage House.

Hume, David. [1779] 1970. *Dialogues Concerning Natural Religion*. Nelson Pike (ed.). Indianapolis: Bobbs-Merrill.

———. [1739–1740] 1888. *A Treatise of Human Nature*, Oxford: Clarendon Press.

Humphreys, Paul. 1985. Why propensities cannot be probabilities. *Philosophical Review* 94: 557–570.

Hutchison, Keith. 1982. What happened to occult qualities in the scientific revolution? *Isis* 73: 233–253.

Jeffrey, Richard. 1968. Probable knowledge. In *The Problem of Inductive Logic*, ed. Imre Lakatos, 166–180. Amsterdam: North-Holland.

———. 1983. Bayesianism with a human face. In *Minnesota Studies in the Philosophy of Science, Vol. 10: Testing Scientific Theories*, ed. John Earman, 133–156. Minneapolis: University of Minnesota Press.

Jeffrey, Richard, ed. 1980. *Studies in Inductive Logic and Probability*, vol. 2. Berkeley, Los Angeles, London: University of California Press.

Jeffreys, Sir Harold. 1939. *Theory of Probability*. Oxford: Clarendon Press.

———. 1957. *Scientific Inference*. Cambridge: Cambridge University Press.

Kemeny, John G., and Paul Oppenheim. 1952. Degree of factual support. *Philosophy of Science* 19: 307–324.

Kitcher, Philip. 1982. *Abusing Science: The Case against Creationism*. Cambridge, Mass.: MIT Press.

Klein, M. J. 1990. The Physics of J. Willard Gibbs in his time. *Physics Today*. Sep. 40–48.

Kuhn, Thomas. 1962. *The Structure of Scientific Revolutions*. Chicago: University of Chicago Press.

———. 1970. *The Structure of Scientific Revolutions*. 2nd ed. Chicago: University of Chicago Press.

———. 1977. Objectivity, value judgment, and theory choice. In *The Essential Tension*, 293–339. Chicago: University of Chicago Press.

Laplace, P. S., Marquis de. [1820] 1951. *A Philosophical Essay on Probabilities*. F. W. Truscott and F. L. Emory (trans.). New York: Dover Publications.

Lewis, Clarence Irving. 1946. *An Analysis of Knowledge and Valuation*. La Salle, Ill.: Open Court.

Lewis, David. 1980. A subjectivist's guide to objective chance. In *Studies in Inductive Logic and Probability*, vol. 2, ed. Richard C. Jeffrey, 263–293. Berkeley, Los Angeles, London: University of California Press.

Locke, John. [1690] 1924. *An Essay Concerning Human Understanding*. Abridged A. S. Pringle-Pattison (ed.). Oxford: Clarendon Press.

Lodge, Sir Oliver Joseph. 1889. *Modern Views of Electricity* (2nd ed. 1892; 3rd ed. 1907). London and New York: Macmillan.

Lucretius. 1951. *The Nature of the Universe*. Ronald Latham (trans.). New York: Penguin.

Machina, Mark J. 1987. Decision-making in the presence of risk. *Science* 236 (1 May): 537–543.

Maher, Patrick. 1986. The irrelevance of belief to practical action. *Erkenntnis* 24: 263–284.

———. 1993. *Betting on Theories*. Cambridge: Cambridge University Press.

260   *References*

Marshall, Eliot. 1986. Feynmann issues his own shuttle report, attacking NASA's risk estimates. *Science* 232 (27 June): 1596.

Mayo, Deborah. 1996. *Error and the Growth of Experimental Knowledge*. Chicago: University of Chicago Press.

McLaughlin, Robert. 1982a. Invention and appraisal. In *What? Where? When? Why?*, 69–100. Dordrecht: Reidel.

McLaughlin, Robert, ed. 1982b. *What? Where? When? Why?* Dordrecht: Reidel.

Mellor, D. H. 1971. *The Matter of Chance*. Cambridge: Cambridge University Press.

Mermin, N. David. 1985. Is the moon there when nobody looks? Reality and the quantum theory. *Physics Today* 38 (4): 38–47.

Miller, Richard W. 1987. *Fact and Method*. Princeton, N.J.: Princeton University Press.

Moffatt, James, 1922. *A New Translation of the Bible Containing the Old and New Testaments*. New York and Harper.

Motley, Michael T. 1985. Slips of the tongue. *Scientific American* 253: 116–123.

Nye, Mary Jo. 1972. *Molecular Reality*. London: Macdonald; New York: American Elsevier.

Nye, Mary Jo, ed. 1984. *The Question of the Atom*. Los Angeles and San Francisco: Tomash Publishers.

Paley, William. 1851. *Natural Theology, or Evidences of the Existence and Attributes of the Deity, Collected from the Appearances of Nature*. Boston: Gould and Lincoln.

Pannekoek, A. 1961. *A History of Astronomy*. New York: Interscience.

Parrini, Paolo. 1994. With Carnap, beyond Carnap: Metaphysics, science, and the realism/ instrumentalism issue. In *Logic, Language, and the Structure of Scientific Theories*, ed. Wesley C. Salmon and Gereon Wolters, 255–278. Pittsburgh and Konstanz: University of Pittsburgh Press and Universitätsverlag Konstanz.

Partington, J. R. 1964. *A History of Chemistry*, vol. 4. London: Macmillan.

Pauling, Linus. 1970. *Vitamin C and the Common Cold*. San Francisco: W. H. Freeman.

Pearson, Karl. 1911. *The Grammar of Science*. 3rd ed., rev. and enl. London: A. and C. Black. Reprint, New York: Meridian Books, 1957.

Peirce, Charles Sanders. 1931. *Collected Papers*, vol. 2. Charles Hartshorne and Paul Weiss (eds.). Cambridge, Mass.: Harvard University Press.

Perrin, Jean. 1908. Mouvement brownien et réalité moléculaire. *Annales de Chemie et de Physique* [8] 18: 1–114.

———. [1908] 1910. Brownian movement and molecular reality. F. Soddy (trans.). Reprinted in *The Question of the Atom*, ed. Mary Jo Nye. Los Angeles and San Francisco: Tomash Publishers, 1984.

———. [1913] 1916. *Atoms*. D. L. Hammick (trans.). London: Constable & Company.

———. 1923. *Atoms*. 2nd rev. English ed. D. L. Hammick (trans.). New York: Van Nostrand.

Popper, Karl R. [1935] 1959. *The Logic of Scientific Discovery*. English translation of *Logik der Forschung* with added appendices. New York: Basic Books.

———. 1963. *Conjectures and Refutations: The Growth of Scientific Knowledge*. London: Rutledge & Regan Paul.

———. 1974. Autobiography. In *The Philosophy of Karl Popper*, ed. Paul Arthur Schilpp, 3–181. La Salle, Ill.: Open Court.

Popper, Karl R., and David Miller. 1983. A proof of the impossibility of inductive probability. *Nature* 302 (5910): 687–688.

Ramsey, F. P. 1950. *Foundation of Mathematics*, ed. R. B. Braithewaite. New York: Humanities Press.

Raynor, David. 1980. Hume's knowledge of Bayes's theorem. *Philosophical Studies* 38: 105–106.

Reichenbach, Hans. 1933. Rudolf Carnap, "Der logische Aufbau der Welt" [review]. *Kantstudien* 38: 199–201. English translation in *Selected Writings, 1909–1953*, trans. Hans Reichenbach, vol. 1, 405–408. Dordrecht: Reidel, 1978.

———. 1935. *Wahrscheinlichkeitslehre*. Leyden: A. W. Sijthoff.

———. 1938. *Experience and Prediction*. Chicago: University of Chicago Press.

———. 1946. *Philosophic Foundations of Quantum Mechanics*. Berkeley and Los Angeles: University of California Press.

———. [1935] 1949. *The Theory of Probability*. English translation with new material. Berkeley and Los Angeles: University of California Press.

———. 1951. The verifiability theory of meaning. *Proceedings of the American Academy of Arts and Sciences* 80: 46–60.

———. 1954. *Nomological Statements and Admissible Operations*. Amsterdam: North-Holland. Reprinted as *Laws, Modalities, and Counterfactuals*. Berkeley and Los Angeles: University of California Press, 1976.

———. 1956. *The Direction of Time*. Berkeley and Los Angeles: University of California Press.

———. 1968–. *Gesammelte Werke in 9 Bänden*. Braunschweig: Friedr. Vieweg & Sohn.

———. [1954] 1976. *Laws, Modalities, and Counterfactuals*. Foreword by Wesley C. Salmon. Berkeley and Los Angeles: University of California Press.

Reif, F. 1965. *Statistical Physics*. New York: McGraw-Hill.

Reisch, George A. 1991. Did Kuhn kill logical empiricism? *Philosophy of Science* 58: 264–277.

Rescher, Nicholas. 1961. Plausible implication. *Analysis* 21: 128–135.

Rescher, Nicholas, and Robert Brandom. 1980. *The Logic of Inconsistency*. Oxford: Blackwell.

Ronchi, Vasco. 1967. The influence of the early development of optics on science and philosophy. In *Galileo: Man of Science*, ed. Ernan McMullin, 195–206. New York: Basic Books.

Russell, Bertrand. 1939. Dewey's new logic. In *The Philosophy of John Dewey*, ed. Paul Arthur Schilpp, 135–156. New York: Tudor.

———. 1948. *Human Knowledge, Its Scope and Limits*. New York: Simon and Schuster.

Salmon, Wesley C. 1951. A modern analysis of the design argument. *Research Studies of the State College of Washington*.

———. 1955. The short run. *Philosophy of Science* 22: 214–221.

———. 1963a. *Logic*. (2nd ed. 1973; 3rd ed. 1984.) Englewood Cliffs, N.J.: Prentice-Hall.

———. 1963b. On vindicating induction. In *Induction: Some Current Issues*, ed. Henry E. Kyburg and Ernest Nagel. Middletown, Conn.: Wesleyan University Press. Also printed in *Philosophy of Science* 30: 252–261.

———. 1963c. Review of "Inductive probability" by John Patrick Day. *Philosophical Review* 72: 392–396.

———. 1965a. Consistency, transitivity, and inductive support. *Ratio* 7: 164–169.

———. 1965b. Symposium on inductive evidence. With S. Barker and H. Kyburg. *American Philosophical Quarterly* 2: 1–16.

———. 1966. Verifiability and logic. In *Mind, Matter and Method: Essays in Honor of Herbert Feigl*, ed. Paul K. Feyerabend and Grover Maxwell, 354–376. Minneapolis: University of Minnesota Press.

———. 1967a. Carnap's inductive logic. *The Journal of Philosophy* 64: 725–739.

———. 1967b. *Foundations of Scientific Inference*. Pittsburgh: University of Pittsburgh Press.

———. 1968. Inquiries into the foundations of science. In *Vistas in Science*, ed. David L. Arm, 1–24. Albuquerque: University of New Mexico Press.

——. 1971. *Statistical Explanation and Statistical Relevance*, with contributions by Richard C. Jeffrey and James G. Greeno. Pittsburgh: University of Pittsburgh Press.

——. 1973a. Confirmation. *Scientific American* 228 (5): 75–83.

——. 1973b. *Logic*. 2nd ed. Englewood Cliffs, N.J.: Prentice-Hall.

——. 1974. Russell on scientific inference, or Will the real deductivist please stand up? In *Bertrand Russell's Philosophy*, ed. George Nakhnikian, 183–208. London: Gerald Duckworth.

——. 1979. Propensities: A discussion-review. *Erkenntnis* 14: 183–216.

——. 1981. Rational prediction. *British Journal for the Philosophy of Science* 32: 115–125.

——. 1984a. *Logic*. 3rd ed. Englewood Cliffs, N.J.: Prentice-Hall.

——. 1984b. *Scientific Explanation and the Causal Structure of the World*. Princeton, N.J.: Princeton University Press.

——. [1989] 1990. *Four Decades of Scientific Explanation*. Minneapolis: University of Minnesota Press. Reprinted from *Minnesota Studies in the Philosophy of Science Volume XIII: Scientific Explanation*, eds. Philip Kitcher and Wesley C. Salmon, 3–219. Minneapolis: University of Minnesota Press.

——. 1994. Carnap on realism: Comments on Parrini. In *Logic, Language, and the Structure of Scientific Theories*, ed. Wesley C. Salmon and Gereon Wolters, 279–286. Pittsburgh and Konstanz: University of Pittsburgh Press and Universitätsverlag Konstanz.

——. 1999. Ornithology in a cubical world: Reichenbach on scientific realism. In *Epistemological and Experimental Perspectives on Quantum Physics*, ed. Daniel Greenberger, Reiter L. Wolfgang, and Anton Zeilinger, 303–315. Dordrecht: Kluwer.

Salmon, Wesley C., and Gereon Wolters, eds. 1994. *Logic, Language, and the Structure of Scientific Theories*. Pittsburgh and Konstanz: University of Pittsburgh Press and Universitätsverlag Konstanz.

Savage, L. J. 1954. *The Foundations of Statistics*. New York: Wiley.

Schilpp, Paul Arthur, ed. 1939. *The Philosophy of John Dewey*. New York: Tudor.

——. 1963. *The Philosophy of Rudolf Carnap*. La Salle, Ill.: Open Court.

Schwartz, Jeffrey H. 1987. *The Red Ape: Orangutans and Human Origins*. Boston: Houghton Mifflin.

Shearer, Mary. 1966. "An investigation of $c^\dagger$." M.A. thesis, Wayne State University, Detroit.

Shimony, Abner. 1970. Scientific inference. In *The Nature and Function of Scientific Theories*, ed. Robert G. Colodny, 79–172. Pittsburgh: University of Pittsburgh Press.

Skyrms, Brian. 1985. Degrees of belief and coherence epistemology. University of Pittsburgh Center for Philosophy of Science Annual Lecture, 8 October.

Smith, Joel. 1987. "The status of inconsistent statements in scientific inquiry." Ph.D. diss., University of Pittsburgh.

Smokler, Howard. 1967. The equivalence condition. *American Philosophical Quarterly* 5: 300–307.

Sobel, Michael. 1987. *Light*. Chicago: University of Chicago Press.

Sober, Elliott. 1988. The principle of the common cause. In *Probability and Causality*, ed. James H. Fetzer, 211–228. Dordrecht: Reidel.

Stuewer, Roger H., ed. 1970. *Minnesota Studies in the Philosophy of Science Volume V: Historical and Philosophical Perspectives of Science*. Minneapolis: University of Minnesota Press.

Suppes, Patrick. 1966. A Bayesian approach to the paradoxes of confirmation. In *Aspects of Inductive Logic*, ed. Jaakko Hintikka and Patrick Suppes, 198–207. Amsterdam: North-Holland.

van Fraassen, Bas C. 1980. *The Scientific Image*. Oxford: Clarendon Press.

———. 1989. *Laws and Symmetry*. Oxford: Clarendon Press.

Velikovsky, Immanuel. 1950. *Worlds in Collision*. New York: Doubleday.

Wallace, John R. 1966. Goodman, logic, induction. *The Journal of Philosophy* 63: 310–328.

Watson, James D. 1969. *The Double Helix*. New York: New American Library.

Weber, Joseph. 1968. Gravitational waves. *Phys. Rev. Lett.* 20, 1307–1308.

Weinberg, Steven. 1972. *Gravitation and Cosmology*. New York: Wiley.

Whitehead, A. N., and B. Russell. 1910–1913. *Principia Mathematica*, vols. 1–3. Cambridge: Cambridge University Press.

Wichmann, Eyvind H. 1967. *Quantum Physics*. New York: McGraw-Hill.

Wilde, Oscar. 1979. *An Ideal Husband*. New York: Penguin Books.

Wolters, G. 1989. Phenomenalism, relativity and atoms: Rehabilitating Ernst Mach's philosophy of science. In *Logic, Methodology and Philosophy of Science, vol. 8: Proceedings of the Eighth International Congress of Logic, Methodology and Philosophy of Science, Moscow*, ed. J. E. Fenstad, I. T. Frolov, and R. Hilpinen, 641–660, 1987. Amsterdam: North-Holland.

Worrall, John. 1990. Scientific revolutions and scientific rationality: The case of the elderly holdout. In *Minnesota Studies in the Philosophy of Science, Volume XIV: Scientific Theories*, ed. C. Wade Savage, 319–354. Minneapolis: University of Minnesota Press.

Wright, Larry. 1976. *Teleological Explanations*. Berkeley, Los Angeles, London: University of California Press.

Wyden, Peter. 1984. *Day One*. New York: Simon and Schuster.

Bibliography of Works
by Wesley C. Salmon

Books

Logic. Englewood Cliffs, N.J.: Prentice-Hall, 1963.
   Spanish translation. Mexico City: Union Tipografica Editorial Hispano Americano, 1965.
   Japanese translation. Tokyo: Baifu Kan, 1967.
   Italian translation. Rome: Societa editrice il Mulino, 1969.
   Portuguese translation. Rio de Janeiro, Brazil: Zahar Editores, 1969.
The Foundations of Scientific Inference. Pittsburgh: University of Pittsburgh Press, 1967.
Zeno's Paradoxes (ed.). Indianapolis: Bobbs-Merrill, 1970. Reprinted (with new preface and
   expanded bibliography): Indianapolis: Hackett, 2001.
Statistical Explanation and Statistical Relevance, with contributions by Richard C. Jeffrey
   and James G. Greeno. Pittsburgh: University of Pittsburgh Press, 1971.
Logic, 2nd ed. Englewood Cliffs, N.J.: Prentice-Hall, 1973.
   Japanese translation. Tokyo: Baifu Kan, 1975.
   German translation. Stuttgart: Philipp Reclam, 1983.
Space, Time, and Motion: A Philosophical Introduction. Encino, Cal.: Dickenson, 1975.
Hans Reichenbach: Logical Empiricist (editor). Dordrecht: Reidel., 1979.
Space, Time, and Motion: A Philosophical Introduction, 2nd ed. Minneapolis: University of
   Minnesota Press, 1981.
Logic, 3rd ed. Englewood Cliffs, N.J.: Prentice-Hall, 1984.
   Japanese translation. Tokyo: Baifu Kan, 1987.
   Portuguese translation. Rio de Janeiro: Prentice-Hall do Brasil, 1993.
Scientific Explanation and the Causal Structure of the World. Princeton, N.J.: Princeton
   University Press, 1984.
The Limitations of Deductivism (co-editor with Adolf Grünbaum). Berkeley: University of
   California Press, 1988.
Scientific Explanation. Minnesota Studies in the Philosophy of Science, Volume XIII (co-
   editor with Philip Kitcher). Minneapolis: University of Minnesota Press, 1989.
Four Decades of Scientific Explanation. Minneapolis: University of Minnesota Press,
   1990.
   Italian translation. *40 anni di spiegazione scientifica.* Padova: Franco Muzzio Editore,
   1992.

Introduction to the Philosophy of Science (co-author with other members of the Department of History and Philosophy of Science). Englewood Cliffs, N.J.: Prentice-Hall, 1992. Greek translation. ΣΤΗ ΦΙΛΟΣΟΦΙΑ ΤΗΣ ΕΠΙΣΤΗΜΗΣ ΠΑΝΕΠΙΣΤΗΜΙΑΚΕΣ ΕΚΔΟΣΕΙΣ ΚΡΗΤΗΣ, ΗΡΑΚΛΕΙΟ, 1998

Logic, Language, and the Structure of Scientific Theories (co-editor with Gereon Wolters). Pittsburgh and Konstanz: University of Pittsburgh Press and Universitätsverlag Konstanz, 1994.

Causality and Explanation. New York: Oxford University Press, 1998.

Logical Empiricism: Historical and Contemporary Perspectives (co-editor with Paolo Parrini and Merrilee H. Salmon). Pittsburgh: University of Pittsburgh Press, 2003.

Reality and Rationality. Phil Dowe and Merrilee H. Salmon (eds.). New York: Oxford University Press, 2005.

Articles

"A Modern Analysis of the Design Argument." Research Studies of the State College of Washington 19, 4 (December 1951): 207–220.

"The Frequency Interpretation and Antecedent Probabilities." Philosophical Studies 4, 3 (April 1953): 44–48.

"The Uniformity of Nature." Philosophy and Phenomenological Research 14, 1 (September 1953): 39–48.

"The Short Run." Philosophy of Science 22, 3 (July 1955): 214–221.

Reply to critic:

"Reply to Pettijohn." Philosophy of Science 23, 2 (April 1956): 150–152.

"Regular Rules of Induction." Philosophical Review 65, 3 (July 1956): 385–388.

"Should We Attempt to Justify Induction?" Philosophical Studies 8, 3 (April 1957): 33–38.

Reprinted (in part or in toto):

In Readings in the Theory of Knowledge, ed. John V. Canfield and Franklin Donnell, 363–367. New York: Appleton-Century-Crofts, 1964.

In Meaning and Knowledge, ed. Ernest Nagel and Richard B. Brandt, 365–369. New York: Harcourt, Brace and World, 1965.

"Should We Attempt to Justify Induction?" Indianapolis: Bobbs-Merrill Reprint, Phil 184, 1969.

In New Readings in Philosophical Analysis, ed. H. Feigl, W. Sellars, and K. Lehrer, 500–510. New York: Appleton-Century-Crofts, 1972.

In Philosophical Problems of Science and Technology, ed. Alex Michalos, 357–373. Boston: Allyn and Bacon, 1974.

"The Predictive Inference." Philosophy of Science 24, 2 (April 1957): 180–190.

" 'Exists' as a Predicate" (with George Nakhnikian). Philosophical Review 66, 4 (October 1957): 535–542.

Spanish translation: " 'Existe' como Predicado." Revista "Universidad De San Carlos" 41: 145–155.

"Psychoanalytic Theory and Evidence." In Psychoanalysis, Scientific Method, and Philosophy, ed. Sidney Hook, 252–267. New York: New York University Press, 1959.

Reprinted:

In Freud: A Collection of Critical Essays, ed. Richard Wollheim, 271–284. Garden City, N.Y.: Anchor/Doubleday, 1974.

In The Freudian Paradigm, ed. Md. Mujeeb ur-Rahman, 187–200. Chicago: Nelson-Hall, 1977.

"Barker's Theory of the Absolute." Philosophical Studies 10, 4 (June 1959): 50–53.

"Vindication of Induction." In *Current Issues in the Philosophy of Science*, ed. Herbert Feigl
and Grover Maxwell, 245–256. New York: Holt, Rinehart and Winston, 1961.
Reply to critic:
"Rejoinder to Barker." Ibid., 260–262.
"On Vindicating Induction." In *Induction: Some Current Issues*, ed. Henry E. Kyburg, Jr.,
and Ernest Nagel, 27–41. Middletown, Conn.: Wesleyan University Press, 1963.
Reply to critic:
"Reply to Black." Ibid., 49–54.
Reprinted:
Philosophy of Science 30, 3 (July 1963): 252–261.
In *Probabilities, Problems, and Paradoxes*, ed. Sidney Luckenbach, 200–212. Encino,
Cal.: Dickenson, 1972.
"Inductive Inference." In *Philosophy of Science: The Delaware Seminar*, vol. 2, ed. Bernard
H. Baumrin, 341–370. New York: Wiley, 1963.
Reprinted:
In *Readings in the Philosophy of Science*, ed. Baruch A. Brody, 597–617. Englewood
Cliffs, N.J.: Prentice-Hall, 1970.
"The Pragmatic Justification of Induction." In *The Justification of Induction*, ed.
Richard Swinburne, 85–97. London: Oxford University Press, 1974.
Spanish translation. "La justificacion pragmatica de la induccion." In *La justificacion
del razonamento inductivo*, ed. Richard Swinburne, 105–118. Madrid: Alianza
Editorial, 1976.
"What Happens in the Long Run?" *Philosophical Review* 74, 3 (July 1965): 373–378.
"The Concept of Inductive Evidence." *American Philosophical Quarterly* 2, 4 (October
1965): 1–6.
Reply to critics:
"Rejoinder to Barker and Kyburg." Ibid.: 13–16.
Reprinted:
"Symposium on Inductive Evidence" (with S. Barker and H. Kyburg). Indianapolis:
Bobbs-Merrill Reprint, Phil 240, 1969.
In *The Justification of Induction*, ed. Richard Swinburne, 48–57. London: Oxford
University Press, 1974.
"Rejoinder to Barker and Kyburg." Ibid.: 66–73.
Spanish translation. "El concepto de evidencia inductiva." In *La justificacion del razo-
namiento inductivo*, ed. Richard Swinburn, 61–74. Madrid: Alianza Editorial, 1976.
"Replica a Barker y Kyburg." Ibid., 85–92.
"The Status of Prior Probabilities in Statistical Explanation." *Philosophy of Science* 33, 2
(April 1965): 137–146.
Reply to critic:
"Reply to Kyburg." Ibid.: 152–154.
"Consistency, Transitivity, and Inductive Support." *Ratio* 7, 2 (December 1965): 164–169.
German translation. "Widerspruchsfreiheit, Transitivität und induktive Bekraftigung."
Ratio 7, 2 (December 1965): 152–157.
"Use, Mention, and Linguistic Invariance." *Philosophical Studies* 17, 1–2 (January/February
1966): 13–18.
"Verifiability and Logic." In *Mind, Matter, and Method*, ed. P. K. Feyerabend and Grover
Maxwell, 347–376. Minneapolis: University of Minnesota Press, 1966.
Reprinted:
In *The Logic of God*, ed. Malcolm L. Diamond and Thomas V. Litzenburg, Jr.,
456–480. Indianapolis: Bobbs-Merrill, 1975.

"The Foundations of Scientific Inference." In *Mind and Cosmos*, ed. Robert G. Colodny, 135–275. Pittsburgh: University of Pittsburgh Press, 1966.
Reprinted:
> Wesley C. Salmon, *The Foundations of Scientific Inference*. Pittsburgh: University of Pittsburgh Press, 1967 (addendum, April 1967).
> "The Problem of Induction." In *Introduction to Philosophy*, ed. John Perry and Michael Bratman, 265–286 (includes 5–27, 40–43, 48–56, 136, from *The Foundations of Scientific Inference*). New York: Oxford University Press, 1986.

"Carnap's Inductive Logic." *Journal of Philosophy* 64, 21 (November 9, 1967): 725–739.
> Italian translation. "La logica induttiva di Carnap." In *L'induzione L'ordine Dell'universo*, ed. Alberto Meotti, 191–192 (translation of small extract). Milano: Edizioni de Comunita, 1978.

"The Justification of Inductive Rules of Inference." In *The Problem of Inductive Logic*, ed. Imre Lakatos, 24–43. Amsterdam: North-Holland, 1968.
Reply to critics:
> "Reply." Ibid., 74–97.
> Spanish translation. "La justificacion de las reglas inductivas de inferencia." In *Problems de la Filosofia*, ed. Luis O. Gomez and Roberto Torretti, 373–385. Editorial Universitaria, Universidad de Puerto Rico, 1974.

"Inquiries into the Foundations of Science." In *Vistas in Science*, ed. David L. Arm, 1–24. Albuquerque: University of New Mexico Press, 1968.
Reprinted:
> In *Probabilities, Problems, and Paradoxes*, ed. Sidney Luckenbach, 139–158. Encino, Cal.: Dickenson, 1972.

"Introduction: The Context of These Essays" (with Adolf Grünbaum; introduction to "A Panel Discussion of Simultaneity by Slow Clock Transport in the Special and General Theories of Relativity"). *Philosophy of Science* 36, 1 (March 1969): 1–4.

"The Conventionality of Simultaneity." *Philosophy of Science* 36, 1 (March 1969): 44–63.

"Partial Entailment as a Basis for Inductive Logic." In *Essays in Honor of Carl G. Hempel*, ed. Nicholas Rescher, 47–82. Dordrecht: Reidel, 1969.

"Statistical Explanation." In *The Nature and Function of Scientific Theories*, ed. Robert G. Colodny, 173–232. Pittsburgh: University of Pittsburgh Press, 1970.
Reprinted:
> Wesley C. Salmon, *Statistical Explanation and Statistical Relevance*, 29–87. Pittsburgh: University of Pittsburgh Press, 1971.
> In *Foundations of Philosophy of Science*, ed. James H. Fetzer, 224–249 (in part). New York: Paragon House, 1993.
> In *Philosophies of Science*, Jennifer McErlean, 72–87. Belmont, Cal.: Wadsworth, 2000.

"Bayes's Theorem and the History of Science." In *Minnesota Studies in the Philosophy of Science Volume V: Historical and Philosophical Perspectives of Science*, ed. Roger H. Stuewer, 68–86. Minneapolis: University of Minnesota Press, 1970.

"Determinism and Indeterminism in Modern Science."* In *Reason and Responsibility*, 2nd and subsequent eds., ed. Joel Feinberg. Encino, Cal.: Dickenson, 1971.

"A Zenoesque Problem." Published without title in Martin Gardner, "Mathematical Games," *Scientific American* 225, 6 (December 1971): 97–99.

"Logic." *Encyclopedia Americana* (1972) 17: 673–687.

"Laws in Science." *Encyclopedia Americana* (1972) 17: 81.

"Numerals vs. Numbers." *De Pauw Mathnews* 2, 1 (March 1973): 8–11.

"Confirmation." *Scientific American* 228, 5 (May 1973): 75–83.

Reply to critic:
"Reply to Bradley Efron." *Scientific American* 229, 3 (September 1973): 8–10.
"Memory and Perception in *Human Knowledge*." In *Bertrand Russell's Philosophy*, ed. George Nakhnikian, 139–167. London: Gerald Duckworth, 1974.
"Russell on Scientific Inference, or Will the Real Deductivist Please Stand Up?" Ibid., 183–208.
Reprinted:
   In *Bertrand Russell: Critical Assessments*, vol. 3, ed. Andrew Irvine. London: Routledge, 1998.
"Dedication to Leonard J. Savage." In *PSA 1972*, ed. Kenneth F. Schaffner and Robert S. Cohen, 101. Dordrecht: Reidel, 1974.
"An Encounter with David Hume." In *Reason and Responsibility*, 3rd and subsequent eds., ed. Joel Feinberg, 190–208. Encino, Cal.: Dickenson, 1975.
Reprinted:
   In *Foundations of Philosophy of Science*, ed. James H. Fetzer, 277–298. New York: Paragon House, 1993.
"Confirmation and Relevance." In *Minnesota Studies in the Philosophy of Science, Vol. 6: Induction, Probability, and Confirmation*, ed. Grover Maxwell and Robert M. Anderson, Jr., 3–36. Minneapolis: University of Minnesota Press, 1975.
Reprinted:
   In *The Concept of Evidence*, ed. Peter Achinstein, 93–123. Oxford: Oxford University Press, 1983.
"Theoretical Explanation."* In *Explanation*, ed. Stephan Körner, 118–145. Oxford: Blackwell, 1975.
Reply to critics:
   "Reply to comments." Ibid., 160–184.
"Clocks and Simultaneity in Special Relativity, or Which Twin Has the Timex?" In *Motion and Time, Space and Matter*, ed. Peter K. Machamer and Robert G. Turnbull, 508–145. Columbus: Ohio State University Press, 1976.
Reprinted:
   Wesley C. Salmon, *Space, Time, and Motion: A Philosophical Introduction*, chap. 4. Encino, Cal.: Dickenson, 1975.
Foreword to *Laws, Modalities, and Counterfactuals* by Hans Reichenbach, vii–xlii. Berkeley and Los Angeles: University of California Press, 1976.
Reprinted (with minor revisions):
   "Laws, Modalities, and Counterfactuals." *Synthese* 35, 2 (June 1977): 191–229.
   In *Hans Reichenbach: Logical Empiricist*, ed. Wesley C. Salmon, 655–696. Dordrecht: Reidel, 1979.
Preface to "Hans Reichenbach: Logical Empiricist," *Synthese* 34, 1 (January 1977): 1–2.
"The Philosophy of Hans Reichenbach." *Synthese* 34, 1 (January 1977): 5–88.
Reprinted:
   In *Hans Reichenbach: Logical Empiricist*, ed. Wesley C. Salmon, 1–84. Dordrecht: Reidel, 1979.
   German translation with revisions. "Hans Reichenbachs Leben und die Tragweite seiner Philosophie" (Einleiting zur Gesamtausgabe). In *Hans Reichenbach: Gesammelte Werke*, vol. 1, ed. Andreas Kamlah and Maria Reichenbach, trans. from English Maria Reichenbach, 5–81. Wiesbaden: Vieweg, 1977.
"Hans Reichenbach." In *Dictionary of American Biography*, Supplement 5, 1951–1955, ed. John A. Garraty, 562–563. New York: Scribner, 1977.
"Hempel's Conception of Inductive Inference in Inductive-Statistical Explanation." *Philosophy of Science* 44, 2 (June 1977): 180–185.

"Indeterminism and Epistemic Relativization." Ibid.: 199–202.

"An 'At-At' Theory of Causal Influence."* Ibid.: 215–224.

"The Curvature of Physical Space." In *Minnesota Studies in the Philosophy of Science, Volume VIII: Foundations of Space-Time Theories*, ed. John S. Earman, Clark N. Glymour, and John Stachel, 281–302. Minneapolis: University of Minnesota Press, 1977.

"The Philosophical Significance of the One-way Speed of Light." *Nous* 11, 3 (September 1977): 353–392.

"A Third Dogma of Empiricism."* In *Basic Problems in Methodology and Linguistics, Part III, Proceedings of the Fifth International Congress of Logic, Methodology, and Philosophy of Science*, ed. Robert E. Butts and Jaakko Hintikka, 149–166. Dordrecht: Reidel, 1977.

"An Ontological Argument for the Existence of the Null Set." In *Mathematical Magic Show*, ed. Martin Gardner, 32–33. New York: Knopf, 1977.

"Objectively Homogeneous Reference Classes." *Synthese* 36, 4 (December 1977): 399–414.

"Religion and Science: A New Look at Hume's *Dialogues*." *Philosophical Studies* 33 (1978): 143–176.

Reply to critic:

    "Experimental Atheism," *Philosophical Studies* 35 (1979): 101–104.

"Unfinished Business: The Problem of Induction." *Philosophical Studies* 33 (1978): 1–19.

"Why Ask, 'Why?'? — An Inquiry Concerning Scientific Explanation."* *Proceedings and Addresses of the American Philosophical Association* 51, 6 (August 1978): 683–705.

Reprinted:

    In *Hans Reichenbach: Logical Empiricist*, ed. Wesley C. Salmon, 403–425. Dordrecht: Reidel, 1979.

    In *Scientific Knowledge*, ed. Janet A. Kourany, 51–64. Belmont, Cal.: Wadsworth, 1987.

    In *Readings in the Philosophy of Science*, ed. Theodore Schick, Jr., 79–86. Mountain View, Cal.: Mayfield, 2000.

"Hans Reichenbach: A Memoir." In *Hans Reichenbach: Selected Writings, 1909 – 1953*, vol. 1, ed. Maria Reichenbach and Robert S. Cohen, 69–77. Dordrecht: Reidel, 1978.

"Alternative Models of Scientific Explanation"* (with Merrilee H. Salmon). *American Anthropologist* 81, 1 (March 1979): 61–64.

"A Philosopher in a Physics Course." *Teaching Philosophy* 2, 2 (1979): 139–146.

Reprinted:

    In *Demonstrating Philosophy*, ed. Arnold Wilson, 133–140. Lanham, Md.: University Press of America, 1988.

"Propensities: A Discussion Review." *Erkenntnis* 14 (1979): 183–216.

"Informal Analytic Approaches to the Philosophy of Science." In *Current Research in Philosophy of Science*, ed. Peter Asquith and Henry E. Kyburg, Jr., 3–15. East Lansing, Mich.: Philosophy of Science Assn., 1979.

"Probabilistic Causality."* *Pacific Philosophical Quarterly* 61, 1–2 (1980): 50–74.

Reprinted:

    In *Causation*, ed. Ernest Sosa and Michael Tooley, 137–153. Oxford: Oxford University Press, 1993.

"Logic and Language." In *The Study of Logic and Language* (in Chinese), 1980, 174–195 (translation of pars. 29–31 of *Logic*, 2nd ed.).

"Wissenschaft, Grundlagen der." In *Handbuch Wissenschaftstheoretischer Begriff*, ed. J. Speck (translated from English by the editor), 752–757. Göttingen: Vandenhoeck and Ruprecht, 1980.

"John Venn's *Logic of Chance*." In *Probabilistic Thinking, Thermodynamics, and the Interaction of the History and Philosophy of Science*, ed. J. Hintikka, D. Gruender, and E. Agazzi, 125–138. Dordrecht: Reidel, 1980.

"Robert Leslie Ellis and the Frequency Theory." Ibid., 139–143.

"Rational Prediction." *British Journal for the Philosophy of Science* 32, 2 (June 1981): 115–125.
Reprinted:
In *The Limitations of Deductivism*, ed. Adolf Grünbaum and Wesley C. Salmon, 47–60. Berkeley, Los Angeles, London: University of California Press, 1988.
In *Philosophy of Science: The Central Issues*, ed. Martin Curd and J. A. Cover, 433–444. New York: Norton, 1998.
French translation. "La Prédiction Rationnelle." In *Karl Popper Cahiers — Science, Technologie, Société*, vol. 8, 108–120. Paris: Editions du Centre National de la Recherche Scientifique, 1985.

"In Praise of Relevance." *Teaching Philosophy* 4, 3–4 (July and October 1981): 261–275.

"Comets, Pollen, and Dreams: Some Reflections on Scientific Explanation."* In *What? Where? When? Why?*, ed. Robert McLaughlin, 155–178. Dordrecht: Reidel, 1982.

"Further Reflections." Ibid., 231–280.

"Autobiographical Note." Ibid., 281–283.

"Causality: Production and Propagation."* In *PSA 1980*, ed. Peter D. Asquith et al., 49–69. East Lansing, Mich.: Philosophy of Science Assn., 1982.
Reprinted:
In *Causation*, ed. Ernest Sosa and Michael Tooley, 154–171. Oxford: Oxford University Press, 1993.

"Causality in Archaeological Explanation."* In *Theory and Explanation in Archaeology*, ed. Colin Renfrew et al., 45–55. New York: Academic Press, 1982.

"Carl G. Hempel on the Rationality of Science." *Journal of Philosophy* 80, 10 (October 1983): 555–562.

"Probabilistic Explanation: Introduction." In *PSA 1982*, ed. Peter Asquith and Thomas Nickles, 179–180. East Lansing, Mich.: Philosophy of Science Assn., 1983.

"Empiricism: The Key Question." In *The Heritage of Logical Positivism*, ed. Nicholas Rescher, 1–21. Lanham, Md.: University Press of America, 1985.

"Conflicting Conceptions of Scientific Explanation" (abstract of contribution to the APA Symposium on Wesley Salmon's *Scientific Explanation and the Causal Structure of the World*). *Journal of Philosophy* 82, 11 (November 1985): 651–654.

"Scientific Explanation: Consensus or Progress?" (abstract of contribution to AAAS Symposium on "Is There a New Consensus in Philosophy of Science"). In *1986 AAAS Annual Meeting: Abstracts of Papers*, compilers Barbara C. Klein-Helmuth and David Savold, 90. Washington, D.C.: American Association for the Advancement of Science, 1986.

"Scientific Explanation: Three Basic Conceptions."* In *PSA 1984*, ed. Peter Asquith and Philip Kitcher, 239–305. East Lansing, Mich.: Philosophy of Science Assn., 1986.

"Van Fraassen on Explanation"* (with Philip Kitcher). *Journal of Philosophy* 84 (1987): 315–330.
Reprinted:
In *The Philosopher's Annual* 10 (1987): 75–90.
In *Explanation*, ed. David-Hillel Ruben, 78–112. Oxford: Oxford University Press, 1993.

"Statistical Explanation and Causality." In *Theories of Explanation*, ed. Joseph Pitt, 75–118. New York: Oxford University Press, 1988. (Contains a brief extract from *Statistical Explanation and Statistical Relevance* and chaps. 5–6 of *Scientific Explanation and the Causal Structure of the World*.)

"Dynamic Rationality: Propensity, Probability, and Credence." In *Probability and Causality*, ed. James H. Fetzer, 3–40. Dordrecht: Reidel, 1988.
Reprinted:
  In *Foundations of Philosophy of Science*, ed. James H. Fetzer, 224–229. New York: Paragon House, 1993.
"Publications: An Annotated Bibliography." In *Probability and Causality*, ed. James H. Fetzer, 271–336. Dordrecht: Reidel, 1988.
"In Memoriam: J. Alberto Coffa." In *The Limitations of Deductivism*, ed. Adolf Grünbaum and Wesley C. Salmon, xxi–xxiv. Berkeley, Los Angeles, London: University of California Press, 1988.
  Spanish translation. *Revista Latinoamericana de Filosofia* 12, 2 (July 1986): 245–247.
"Introduction. " In *The Limitations of Deductivism*, ed. Adolf Grünbaum and Wesley C. Salmon, 1–18. Berkeley, Los Angeles, London: University of California Press, 1988.
"Deductivism Visited and Revisited."* Ibid., 95–127.
"Intuitions: Good and Not-So-Good."* In *Causation, Chance, and Credence*, ed. William Harper and Brian Skyrms, 51–71. Dordrecht: Reidel, 1988.
  Italian translation. "Intuizioni. Sulla teoria della causalità di I. J. Good." In *Epistemologia ed economica*, ed. Maria Carla Galavotti and Guido Gambetta, 189–209. Bologna: CLUEB, 1988.
Extract from *Foundations of Scientific Inference*. In Alec Fisher, *The Logic of Real Arguments*, 178–180. Cambridge: Cambridge University Press, 1988.
"Statistical Explanation and Its Models" (reprint of chap. 2 of *Scientific Explanation and the Causal Structure of the World*). In *Readings in the Philosophy of Science*, 2nd ed., ed. Baruch A. Brody and Richard E. Grandy, 167–183. Englewood Cliffs, N.J.: Prentice-Hall, 1989.
"The Twin Sisters: Philosophy and Geometry" (reprint of chap. 1 of *Space, Time, and Motion: A Philosophical Introduction*). Ibid.: 450–471.
"Four Decades of Scientific Explanation" (reprinted as a separate book). In *Minnesota Studies in Philosophy of Science, Volume XIII: Scientific Explanation*, ed. Philip Kitcher and Wesley C. Salmon, 3–219. Minneapolis: University of Minnesota Press, 1989.
"Causal Propensities: Statistical Causality vs. Aleatory Causality."* *Topoi* 9 (1990): 95–100.
"Rationality and Objectivity in Science, or Tom Kuhn Meets Tom Bayes." In *Minnesota Studies in the Philosophy of Science, Volume XIV: Scientific Theories*, ed. C. Wade Savage, 175–204. Minneapolis: University of Minnesota Press, 1990.
Reprinted:
  In *Philosophy of Science: The Central Issues*, ed. Martin Curd and J. A. Cover, 551–583. New York: Norton, 1998.
  Italian Translation. "Razionalità e oggettività nella scienza ovvero Tom Kuhn incontra Tom Bayes." *Iride* 1 (July–December 1988): 21–52.
"Scientific Explanation: Causation *and* Unification."* *Critica* 22 (December 1990): 3–21.
"A Contemporary Exposition of Zeno's Paradoxes" (reprint of portions of chap. 2 of *Space, Time, and Motion: A Philosophical Introduction*). In *Metaphysics: Classic and Contemporary Readings*, ed. Ronald C. Hoy and L. Nathan Oaklander, 11–13. Belmont, Cal.: Wadsworth, 1991.
"Philosophy and the Rise of Modern Science." *Teaching Philosophy* 13 (September 1990): 233–239.
"Propensioni causali: Causalità statistica e causalità aleatoria," *Rivista di filosofia* 81 (August 1990): 167–180.
"Hans Reichenbach's Vindication of Induction." *Erkenntnis* 35 (July 1991): 99–122.

"Ragionamento informale e regole formali." In *Nuova Civiltà delle Macchine*, Anno 10, N. 3–4 (1992): 27–37.

"Explanation in Archaeology: An Update."* In *Metaarchaeology*, ed. Lester Embree, 243–253. Dordrecht: Kluwer, 1992.

"Carnap, Rudolf." In *A Companion to Epistemology*, ed. Jonathan Dancy and Ernest Sosa, 56–57. Oxford: Blackwell, 1992.

"Explanation." Ibid., 129–132.

"Geometry." Ibid., 155–157.

"Methodology." Ibid., 279.

"Natural Science, Epistemology of." Ibid., 292–296.

"Reichenbach, Hans." Ibid., 426–427.

"Theory." Ibid., 506.

"The Value of Scientific Understanding." *Philosophica* 51 (1993): 9–19.

"Scientific Explanation and the Causal Structure of the World" (selections from *Scientific Explanation and the Causal Structure of the World*). In *Explanation*, ed. David-Hillel Ruben, 72–112. Oxford: Oxford University Press, 1993.

"On the Alleged Temporal Anisotropy of Explanation."* In *Philosophical Problems of the Internal and External Worlds*, ed. John Earman et al., 229–248. Pittsburgh and Konstanz: University of Pittsburgh Press and Universitätsverlag Konstanz, 1993.

"Zeno's Achilles" (brief selection from *Zeno's Paradoxes*). In Marilyn von Savant, *The World's Most Famous Math Problem*, 53–55. New York: St. Martin, 1993.

"Causality Without Counterfactuals."* *Philosophy of Science* 61 (1994): 297–312.

"Carnap, Hempel, and Reichenbach on Scientific Realism." In *Logic, Language, and the Structure of Scientific Theories*, ed. Wesley Salmon and Gereon Wolters, 237–254. Pittsburgh and Konstanz: University of Pittsburgh Press and Universitätsverlag Konstanz, 1994.

"Reichenbach, Hans," In *A Companion to Metaphysics*, ed. Ernest Sosa and Jaegwon Kim, 442–443. Oxford: Blackwell, 1995.

"Zeno of Elea." Ibid., 518–519.

"La comprensión científica en el siglo veinte." *Revista Latinoamericana de Filosofía* 21, 1 (Otoño 1995): 3–21.

"La causalità." *Iride*, 8, 14 (Aprile 1995): 187–196.

"Confirmation." In *Cambridge Dictionary of Philosophy*, ed. Robert Audi, 150–151. Cambridge: Cambridge University Press, 1995.

"Induction, Problem of." Ibid., 651–652.

"Reichenbach, Hans." Ibid., 683–684.

"Theoretical Term." Ibid., 796–797.

"Explanation, Theories of." In *Supplement to the Encyclopedia of Philosophy*, ed. Donald M. Borchert, 165–167. New York: Simon and Schuster, 1996.

"Carl G. Hempel." Obituary (London) *Times*, 10 December 1997. (Mutilated by the editor.)

"Causality and Explanation: A Reply to Two Critiques." *Philosophy of Science* 64 (1997): 461–477.

"Reichenbach, Hans," In *Routledge Encyclopedia of Philosophy*, ed. Edward Craig, 161–171. London and New York: Routledge, 1998.

"A Contemporary Look at Zeno's Paradoxes" (chap. 2 of *Space, Time, and Motion*.) In *Metaphysics: The Big Questions*, ed. Peter van Inwagen and Dean W. Zimmerman, 129–149. Oxford: Blackwell, 1998.

"Scientific Explanation: How We Got from There to Here" (Reprinted from *Causality and Explanation*). In *Introductory Readings in the Philosophy of Science*, ed. E. D. Klemke et al., 241–263. Amherst, N.Y.: Promethius Books, 1998.

"The Problem of Induction" (excerpts from *Foundations of Scientific Inference*). In *Introduction to Philosophy*, 3rd ed., ed. John Perry and Michael Bratman, 230–251. New York: Oxford University Press, 1999.

"Carnap, Rudolf." In *American National Biography*, vol. 4, ed. John A. Garraty and Mark C. Carnes, 407–408. New York: Oxford University Press, 1999.

"Der Schluß auf die beste Erklärung." In "Argumentation." *Dialektik: Enzyklopädische Zeitschrift für Philosophie und Wissenschaften*, ed. Geert-Lueke Lueken and Pirmin Stekeler-Weithofer, 115–139. Hamburg: Felix Meiner Verlag, 1999/1991. (German translation of "Inference to the Best Explanation," presented orally at Leipzig in 1996.)

"The Spirit of Logical Empiricism: Carl G. Hempel's Role in Twentieth-Century Philosophy of Science." *Philosophy of Science* 66 (1999): 333–350.
Reprinted:
    In *Science, Explanation and Rationality*, James H. Fetzer, ed. 309–324. Oxford: Oxford University Press, 2000.

"Ornithology in a Cubical World: Reichenbach on Scientific Realism." In *Epistemological and Experimental Perspectives on Quantum Physics*, ed. D. Greenberger, R. L. Wolfgang, and A. Zeilinger, 305–315. Dordrecht: Kluwer, 1999.

"Logical Empiricism." In *Blackwell's Companion to Philosophy of Science*, ed. W. H. Newton-Smith, 233–242. Oxford: Blackwell, 2000.

"Statistical Explanation" (with Christopher Hitchcock). Ibid., 470–479.

"Quasars, Causality, and Geometry: A Scientific Controversy That Should Have Happened but Didn't." In *Scientific Controversies: Philosophical and Historical Perspectives*, ed. Peter Machamer, Marcello Pera, and Aristides Baltas, 254–270. New York: Oxford University Press, 2000. (Also published in W. Salmon, *Causality and Explanation*. New York: Oxford University Press, 1998.)

"Scientific Understanding in the Twentieth Century." In *John von Neumann and the Foundations of Quantum Physics*, ed. M. Redei and M. Stöltzner, 289–304. Dordrecht: Kluwer, 2001.

"Explanation and Confirmation: A Bayesian Critique of Inference to the Best Explanation." In *Explanation: Theoretical Approaches and Applications*, ed. Giora Hon and Sam S. Rakover, 61–91. Dordrecht: Kluwer, 2001.

"Reflections of a Bashful Bayesian: A Reply to Peter Lipton." Ibid., 121–136.

"Explaining Things Probabilistically." *The Monist* 84, 2 (April 2001): 208–217.
Reprinted:
    In *Probability Is the Very Guide of Life*, ed. Henry E. Kyburg, Jr., and Miriam Thalos, 349–358. Chicago and LaSalle, Ill.: Open Court, 2003.

"A Realistic Account of Causation." In *The Problem of Realism*, ed. Michele Marsonet, 106–134. Hampshire, Eng., and Burlington, Vt.: Ashgate, 2002.

"Causation." In *The Blackwell Guide to Metaphysics*, ed. Richard M. Gale, 19–42. Oxford: Blackwell, 2002.

"Commit It Then to the Flames. . . ." In *Logical Empiricism: Historical and Contemporary Perspectives*, ed. Paolo Parrini, Wesley C. Salmon, and Merrilee H. Salmon, 375–387. Pittsburgh: University of Pittsburgh Press, 2003.

Forthcoming:

"Explanatoriness: Cause vs. Craig" (with Jukka Keranin). *Synthese*.

"Grover Maxwell: In Memoriam." In *Collected Works of Grover Maxwell*. Minneapolis: University of Minnestoa Press.

"Ancient Geometry and Modern Epistemology." *Poznan Studies in Philosophy*

Brief Comments on Works of Other Authors:

"Comments on Barker's 'The Role of Simplicity in Explanation.'" In *Current Issues in the Philosophy of Science*, ed. Herbert Feigl and Grover Maxwell, 274–276. New York: Holt, Rinehart and Winston, 1961.

"Empirical Statements about the Absolute" (reply to an article by George Schlesinger). *Mind* 76, 303 (July 1967): 430–431.

"Who Needs Inductive Acceptance Rules?" (comments on an article by Henry Kyburg). In *The Problem of Inductive Logic*, ed. Imre Lakatos, 139–144. Amsterdam: North-Holland, 1968.

"Comment" (on R. M. Martin). In *Fact and Existence*, ed. Joseph Margolis, 95–97. Oxford: Blackwell, 1969.

"Induction and Intuition: Comments on Ackermann's 'Problems'." In *Philosophical Logic*, ed. J. W. Davis et al., 158–163. Dordrecht: Reidel, 1969.

"Foreword to the English Edition." In Hans Reichenbach, *Axiomatization of the Theory of Relativity*, v–x. Berkeley and Los Angeles: University of California Press, 1969.

"Discussion Remarks" (edited without my approval). In *Minnesota Studies in the Philosophy of Science Volume IV: Analysis of Theories and Methods of Physics and Psychology*, ed. Michael Radner and Stephen Winokur, 225f., 227, 243, 248f., 251. Minneapolis: University of Minnesota Press, 1970.

"Postscript to 'Probabilities and the Problem of Individuation' by Bas van Fraassen" (paraphrased by van Fraassen with my approval). In *Probabilities, Problems, and Paradoxes*, ed. Sidney Luckenbach, 135–138. Encino, Cal.: Dickenson, 1972.

"Explanation and Relevance: Comments on James G. Greeno's 'Theoretical Entities in Statistical Explanation'." In *PSA 1970*, ed. Roger C. Buck and Robert S. Cohen, 27–39. Dordrecht: Reidel, 1971.

"Reply to Lehman," *Philosophy of Science* 40, 3 (September 1973): 397–402.

"Comments on 'Hempel's Ambiguity' by J. Alberto Coffa." *Synthese* 28, 4 (October 1974): 165–170.

"A Note on Russell's Anticipations." *Russell*, 17 (Spring 1975): 29.

"Comment on Mühlhölzer: Reichenbach on Realism." In *Logic, Language, and the Structure of Scientific Theories*, ed. Wesley C. Salmon and Gereon Wolters, 139–146. Pittsburgh and Konstanz: University of Pittsburgh Press and Universitätsverlag Konstanz, 1994.

"Comment on Parrini: Carnap on Realism." Ibid., 279–286.

"Logical Positivism, Logical Empiricism, and Thomas Kuhn: Comment on Robert Butts." In *Experience, Reality and Scientific Explanation*, ed. Maria Carla Galavotti and Alessandro Pagnini, 209–212. Dordrecht: Kluwer, 1999.

"Empiricism and the A Priori: Comment on Paolo Parrini." Ibid., 212–216.

"Probabilistic Explanation and Inference: Comment on Maria Carla Galavotti." Ibid., 216–218.

"Objective Bayesianism: Comment on Roberto Festa." Ibid., 219–223.

"Hume and Bayes: Comment on Alessandro Pagnini." Ibid., 226–229.

"Conventionality of Simultaneity Revisited: Comment on Martin Carrier." Ibid., 229–231.

"Quantum Realisms: Comment on Michael Stöltzner." Ibid., 231–234.

"Words of Appreciation." Ibid., 241.

Reviews

Sheldon J. Lachman, *The Foundations of Science*. *Philosophy of Science* 24, 4 (October 1957): 358–359.

Sir Harold Jeffreys, *Scientific Inference. Philosophy of Science* 24, 4 (October 1957): 364–366.

James T. Culbertson, *Mathematics and Logic for Digital Devices. Mathematical Reviews* 19, 11 (December 1958): 1200.

Hakan Tornebohm, "On Two Logical Systems Proposed in the Philosophy of Quantum Mechanics." *Mathematical Reviews* 20, 4 (April 1959): #2278.

Stephen F. Barker, *Induction and Hypothesis. Philosophical Review* 68, 2 (April 1959): 247–253.

G. H. von Wright, *The Logical Problem of Induction*, 2nd ed. *Philosophy of Science* 26, 2 (April 1959): 166.

Richard von Mises, *Probability, Statistics and Truth*, 2nd English ed. *Philosophy of Science* 26, 4 (October 1959): 387–388.

Hans Reichenbach, *The Philosophy of Space and Time. Mathematical Reviews* 21, 5 (May 1960), #3247.

Hans Reichenbach, *Modern Philosophy of Science. Philosophical Review* 69, 3 (July 1960): 409–411.

J. L. Destouches, "Physico-logical Problems." *Mathematical Reviews* 21, 10 (November 1960): #6319.

Rom Harré, *An Introduction to the Logic of Science*; Greenwood, *The Nature of Science*; Smith, *The General Science of Nature. Isis* (1962): 234–235.

John Patrick Day, *Inductive Probability. Philosophical Review* 72 (1963): 392–396.

I. J. Good, "A Causal Calculus, I, II." *Mathematical Reviews* 26, 1 (July 1963): #789 a–b.

Letter to the Editor, *New York Review of Books*, May 14, 1964. (Review of Stephen Toulmin's review of Adolf Grünbaum, *Philosophical Problems of Space and Time*.)

Robert G. Colodny, ed., *Frontiers of Science and Philosophy. Isis* 55, 2 (June 1964): 215–216.

Adolf Grünbaum, *Philosophical Problems of Space and Time. Science* 147 (12 February 1965): 724–725.

Rudolf Carnap, *Philosophical Foundations of Physics. Science* 155 (10 March 1967): 1235.

Henry E. Kyburg, Jr., *Probability and the Logic of Rational Belief. Philosophy of Science* 34, 3 (September 1965): 283–285.

Adolf Grünbaum, *Modern Science and Zeno's Paradoxes. Ratio* 12, 2 (December 1970): 178–182. (Also published in German ed. of *Ratio* in German trans.)

Wesley C. Salmon, *Zeno's Paradoxes* (author's abstract-review). *The Monist* 56, 4 (October 1972): 632.

John Earman, *Bayes or Bust. American Scientist* 82, 1 (January-February 1994): 91–92.

Marcello Pera, *The Ambiguous Frog* (brief review). *Library Review* 3, 3 (Summer 1994): 5.

Marcello Pera, *The Ambiguous Frog. Philosophy of Science* 62, 1 (March 1995): 164–166.

*All items marked with an asterisk can also be found in Wesley C. Salmon, *Causality and Explanation* (New York: Oxford University Press, 1998).

Index

Printed in the United States
by Blackmont

Printed in the United States
By Bookmasters